Lifetime Reproduction in Birds

Lifetime Reproduction in Birds

Edited by

Ian Newton
Monks Wood Experimental Station
Huntingdon, Cambridgeshire

ACADEMIC PRESS
Harcourt Brace Jovanovich, Publishers
London San Diego New York Berkeley Boston Sydney Toronto Tokyo

This book is printed on acid-free paper. ⊗

ACADEMIC PRESS LIMITED.
24/28 Oval Road
London NW1 7DX

United States Edition published by
ACADEMIC PRESS INC.
San Diego, CA 92101

Copyright © 1989 by
ACADEMIC PRESS LIMITED

British Library Cataloguing in Publication Data
Lifetime reproduction in birds.
  1. Birds. Reproduction
  I. Newton, Ian
  598.2′16

  ISBN 0–12–517370–9

Typeset by Photo·Graphics, Honiton, Devon
and printed in Great Britain by TJ Press Ltd, Padstow, Cornwall

Contents

Editor's acknowledgements

I would like to thank T.H. Clutton-Brock for helpful advice at the start of this book and P.J. Bacon and J.P. Dempster for helpful discussion of particular issues. For refereeing the various chapters, I am grateful to the contributors themselves, and to T.H. Clutton-Brock, J.C. Coulson, C. Cummins, J.P. Dempster, M.P. Harris, G. Hogstedt, D. Jenkins, A. Lundberg, M. Marquiss, M. Mountford, D. Scott, I.R. Taylor and A. Village. Finally, it is a pleasure to thank A. Richford, of Academic Press, for encouragement and interest throughout.

List of Contributors

Pelle Andersen-Harild, *Nature Conservancy, Ministry of Environment, Slotsmarken 13, DK 2970, Horsholm, Denmark.*

Peter Arcese, *Serengeti Wildlife Research Center, PO Box 3134, Arusha, Tanzania.*

Philip J. Bacon, *Institute of Terrestrial Ecology, Merlewood Research Station, Grange-over-Sands, Cumbria LA11 6JU, UK.*

Les D. Beletsky, *Department of Zoology, NJ-15 University of Washington, Seattle, Washington, 98195, USA.*

T.R. Birkhead, *Department of Animal Biology, University of Sheffield, Sheffield S10 2TN, UK.*

Jeffrey M. Black, *The Wildfowl Trust, Slimbridge, Glos GL2 7BT, UK.*

J.S. Bradley, *Biological Sciences, Murdoch University, Western Australia 6150.*

David M. Bryant, *School of Molecular & Biological Sciences, University of Stirling, Stirling FK9 4LA, Scotland, UK.*

Margret Bunzel, *Mester-Godert-Weg 8, D-4770 Soest, FRG.*

André A. Dhondt, *Department of Biology, Universiteitsplein 1-B-2610, Antwerpen (Wilrijk), Belgium.*

Joachim Drüke, *Hammer Str. 12a, D-4400 Münster, FRG.*

John W. Fitzpatrick, *Department of Zoology, Field Museum of Natural History, Chicago, Illinois 60605, USA.*

Frederick R. Gehlbach, *Department of Biology, Baylor University, Waco, Texas, USA.*

S.F. Goodburn, *Department of Animal Biology, University of Sheffield, Sheffield S10 2TN, UK.*

Lars Gustafsson, *Department of Zoology, Uppsala University, Box 561, S-751 22 Uppsala, Sweden.*

Wesley M. Hochachka, *Department of Zoology, 6270, University Boulevard, Vancouver, BC V6T 2A9 Canada.*

Hermann Hötker, *Institut für Haustierkunde, University of Kiel, Olshausenstrasse 40, 2300 Kiel, FRG.*

J. David Ligon & Sandra H. Ligon, *Department of Biology, University of New Mexico, Albuquerque, NM 87131, USA.*

R.H. McCleery, *Edward Grey Institute, Department of Zoology, University of Oxford, South Parks Road, Oxford OX1 3PS, UK.*

J.A. Mills, *Department of Conservation, PO Box 10 420, Wellington, New Zealand.*

Ian Newton, *Monks Wood Experimental Station, Abbots Ripton, Huntingdon, Cambs PE17 2LS, UK.*

Gordon H. Orians, *Department of Zoology, NJ-15, University of Washington, Seattle, Washington, 98195, USA.*

Myrfyn Owen, *The Wildfowl Trust, Slimbridge, Glos GL2 7BT, UK.*

Linda Partridge, *Department of Zoology, Edinburgh University, West Mains Road, Edinburgh, EH9 3JT, UK.*

Robert B. Payne, *Museum of Zoology, University of Michigan, Ann Arbor, Michigan 48109, USA.*

C.M. Perrins, *Edward Grey Institute, Department of Zoology, University of Oxford, South Parks Road, Oxford, OX1 3PS, UK.*

Sergej Postupalsky, *Department of Wildlife Ecology, University of Wisconsin, Madison, Wisconsin 53706, USA.*

Ian Rowley, *CSIRO Division of Wildlife and Ecology, LMB No.4, PO Midland, Western Australia 6056.*

Eleanor Russell, *CSIRO Division of Wildlife and Ecology, LMB No.4, PO Midland, Western Australia 6056.*

Pertti Saurola, *Zoological Museum, University of Helsinki, P. Rautatiekatu 13, SF-00100 Helsinki, Finland.*

D.L. Serventy, *CSIRO, Division of Wildlife & Rangelands Research, LMB 4, PO Midland, Western Australia 6056.*

I.J. Skira, *Department of Lands, Parks & Wildlife, Hobart, Tasmania, Australia 7001.*

James N.M. Smith, *Department of Zoology 6270, University Boulevard, Vancouver, BC 16T 2A9, Canada.*

Helmut Sternberg, *Ornithologische Arbeitsgemeinschaft für Populations forschung, Im Schapenkamp 11, D-3300 Braunschweig, FRG.*

Glen E. Woolfenden, *Department of Biology, University of South Florida, Tampa, Florida 33620, USA.*

R.D. Wooller, *Biological Sciences, Murdoch University, Western Australia 6150.*

Amotz Zahavi, *Institute for Nature Conservation Research, Tel-Aviv University, Ramat Aviv, 69 978, Tel-Aviv, Israel.*

1. Introduction

IAN NEWTON

For more than half a century detailed field studies have been made of
the breeding ecology of birds, but most have been concerned with the
numbers of young raised in individual attempts or in individual years.
Such studies can be described as "cross sectional", in that they involve
the collection of data at specific points in time, mostly from different and
unknown individuals. It is only recently, as a result of long-term studies
of marked individuals, that it has become possible to track the breeding
performance of particular birds throughout their lives. This has enabled
the measurement of lifetime reproductive success (LRS): that is, the total
number of young raised by recognizable individuals during their lifespans.
Interest in such "longitudinal" studies has grown rapidly with the realization
that lifetime reproductive rates could provide good approximations of
biological fitness: that is, of the contributions that particular types of
individuals make to future gene pools. Despite the great theoretical
importance of fitness, which is discussed in almost every modern textbook
on evolution, it has remained one of the most elusive measures in biology
(Lewontin 1974).

The aim of this book is to bring together most of the major avian
studies of lifetime reproduction made to date, and find what general
patterns emerge. Most of the book consists of 23 chapters devoted to
field studies of individual species, written by the researchers themselves.
Eight of these studies had previously been reported elsewhere, but in
each case the authors have brought out important new points in a reanalysis of enlarged samples. The remaining 15 species chapters present
new findings not previously published. The book ends with two general
chapters, concerned with life-history evolution and conclusions.

A wide array of species is represented: small and large, short-lived and
long-lived, landbirds and seabirds, monogamists and polygynists, solitary,
group-living and colonial types. They range from the Kingfisher *Alcedo
atthis*, which seldom lives for more than four years, breeds at less than
one year old and can raise more than 20 young in a year; to the Short-

LIFETIME REPRODUCTION IN BIRDS
ISBN 0-12-517370-9

tailed Shearwater *Puffinus teniurostris*, which can live for more than 30 years, starts breeding at 4–15 years old and raises no more than a single chick each year.

Birds are ideal subjects for work of this type. Not only are most species conspicuous and diurnal, but they can be readily trapped and marked with leg bands or tags, and their performance can be measured accurately at every breeding attempt. After a distinct period of growth and development, they have regular breeding seasons, and their eggs (equivalent to zygotes) and young can be counted with relative ease. Moreover, the life cycle of birds follows a simple pattern, with successive stages which have equivalents in most other organisms that reproduce sexually. Thus the life of any bird begins with the fertilization, development and subsequent hatching of an egg. There then follows a period of growth and further development, leading eventually to sexual maturity, which is marked by the presence of functional gonads. In some bird species, however, as in some mammals, physiological maturity may precede the first breeding attempt by up to several years. There then follows the reproductive phase, during which, in successive breeding attempts, the individual is able to contribute to zygotes, some of which may develop into independent offspring. The reproductive phase may be followed by a senescent phase, but as deaths occur at all stages, only a minority of individuals reach old age.

Because of the advantages that birds offer, they have been subject to more detailed and long-term studies than any other types of animals. Despite this, the first publications on lifetime reproductive success in birds did not appear until the 1980s (Woolfenden & Fitzpatrick 1984, Coulson & Thomas 1985, Newton 1985). These pioneer studies were quickly followed by others, including several which were reported in the book *Reproductive Success*, edited by T.H. Clutton-Brock (1988) and covering a wide range of animals, not just birds.

Advantages and uses of LRS measures

Studies of lifetime reproductive success have several advantages over traditional cross-sectional studies. The most obvious is that they combine the two key measures of individual performance, namely survival and success in particular breeding attempts, into a single overall measure of performance. Moreover, this single measure takes account of any trade-offs between reproduction and survival that might occur. Secondly, for any population, lifetime studies enable us to assess more accurately than before the full extent of individual variation in reproductive success.

Barely perceptible differences in the annual success of individuals can become substantial when repeated over whole lifespans of variable duration. Lifetime studies also provide a comparison on equal terms between the sexes of species which have polygamous mating systems, and in which reproduction may be confined to a smaller part of the lifespan in one sex than in the other. Hitherto, most studies of recognizable individuals have either lasted less than the lifetimes of the birds concerned, or have fused their individual breeding rates into population averages. Yet it is the very differences between individual genotypes in the number of zygotes to which they contribute, in the survival and breeding success of offspring, and the long-term continuation of lineages, which together determine the fitness of the genotypes concerned (Kemp 1985).

Thirdly, lifetime measures of reproductive success are less affected by short-term changes in environment or individual performance than are annual measures. A poor year for breeding may appear disastrous for many individuals at the time, but have little impact on their longer-term performance. Similarly, an individual whose breeding fails in one year because of youthful inexperience may live to breed successfully in later years. In theory, then, lifetime studies might reduce the proportion of variance in breeding success that is caused by short-term environmental and individual circumstances. This in turn can make it easier to measure the effects of phenotype on breeding success.

In the wider arena, therefore, the main applications of data on lifetime performance are in the fields of population biology, life history and evolution. Their relevance in evolutionary studies is immediately obvious, because it is the relative contributions of individuals of different genotypes which determine the composition of future gene pools. Selection at the level of the individual genotype, which is manifest in differential reproductive success, is the process by which adaptations arise and populations evolve. Studies of LRS enable us to investigate which components of life cycle, environment, phenotype and genotype contribute most importantly to variance in overall breeding success. They thus allow the identification of those individual attributes and individual circumstances that could contribute most importantly to fitness. Moreover, in providing a measure of 'fitness' over the whole lifespan for different categories of individuals, they have an advantage over the many studies which have assessed fitness over only a part of a lifespan, representing a single episode of selection (Endler 1986). Some behavioural or morphological traits can affect several components of fitness. They may appear beneficial at one stage of a life cycle yet detrimental at another (Smith 1988), so only from studies over the whole lifespan can the net effects of such traits be assessed. Such matters of selection and adaptation provided major themes in the Clutton-Brock (1988) volume, and in the present book, by contrast, emphasis is placed on population and life history issues.

One incidental result of lifetime studies is that they reveal changes in survival and reproduction associated with age. An understanding of the effects of age, and the extent to which they vary with environmental circumstances, is relevant in studies of population demography (Caughley 1977), as well as in understanding the evolution of life histories (Pianka 1978). Longitudinal studies, that can compare the breeding success of the same individuals at different ages, have an advantage over the usual practice of comparing the breeding of separate samples of birds observed at different ages. In the former, we can compare changes in the same individuals, but in the latter it is difficult to separate changes in individuals from changes in the composition of the sample, as different birds start and end their reproductive careers at different ages.

While studies of LRS represent a genuine advance, which should enhance our understanding of birds and other animals, they have their limitations. Like any other measure of individual performance, they are open to error, and for some purposes require careful interpretation, as Grafen (1988) and others have stressed.

Problems in measuring LRS

Despite the advantages that birds offer, lifetime reproduction is not always easy to record accurately. One problem results from the duration of study in long-lived species. In all birds so far studied, lifespan has emerged as a major determinant of lifetime reproductive success. Hence, for realistic findings, any study should include a representative cross-section of lifespans, and should not be weighted towards short-lived individuals. With few exceptions, however, most studies of long-lived birds have not progressed long enough to include the complete lifespans of the longest-lived individuals, which sometimes exceed 30 years. Although such long-lived birds form a tiny minority of the population, they are important because they represent extremes in success. The extent to which recorded lifespans are truncated by the duration of the study, or biased by over-representation of short-lived individuals, varies between the studies reported, as do the methods used to circumvent these biases.

A second constraint results from inevitable gaps in records. In some species it is often difficult to keep track of a large number of individuals throughout their breeding lives, and record their every nesting attempt. In certain species movements are a problem, because some individuals may change breeding localities during their lives. Such birds mostly move over short distances, but can sometimes cross study area boundaries, leading to underestimates of their LRS values. Birds which leave study areas undetected must generally be assumed to have died, along with all other individuals which disappear without trace. The severity of this potential under-recording of LRS also varies with the species, as does its effects on the findings, as discussed in the ensuing chapters.

An assumption necessary in most field studies, whether of annual or of lifetime success, is that the adults attending a brood are the true biological parents of that brood. In monogamous species this assumption is probably often justified, but in some such species copulations outside the pair bond are not uncommon. They may be solicited by the female or forced upon her. Extra-pair coition has been recorded in a wide range of species, including various passerines, raptors, waterfowl, seabirds and others, so that true paternity becomes uncertain (McKinney *et al.* 1983, Birkhead *et al.* 1987, Moller 1987, Westneat 1987a,b). In regular polyandrous or promiscuous species, the female will by definition copulate with more than one male, so that it is usually impossible to know by observation alone how many of her offspring to attribute to any particular male.

Nest parasitism (or egg dumping) where females lay in nests of conspecifics, is also regular in certain birds, and gives further scope for error in LRS values. It occurs commonly in some waterfowl and gallinaceous birds, and also in other species, such as hirundines, various ploceids and Starling *Sturnus vulgaris* (Yom-Tov 1980, Moller 1987). Cliff Swallows *Hirundo pyrrhonota* not only lay in other nests, but also transfer eggs in their beaks, so that more than 6 % of nests in one large colony were parasitized (Brown & Brown 1988). In addition, polygyny, in which the male mates with more than one female, is known to be regular in some bird species, but can occur as an occasional and unexpected event in others (Chapter 2), where it is easily missed. If the male concerned is not recorded with his secondary mate(s), or at his secondary nest(s), his breeding success is underestimated.

The consequences of extra-pair copulations (EPC) and egg dumping for the assignment of parentage have been assessed in a few species using at least four different techniques, namely (a) genetic plumage markers, (b) morphological correlates, (c) enzyme polymorphisms, and (d) genetic fingerprinting (Table 1.1). In these species, the percentage of offspring attributable to extra-pair copulations (i.e. a mismatch between the characters of the offspring and the putative father) varied between 1 and 33 % (though estimates resulting from enzyme electrophoresis may have been underestimated, and revised values could have been as high as 42 %). Certain of these estimates are higher than would be expected in

Table 1.1 Percentage of offspring in different species attributable to extra-pair paternity.

Method of study	Species	% offspring attributed to extra-pair paternity	Reference
1. Genetic plumage marker	Zebra Finch	6	Birkhead *et al.* 1988
	Mallard (captive) *Anas platyrhynchos*	8	Burns *et al.* 1980
2. Morphological characters	Pied and Collared Flycatchers *Ficedula hypoleuca & F.collaris*	24	Alatalo *et al.* 1984
	Swallow *Hirundo rustica*	26	Moller 1987
3. Enzyme electrophoresis	Indigo Bunting *Passerina cyanea*	14(42)[a]	Westneat 1987a,b
	Mountain White-crowned Sparrow *Zonotrichia leucophrys*	14(38)[a]	Sherman & Morton, 1989
	Acorn Woodpecker *Melanerpes formicivorus*	33 2	Joste *et al.* 1985 Mumme *et al.* 1985
	Eastern Blue Bird *Sialis sialis*	5	Gowaty & Karlin 1984
	White-fronted Bee-eater *Merops bullockoides*	1	Wrege & Emlen 1987
	Bobolink *Dolichonyx oryzivorus*	17[b]	Gavin & Bollinger 1985
	Mallard *Anas platyrhynchos*	>4	Evarts & Williams 1987
	Starling *Sturnus vulgaris*	2–8	Hoffenberg *et al.* 1988
4. DNA fingerprinting	House Sparrows *Passer domesticus*	15	Wetton *et al.* 1987 Burke & Bruford 1987
	Dunnock *Prunella modularis*	1	Burke *et al.* 1989

most birds: in some of the species studied individuals nest in social groups or in colonies, giving good opportunities for both EPC and nest parasitism, while in others "rapes" appear unusually frequent. Only two of the studies found evidence for egg dumping (a mismatch between offspring and putative mother, or both parents) but none of the species most involved in nest parasitism has been studied in this way.

Together with the observational evidence (Yom-Tov 1980, Birkhead *et al.* 1987), such studies indicate that many "monogamous" birds, far from adhering strictly to monogamy, may pursue mixed reproductive strategies, in which EPC or egg dumping figure prominently, at least in part of the population. For the male, the advantage of EPC is obvious, as it may provide the chance to father additional offspring, at little or no extra cost. On the other hand, any male who is himself cuckolded wastes time and energy rearing another male's offspring rather than his own. Not surprisingly, males in many birds show behaviour thought to minimize the risk of being cuckolded, including mate guarding during the fertile period, or frequent copulation which may devalue inseminations from EPC (Birkhead 1987). For the female, EPC may provide a hedge against infertility in her own mate and, in some cases, the opportunity to mate with a male superior to her own. There may in addition be other benefits, such as help in parental care or the opportunity to acquire a superior mate or territory for later breeding attempts. The costs to the female might include injury and (if she is discovered) a lower contribution to parental care by her own mate, or the abandonment of the current nesting attempt by her own mate (Birkhead & Biggins 1987). In forced copulations the female may also be injured.

Genetic fingerprinting has occasionally revealed the mismatch of a whole brood with the putative father (D. Parkin pers. comm.). This may mean that the whole brood resulted from EPC, that mate change occurred during the female's fertile period, or more frequently that a male seen at the nest site was mistakenly identified as the mate of the resident female. In most bird species, this type of observer error is much more likely for

---

[a] This technique underestimates the extent of mixed paternity, and figures in parentheses show maximum likely values.

[b] All values refer to numbers of young, except for Bobolinks, where they refer to numbers of broods in which mixed or incorrect paternity was detected.

Methods 1, 2 and 3 will not usually distinguish between the effects of egg dumping and extra-pair copulation, and additional behavioural observations are needed to indicate the likelihood of each.

male birds than for females, because males spend less time at their own nests than do females.

In egg dumping, the advantage to the parasitic female is that she might produce (extra) young without the cost of parental care. Her success per egg may be lower than that of the average non-parasitic female, but for some individuals parasitism may be the only alternative in some years to not breeding. Females which lay in more than one nest may also spread the risks in the event of total nest failures. The cost to the host female is a reduced success, because with additional eggs, some of her own eggs may get displaced from the nest or incubated less efficiently, resulting in a poorer hatch (Yom-Tov 1980). Moreover, in nidicolous species, with extra mouths to feed, the host's offspring may survive less well than otherwise. Indeed there are no obvious advantages to the host female, and in species with regular egg dumping, females often guard their nests and fight to keep other females away (Moller 1987, Brown & Brown 1988).

The existence of EPC, egg dumping and undetected polygyny means that observed LRS values—like any other measures of observed performance—are open to error. EPC, undetected polygyny and misidentification of the male parent affect the assessment of male LRS, whereas only egg dumping affects the assessment of female or both male and female LRS. In most species, therefore, female LRS values will be more accurate than male values. Moreover, the behavioural events leading to wrongly assigned parentage are unlikely to fall randomly across a population: dominant (old) males are likely to achieve more EPC, while submissive (young) ones are more likely to be cuckolded (Fujioka & Yamagishi 1981, Moller 1985). In effect this means that, in certain species, the true variance in male LRS values is likely to be greater than the recorded variance, as measured from the numbers of apparent offspring. Conversely, submissive (often young) birds may be more likely to practice brood parasitism, while dominant ones are more likely to be the unwitting recipients. This would result in real variance in LRS being less than recorded variance. As yet, little can be done to eliminate these sources of confusion, which of course apply as much to annual measures of breeding success as to lifetime measures.

Another opportunity for error in LRS values arises in certain waterfowl, and perhaps other nidifugous species, from the tendency for recently hatched young to change broods or for broods to amalgamate. In some crowded populations this mixing is so frequent as to render futile any attempt to allocate parentage solely from brood counts of large young. However, this was not a problem in the two waterfowl species discussed in this book.

These, then, are some of the problems encountered in measuring LRS in an unbiased manner in the field, and which should be borne in mind in the following chapters.

In some studies LRS is recorded as the number of young (usually fledglings) produced in a lifetime, whereas in others it is recorded as the number of recruits to future breeding populations. Both measures have merits and drawbacks. Production to the fledging stage is primarily determined by parental attributes, because until then, the young are partly or wholly dependent on their parents, according to species. This period covers only a small part of the potential lifespan of the offspring, but can usually be measured "accurately" for every individual in a study area. LRS measured by numbers of recruits gives a measure of individual contributions to future breeding populations, and incorporates survival throughout the whole pre-breeding part of the life-cycle, not just to the stage of nest-leaving or independence. However, it confounds the contributions to offspring survival of both the parents (pre-independence) and the offspring themselves (post-independence). Moreover, it can generally include only recruits to the local study area, because more distant settlers are not detected. In studies where both fledgling and local recruit productions were available, however, the two measures were related, with correlation coefficients in the range 0.4–0.8 (Chapters, 5, 8, 9, 13, 15, 17, 20).

The contribution that an individual makes to a future population is not dependent solely on the number of offspring produced, whether fledglings or recruits, but also on the population status and trend. A given number of offspring, if they survive, will form a greater proportion of the next generation in a low or declining population than in a large or increasing one (Fisher 1930). Similarly, in expanding populations, where offspring more than replace parents, an offspring produced early in the lifespan is more valuable than one produced later, because it forms a higher proportion of its cohort; in a decreasing population the opposite is true.

In the species accounts which follow, each author was asked to include some details on: (a) annual breeding success, survival and longevity; (b) changes in survival and breeding success associated with age; and (c) lifetime success for a representative sample of individuals. The author was also asked to investigate, where possible, the effects of individual life history, environment and phenotype on lifetime success. Such knowledge is basic to all attempts to understand the ecological needs of different species, and the factors that affect individual performance in the wild.

References

Alatalo, R.V., Gustafsson, L. & Lundberg, A. 1984. High rate of cuckoldry in Pied and Collared Flycatchers. *Oikos* **42**, 41–7.

Birkhead, T.R. 1988. Behavioural aspects of sperm competition in birds. *Advances in the Study of Behaviour* **18**: 35–72.

Birkhead, T.R. & Biggins, J.D. 1987. Reproductive synchrony and extra-pair copulation in birds. *Ethology* **74**: 320–34.

Birkhead, T.R., Atkin, L. & Moller, A.P. 1987. Copulation behaviour of birds. *Behaviour* **102**: 101–38.

Birkhead, T.R., Pellatt, J. & Hunter, F.M. 1988. Extra-pair copulation and sperm competition in the Zebra Finch. *Nature* **334**: 60–2.

Brown, C.R. & Brown, M.B. 1988. The new form of reproductive parasitism in Cliff Swallows. *Nature* **331**: 66–8.

Burke, T. & Bruford, M.W. 1987. DNA fingerprinting in birds. *Nature* **327**: 149–52.

Burke, T., Davies, N.B., Bruford, M.W. & Hatchwell, B.J. 1989. Parental care and mating behaviour of polyandrous Dunnocks *Prunella modularis* related to paternity by DNA fingerprinting. *Nature* **338**: 249–51.

Burns, J.T., Cheng, K.M. & McKinney, F. 1980. Forced copulation in captive Mallards. 1. Fertilisation of eggs. *Auk* **97**: 875–9.

Caughley, G. 1977. *Analysis of Vertebrate Populations*. New York: Wiley.

Clutton-Brock, T.H. (ed.) 1988. *Reproductive Success*. Chicago: University Press.

Coulson, J.C. & Thomas, C. 1985. Differences in the breeding performance of individual Kittiwake Gulls *Rissa tridactyla* (L.). In *Behavioural Ecology*, ed. R.M. Sibly & R.H. Smith, pp. 489–503. Oxford: Blackwells.

Endler, J.A. 1986. *Natural Selection in the Wild*. Princeton: University Press.

Evarts, S. & Williams, C.J. 1987. Multiple paternity in a wild population of Mallards. *Auk* **104**: 597–602.

Fisher, R.A. 1930. *The Genetical Theory of Natural Selection*. Oxford: University Press.

Fujioka, M. & Yamagishi, S. 1981. Extra-marital and pair copulations in the Cattle Egret. *Auk* **98**: 134–44.

Gavin, T.A. & Bolinger, E.K. 1985. Multiple paternity in a territorial passerine: the Bobolink. *Auk* **102**: 550–5.

Gowaty, P.A. & Karlin, A.A. 1984. Multiple maternity and paternity in single broods of apparently monogamous Eastern Bluebirds *Sialis sialis*. *Behav. Ecol. Sociobiol.* **15**: 91–5.

Grafen, A. 1988. On the uses of data on lifetime reproductive success. In *Reproductive Success*, ed. T.H. Clutton-Brock, pp. 454–71. Chicago: University Press.

Hoffenberg, A.S., Power, H.W., Lombardo, L.G., Lombardo, M.P. & McGuire, T.R. 1988. The frequency of cuckoldry in the European Starling. *Wilson Bull.* **100**: 60–9.

Joste, N., Ligon, J.D. & Stacey, P.B. 1985. Shared paternity in the Acorn Woodpecker *Melanerpes formicivorus*. *Behav. Ecol. Sociobiol.* **17**: 39–41.

Kemp, A.C. 1984. Individual and population expectations related to species and speciation. In *Species and Speciation*, ed. E.S. Vrba, pp. 59–69. Transvaal Museum Monograph No. 4. Pretoria: Transvaal Museum.

Lewontin, R.C. 1974. *The Genetic Basis of Evolutionary Change*. New York:

Columbia University Press.

McKinney, F., Derrickson, S.R. & Mineau, P. 1983. Forced copulation in waterfowl. *Behaviour* **86**: 250–94.

Moller, A.P. 1985. Mixed reproductive strategy and mate guarding in a semi-colonial passerine, the Swallow *Hirundo rustica*. *Behav. Ecol. Sociobiol.* **17**: 401–8.

Moller, A.P. 1987. Behavioural aspects of sperm competition in Swallows. *Behaviour* **100**: 92–104.

Moller, A.P. 1987. Intraspecific nest parasitism and anti-parasite behaviour in Swallows, *Hirundo rustica*. *Anim. Behav.* **35**: 247–54.

Mumme, R.L., Koenig, W.D., Zinc, R.M. & Marten, J.A. 1985. Genetic variation and parentage in a California population of Acorn Woodpeckers. *Auk* **102**: 305–12.

Newton, I. 1985. Lifetime reproductive output of female Sparrowhawks. *J. Anim. Ecol.* **54**: 241–53.

Pianka, E.R. 1978. *Evolutionary Ecology*, 2nd edn. New York: Harper & Row.

Sherman, P.W. & Morton, M.L. 1989. Extra-pair fertilisations in Mountain White-crowned Sparrows. *Behav. Ecol. Sociobiol.* **22**: 413–20.

Smith, J.N.M. 1988. Determinants of lifetime reproductive success in the Song Sparrow. In *Reproductive Success*, ed. T.H. Clutton-Brock, pp. 154–72. Chicago: University Press.

Westneat, D. 1987a. Extra-pair copulations in a predominantly monogamous bird: observations of behaviour. *Anim. Behav.* **35**: 865–76.

Westneat, D. 1987b. Extra-pair fertilisations in a predominantly monogamous bird: genetic evidence. *Anim. Behav.* **35**: 876–86.

Wetton, J.H., Carter, R.E., Parkin, D.T. & Walters, D. 1987. Demographic study of a wild House Sparrow population by DNA fingerprinting. *Nature* **327**: 147–9.

Woolfenden, G.E. & Fitzpatrick, J.W. 1984. *The Florida Scrub Jay; Demography of a Co-operative Breeding Bird*. Princeton: University Press.

Wrege, P.H. & Emlen, S.T. 1987. Biochemical determination of parental uncertainty in White-fronted Bee-eaters. *Behav. Ecol. Sociobiol.* **20**: 153–60.

Yom-Tov, Y. 1980. Intraspecific nest parasitism in birds. *Biol. Rev. Cambridge Philos. Soc.* **55**: 93–108.

Part I. Short-lived Hole Nesters

This first section includes chapters on four hole-nesting passerines that breed commonly in European woodlands, namely the Blue Tit *Parus caeruleus*, Great Tit *P. major*, Pied Flycatcher *Ficedula hypoleuca* and Collared Flycatcher *F. collaris*. As all four species take readily to nestboxes, they have been subject to several long and detailed studies, and now rank among the best known birds in the world (for earlier accounts of lifetime reproduction in *P. major*, see van Noordwijk & van Balen 1987, 1988, McCleery & Perrins 1988). All four species are short-lived and show high breeding success, benefitting from the relative security from predation that hole-nesting confers.

In his study of Blue Tits in Belgium, Dhondt shows how survival and lifetime reproductive success can vary widely between areas; while in Great Tits in southern England, McCleery & Perrins show how lifetime reproductive success (LRS) can vary greatly over time. Birds whose breeding happens to coincide with a good beech crop have higher LRS than other individuals because their offspring survive well in those years.

Both the tit species are year-round residents in the areas where they were studied; both are essentially monogamous, and show only slight sexual dimorphism in size and colour. The two flycatchers, in contrast, are summer migrants to Europe, are often polygynous and show striking sexual dichromatism. In Chapter 4, Sternberg reports the findings from a long-term study of Pied Flycatchers in Germany, showing clearly the benefits for breeding males of polygyny; while in Chapter 5 Gustafsson reports a study of Collared Flycatchers on the Swedish Island of Gotland, where Pied Flycatchers also occur. This study shows the effects of inter-specific competition and hybridization on LRS values, and assesses the heritability of different reproductive traits.

The remaining two species in this section do not nest in tree holes, but in other protected sites, so they too enjoy relative freedom from predation. The House Martin *Delichon urbica* is a familiar summer migrant to Europe, constructing its mud nests colonially under the eaves of buildings.

Its dependence on aerial insects means that its breeding success is highly dependent on weather. In Chapter 6, Bryant also shows how male phenotype (body size) can influence reproductive success.

The last species, the Kingfisher *Alcedo atthis*, is found on rivers and other water bodies throughout much of the Palearctic region, feeding on small fish and nesting in self-made tunnels in earth banks. It is the most extreme r-selected bird species in this book, being short-lived and highly prolific. In their study in Germany, Bunzel & Druke found that few Kingfishers lived longer than four years, but some raised more than 20 young (in three broods) in a single year. The species is highly vulnerable to cold winters, but with its high breeding rate, it can soon recover in numbers again.

References

McClerery, R.H. & Perrins, C.M. 1988. Lifetime reproductive success in the Great Tit. In *Reproductive Success*, ed. T.H. Clutton-Brock, pp. 136–53. Chicago: University Press.

Noordwijk, A.J. van, Balen, J.H. van & Visser, J. 1987. Lifetime reproductive success and recruitment in two Great Tit populations. *Ardea* **75**: 1–11.

Noordwijk, A.J. van & Balen, J.H. van. 1988. The Great Tit, *Parus major*. In *Reproductive Success*, ed. T.H. Clutton-Brock, pp. 119–35. Chicago: University Press.

2. Blue Tit

ANDRÉ A DHONDT

The Blue Tit *Parus caeruleus* is a small (11 g), short-lived, hole-nesting passerine, which breeds in wooded habitats over much of the Palearctic region. It reaches its highest breeding densities in mature oak woodland and feeds mainly on arthropods. During the breeding season caterpillars form the bulk of the food brought to nestlings. Although males are somewhat brighter blue, it takes an experienced observer to tell the difference between the sexes.

In northern Belgium the breeding population is usually resident, although a small proportion of adult females moves over several kilometres between years (Dhondt & Eyckerman 1980a). Post-fledging dispersal also is important, especially in young females. This makes it practically impossible to trap all surviving offspring from known breeding pairs. As in other species, therefore, lifetime reproductive success has to be estimated indirectly from lifetime fledgling production (= the total number of fledglings produced in the lifetime of an individual) and from lifetime recruitment (= number of young known to have survived at least one year that were produced during a bird's lifetime). Most Blue Tits in northern Belgium are single brooded and, in successful nests, on average about 9.5 young fledge from around 11 eggs. The largest clutch recorded was 16 eggs. The first egg is laid on average on 20 April.

Because the eggs are not incubated until the clutch is complete, incubation takes 13 days and the young stay in the nest until they are 20 days old, most young fledge at the end of May. Failed first clutches are not always replaced by a repeat clutch, so that in some years some breeding pairs raise no young at all. In certain years a small proportion of breeding adults starts a second clutch after a successful first clutch. In the population as a whole, the contribution of young from repeat and second broods is small. Over the period 1979–1987 only 1.7 % of 3,219 fledglings in one study area (Plot C) and 4.8 % of 2,297 fledglings in

LIFETIME REPRODUCTION IN BIRDS
ISBN 0-12-517370-9

another (Plot B) flew from non-first broods. Blue Tits are normally monogamous but some 4–10 % of the males are polygynous, having up to three partners simultaneously (Dhondt 1987a,b).

The breeding density of the Blue Tit is lowered by competition from the Great Tit *Parus major*, mainly over roosting sites in winter (Dhondt & Eyckerman 1980a,b). The effect of this competition is manifest especially on juvenile dispersal (very low local recruitment and high immigration in Blue Tit populations subject to intense competition by Great Tit), and perhaps also on adult survival (see below).

Reproduction of Blue Tit females improves from age one to two years, remains "good" at ages two and three, and worsens rapidly thereafter. Females of four years and older fledge fewer young, and the survival of these young is lower than that of young produced by younger females (Dhondt, 1989). No similar effect has yet been demonstrated in males.

The study plots

The Antwerp study started in 1979 and is still continuing. It uses five deciduous plots. In order to manipulate the degree of interspecific competition, the types of nestboxes differ between plots. When small-holed nestboxes (entrance hole diameter of 26 mm, which is too small for Great Tits) are present, Blue Tits use these for roosting in winter and breeding density is higher than when only large-holed boxes are present (cf. Dhondt et al. 1982). I will present data from two plots (Plots B and C) only, where the Blue Tit population is of sufficient size (Plot B: 20–40 pairs, Plot C: 28–51 pairs over nine years), and where the experimental nestbox situation has remained unchanged throughout the study.

Plot B covers 12.5 ha of optimal oakwood habitat well inside a large (ca. 150 ha) wooded estate, the Peerdsbos. Fifty-nine small-holed and 118 large-holed nestboxes (entrance hole diameter 32 mm, suitable both for Blue Tit and Great Tit) were present throughout, so that Great Tits also bred at a high density (see Dhondt & Schillemans 1983).

Plot C, 17 ha, is in an isolated estate, about 2 km south of the Peerdsbos. The vegetation is mainly deciduous, but not quite as rich (fewer mature oak trees) as that in Plot B. One hundred small-holed nestboxes were present, but large-holed boxes were absent outside the breeding season. Each year, at the beginning of April, and for the duration of the breeding season, the entrance hole of 10 boxes was enlarged to 32 mm. Most of these were readily occupied by Great Tits.

About half of the edge of this Plot is adjacent to housing or gardens, in which colour-ringed Blue Tits are seen in winter.

The plots thus differ in four respects: (a) degree of isolation from other tit populations: high in Plot C, lacking in Plot B; (b) Great Tit density: high in Plot B, low in Plot C; (c) vegetation: somewhat better (more mature oak) in Plot B, somewhat poorer, but still good for Blue Tits, in Plot C; (d) degree of isolation from human habitation: well isolated in Plot B, less so in Plot C. Three effects of the position of Plot C were identified: (1) laying in Plot C was about five days earlier than in Plot B, possibly through an earlier overall phenology; (2) nestling weight at 15 days was about 0.5 g lower in Plot C in each year (cf. Dhondt 1988) and nestling mortality was somewhat higher (see later); (3) adult local survival was much higher in Plot C than in Plot B.

Birds excluded from the analyses

Four groups of birds were excluded from the analyses of lifetime success:

(1) Birds of unknown age: Blue Tits can be aged, on plumage, until the age of about 14 months (after their first complete moult). Birds first trapped in yearling plumage can thus be assigned to a year class. Other birds, however, were first trapped in "adult" plumage. Because their exact age and previous history were unknown they were excluded from the analyses, except when explicitly mentioned.

(2) The oldest known Blue Tit in Belgium, comes from my population and lived for eight years. I have excluded all birds first breeding in 1984 or later. Including birds first breeding in 1983 thus produces a small bias against the estimate of lifetime reproduction in one or two birds that could have survived beyond 1987, the cut-off date for the data.

(3) Birds of known age first trapped breeding when older than one year.

(4) Birds not fledging any young in their lifetime; these birds are discussed separately, and are therefore excluded from the general analyses.

The samples left for analysis are: males Plot C: 77; females Plot C: 87; males Plot B: 67; females Plot B: 81 individuals.

Biases in the data

ADULT DISPERSAL

Considering all five Antwerp study plots, at least 4.3 % of 185 females surviving two years or more and breeding in one of the plots moved out of her breeding plot to a non-adjacent one. This is a minimum estimate for adult dispersal, since the chance of recapture after dispersing is relatively small. None of 167 males was known to move. Dhondt & Eyckerman (1980a) found similar (3.3 % of 151 individuals) values for movements of Blue Tit females between study sites in the Ghent area, and also found no movements among 107 males. Differential dispersal of adults according to sex therefore causes a bias against females in the estimation of lifetime reproduction, but probably not against males.

POLYGYNY

Polygyny was observed in both plots, but about half of presumed secondary females were deserted, so that the identity of their partner was unknown (Dhondt 1987a). Some of these nests still produced young, leading to an underestimate of lifetime reproduction of some polygynous males. For that reason, known polygynous males will be discussed separately.

UNIDENTIFIED ADULTS AND AGE AT FIRST BREEDING

A small number of breeding pairs in Plot B, and perhaps in Plot C, breed in natural cavities, so were unrecorded. Other birds were present in the breeding season, but did not nest. They could not be counted, but I know from anecdotal field observations that they included some of each sex.

   The age at first breeding of known age birds that were most probably present in the area, i.e. birds ringed in the nest in the more isolated Plot C and later found breeding there, is shown in Table 2.1 over the period 1980–1987: only 81 % of males and females were first trapped breeding at the age of one year. This cannot be due solely to the fact that some breeders were not captured. Some 95.7 % (331/346) of all females attempting to breed were trapped in Plot C. The 5/26 females first breeding when more than one year old suggests that some of them really did not breed in their first year (G = 6.88, 1 df, $P < 0.01$). Among the males, and assuming all missed males were monogamous, 87.7 % (293/334) were trapped in Plot C, whereas only 80.8 % were first breeding at

Table 2.1 Age at first breeding (Age) of nestlings born and breeding in Plot C (first breeding 1980–1987). The data show that some Blue Tits do not breed as yearlings.

Age	Males	Females
1	59	21
2	10	4
3	3	1
4	1	–
sum	73	26

age 1. Again this suggests that some males did not breed at age 1, although the difference is not statistically significant (G = 2.26, 1 df, ns). The data are not presented in detail for Plot B, because only six local born females were later retapped breeding in that plot (all at age 1) and only 21 males were (19 at age 1, and two at age 2). This low number not breeding at age 1 could in part result from the lower adult survival in Plot B.

The bias in the estimation of lifetime reproduction caused by the failure to trap all breeding adults is very small in females, because only 0.75 % of females in successful nests remained unidentified (Dhondt 1987a). Although in males 6.5 % remained unidentified, the bias will again be small because about half will have been yearlings, that will thus have been excluded from the analysis because they were missed in their first season, and many were probably polygynous, and these are discussed separately.

The use of lifetime fledgling production as an estimate of LRS

Only about 6 % of local nestlings were recovered after a minimum of three months, and only 3–4 % after a minimum of one year. This is insufficient to replace the adult losses. Furthermore, a large proportion of the breeding adults were immigrants. Although one cannot, therefore, calculate total Lifetime Reproductive Success (LRS) directly, significant correlations exist between the number of young fledged, the number of breeding seasons and the number of young subsequently recovered. This is true both after three months and after one year, both within one season (unpublished data) and over a bird's lifetime, and this held in each of the study plots (Table 2.2). I have calculated these correlations also for

Table 2.2 Spearman Rank Correlation Coefficients between lifetime fledgling production (LFP), number of breeding seasons (NBS), lifetime reproductive success measured through the number of juveniles surviving three months (3MOSU) and the number surviving one year (1YRSU). All correlation coefficients are statistically significant ($P < 0.001$) (C and B = Plot C and Plot B).

		NBS	3MOSU	1YRSU	n
Males					
LFP	C	0.90	0.44	0.46	77
	B	0.78	0.55	0.48	67
	B+C	0.86	0.48	0.47	144
NBS	C		0.36	0.39	77
	B		0.47	0.36	67
	B+C		0.38	0.39	144
3MOSU	C			0.92	77
	B			0.74	67
	B+C			0.83	144
Females					
		NBS	3MOSU	1YRSU	n
LFP	C	0.82	0.51	0.50	87
	B	0.77	0.50	0.44	81
	B+C	0.80	0.49	0.49	168
NBS	C		0.38	0.38	87
	B		0.39	0.39	81
	B+C		0.36	0.42	168
3MOSU	C			0.92	87
	B			0.69	81
	B+C			0.80	168

each cohort (both plots, 1978–1982) separately, and found significant correlations in all but two of them (males 1982, $P = 0.08$, females 1981, $P = 0.10$), thus confirming the overall analysis. Lifetime fledgling production, therefore, can be used to estimate LRS. One can assume that factors causing variation in lifetime fledgling production will influence, in a similar fashion, lifetime production of breeders.

Annual reproduction of Blue Tits in Plots C and B

In both areas breeding density was high, and breeding performance similar (Table 2.3). The tarsus length of the pulli was the same in the

Table 2.3 Basic information on Blue Tit populations (average 1979–1987).

Statistic	Plot C	Plot B
Plot size (ha)	17	12.5
Breeding density (pairs per ha)	2.44	2.43
% second broods	2.2	5.7
% repeat broods	9/37(24.3 %)	12/38(31.6 %)
Clutch-size 1st broods (*)	11.53	10.60
Brood-size 1st broods (**)	9.56	9.43
Fledglings per egg 1st brood (**)	0.826	0.890
Reproductive rate (all broods)	8.61	8.38
Nestling weight at 15 days (1st broods)	10.74 ± SD 0.744	11.16 ± 0.53
Nestling tarsus at 15 days (1st broods)	14.83 ± 0.393	14.85 ± 0.311

% second broods: number of second broods as a percentage of first broods which produced fledglings.
% repeat broods: number of repeat broods as a percentage of first broods which produced no fledglings.
(*) complete clutches only
(**) successful broods only
Reproductive rate: total number young fledged/number breeding pairs, including failed broods.
Nestling weight and tarsus: mean of nest means: 298 nests in Plot C and 207 nests in Plot B.

two populations, even though nestlings were heavier in Plot B. To illustrate the extent of annual variations in clutch size, number of young fledged (brood size), and number of young surviving to one year, data from first broods of successful yearling females only were used, so that possible age effects were excluded. The extent of the variation is illustrated by dividing the maximum value by the minimum value. The "between-year" variation is small for clutch size (1.16 and 1.20:1 for Plots B and C respectively), and for brood size (1.23 and 1.30:1), but is much larger for the number of recruits (6.33 and 3.41:1). This variation seems to be related to breeding density (mainly in Plot C) and winter cold (mainly in Plot B) (unpublished data).

Annual adult survival

In general Blue Tits lived longer in Plot C than in Plot B (Table 2.4). A Kruskal–Wallis ANOVA indicates that the differences between the plots were highly significant, but that there was no difference between the

Table 2.4 Maximum age of Blue Tits first breeding as yearlings in the breeding seasons 1979–1983 by Sex and Plot. Survival did not differ between sexes within any one plot, but was significantly different between plots.

Plot	Sex	1	2	3	4	5	6	7	Sum	Annual survival
C	M	40	15	9	7	4	1	1	77	0.51
C	F	37	24	16	8	2	0	0	87	0.50
B	M	48	11	6	1	1	0	0	67	0.31
B	F	60	14	5	2	0	0	0	81	0.27

sexes within plots. The average adult annual survival rate over the birds' lifetime was 0.31 for 67 males in Plot B and 0.27 for 81 females in Plot B; and 0.51 for 77 males in Plot C and 0.50 for 87 females in Plot C. No differences existed between the sexes within any one plot (Plot B: G = 0.383, 1 df, ns; Plot C: G = 0.032, 1 df, ns), but for each sex were highly significant between the plots (males: G = 10.30, 1 df, $P < 0.01$; females: G = 15.55, 1 df, $P < 0.001$).

To calculate the annual survival of birds first breeding when at least two years old, I combined males and females from each plot, to obtain sufficient sample sizes. The values of 0.43 for 53 adults in Plot C and of 0.34 for 45 adults in Plot B, were very similar to those calculated for birds first breeding as yearlings, showing that delayed breeding did not enhance subsequent survival.

These differences between the plots must reflect real differences in mortality, and not only differences in local survival, because in males which had bred, mobility was very low and similar values were obtained for males and females on each plot. Since the difference in survival between the sexes is small and not statistically significant, I will use as an estimate of adult survival the average from the two sexes, that is 0.288 for Plot B and 0.508 for Plot C.

Possible causes of the differences in adult survival between plots could be: (a) differing intensity of competition from Great Tits; (b) differing incidence of predation (Sparrowhawks *Accipiter nisus* were seen regularly in Plot B, but only rarely in Plot C); (c) a better overwinter survival in Plot C, because Plot C is bordered on two sides by gardens where some colour-banded birds fed in winter. Plot B lies completely inside the wood, and birds would have to move over at least 800 m to reach gardens.

Lifetime reproduction of zero

Three groups of birds produced no offspring in their lives: birds which did not survive until the first breeding season; birds which did not breed

Table 2.5 Estimate of proportion of surviving young of each sex recovered after one year ($T/(0.5P)$), survival of young for seasons 1979–86; origin of breeding population for seasons 1980–1987).

Plot	P/2	A	B	C	T	T/ (0.5P)	B/T
C MM	1432.5	74	77	36.50	203	14.2	37.9
C FF	1432.5	26	31	10.08	258	18.0	12.0
B MM	1027	24	42	12.96	185	18.0	22.7
B FF	1027	6	24	2.54	236	23.0	10.2

P: number of fledglings 1979–1986 in plot; assuming the sex ratio at fledging to be equal (my data suggest this is true, unpublished data) P/2 is the number of birds of each sex which fledged.
A: number of local recoveries after one year (local recruits)
B: number of total recoveries after one year (total recruits)
C: % breeding birds born locally
T: calculated total number of fledglings surviving; assuming emigration equalled immigration the total number of surviving young, T, was $100*A/C$ (my data on exchange of young between plots do not show that exchange between plots was mainly unidirectional).
T/(0.5P): calculated survival of juveniles after one year
B/T calculated % of surviving juveniles actually recovered; $1 - B/T$ indicates to what extent LRS is underestimated. This clearly differed between plots, because the dispersal of the juveniles was higher from Plot B than from Plot C. Dispersal was also more important in females than in males.

although alive in a breeding season and birds which attempted to breed, but failed to fledge any young. The numbers in each group are estimated below.

AN ESTIMATE OF THE NUMBER OF YOUNG SURVIVING AT LEAST ONE YEAR

In Table 2.5 the data used in the calculation of the proportion of fledglings which survived at least until the next breeding season are summarized. The total survival of fledgling Blue Tits, both sexes combined, estimated in Table 2.5 was 16.1 % for Plot C and 20.5 % for Plot B. This means that the estimated percentage of the fledglings which never produced any young because they did not survive until the first breeding season was 83.9 % in Plot C and 79.5 % in Plot B. This is a rough estimate because: (a) unverified assumptions were made (e.g. emigration = immigration), (b) numbers on which the calculations were made are small, and (c) data over all years were combined, although the size of the breeding population fluctuated strongly from year to year. It gives, however, a value which is plausible, and more than sufficient to replace adult losses.

AN ESTIMATE OF THE BIRDS PRESENT WHICH DID NOT BREED

Assuming that some one-year olds did not breed and that (in Plot C) their annual survival was 50 %, each bird first breeding at age 2 would then represent two birds which were present at age 1 but did not breed; each bird first breeding at age 3 would represent four such birds, and a bird first breeding at age 4 would represent eight birds present at age 1 which did not breed (after Newton 1985).

Applying these values to the data of Plot C (Table 1) yields the following results:

Males: number present at age 1: 59 + 2*10 + 4*3 + 8*1 = 99
         proportion breeding at least once: 73/99 = 73.7 %
         correction for % adults trapped: (73.7 * 100)/87.7 = 84.1 %

Females: number present at age 1: 21 + 2*4 + 4*1 = 33
         proportion breeding at least once: 26/33 = 78.8 %
         correction for % adults trapped: (78.8 * 100)/95.7 = 82.3 %

Based on these calculations about one fifth of the Blue Tits of each sex surviving to age 1 produced no offspring during their lives because they never bred.

FAILED BREEDERS

About two-thirds of the failed first broods were not later replaced by a successful repeat brood (Table 2.3), and, on the whole, replacement clutches did very poorly.

In Plot C, 37 out of 374 first clutches produced no fledglings over the 1979–1987 period. Nine replacement clutches were laid, seven of which produced fledglings (mean 3.6). Thus only seven out of 37 (18.9 %) pairs which failed at their first attempt did eventually produce fledglings in that season and 30 pairs out of 374 (8 %) which attempted to breed produced no fledglings in that season. Assuming the birds which deserted were still alive, and knowing that the chance of surviving another season is about 50 % in Plot C, we can calculate that about 5 % of birds which attempted to breed had a lifetime production of zero.

In Plot B in the period 1979–1987, 38 first brood attempts produced no fledglings and 12 replacement clutches were produced, nine of which were successful (mean 7.0 young). Thus 9/38 (23.7 %) of the failed first attempts produced young in that season and 29 out of 270 (10.7 %) pairs which attempted a first clutch fledged no young in that season. Using an adult survival rate of 29 %, we can estimate that about 7 % of Blue Tits which attempted to breed in Plot B never fledged any young.

Summing up, the chances of a fledged young producing fledglings in its own lifetime is: (the chance of surviving to one year) * (the chance of breeding) * (the chance of breeding successfully). In Plot C this is: 0.161 * 0.83 * 0.95 = 0.13. Assuming that in Plot B the proportion of birds which survived to one year and bred was the same as in Plot C, the equivalent figures are: 0.205 * 0.83 * 0.93 = 0.16. Thus, taking data from both areas, about 14 % of the fledglings eventually reproduced successfully.

Lifetime recruitment of zero in successful birds

In the previous paragraphs I considered the chance that an individual which fledged would become a reproducing adult. In this section I ask what chance such a successful individual has to contribute to the next generation. When one considers the distribution of the recoveries (Fig. 2.1), for a very high proportion of individuals, recruits were never observed. This is in part caused by the fact that only a small proportion of surviving young were ever retrapped. But what would one find if all young surviving to breed were recovered?

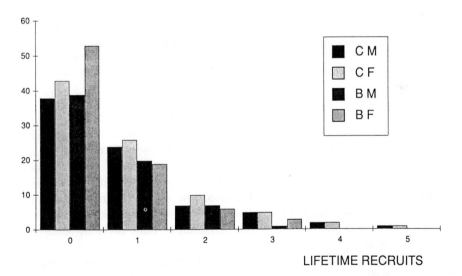

Figure 2.1 Frequency distribution of lifetime recruitment per sex and plot. Note that these distributions deviate significantly from a Poisson distribution in that more birds produced no recruits, and fewer produced one recruit than expected. M: males, F: females; B or C: study plots.

Assuming that young survived according to a Poisson distribution (but see Fig. 2.1 and below), one can calculate for the data in the sample how many individuals would have no recruits surviving until the next year. For this calculation I assumed that 13.7 % of the young in Plot B and 10.7 % in Plot C reproduced successfully (see above). The total number of individuals not producing any young which survive to one year is given in Table 2.6: 20–25 % of the parents never become grandparents. Another way to express this is by calculating the number of offspring required to have a 95 % (22–28) or a 99 % (33–43) chance to produce one recruit. This exercise shows that, if all birds were equal and lived in the same circumstances, a Blue Tit needs two to three successful breeding seasons to have a 95 % chance of producing one recruit in its lifetime.

Table 2.6 Percentage of successfully breeding adults that would never recruit any young, assuming a Poisson distribution of recruitment, and lifetime fledgling production giving a 95 and 99 % chance to have one recruit.

Plot	Sex	N1	N2	%	95 %	99 %
B	M	70	14.4	20.6	22	33
B	F	105	25.0	23.8	22	33
C	M	84	21.0	25.0	28	43
C	F	89	19.0	21.3	28	43

Notes: *N1*: number parents; *N2*, %: calculated number, % of parents not producing any recruits; 95 %, 99 %: lifetime fledgling production needed to have a 95 % or 99 % chance to produce one recruit.

The incorrect assumption in these calculations is that all young have the same chance of surviving. An analysis of the number of surviving offspring per individual shows that it was not Poisson distributed for any of the groups. The distribution deviates significantly from a Poisson distribution in that more birds than expected did not recruit any offspring, and fewer birds than expected recruited one (Fig. 2.1). In consequence the proportion of adults that did not contribute to the next generation is larger than estimated above, and 35 % rather than 25 % of successful parents would produce no recruits.

Several factors contributed to inter-individual variations in survival of young: laying date, possibly nestling weight at fledging, study plot and cohort. Inter-year variations in survival seemed to be related to breeding density (mainly in Plot C) and to winter climatic conditions (mainly in Plot B) (unpublished data). An overall analysis did not show that larger adults produced more surviving offspring than smaller ones (all eight comparisons of tarsus length of adults which did or did not produce at least one surviving offspring after three months or after one year per sex and plot: two-tailed, $P > 0.05$). The single factor which clearly stands out as being important to explain inter-individual variations in lifetime fledgling production and in number of recruits is the number of seasons a bird breeds (see Table 2.2), as discussed in the next section.

Variation in lifetime reproduction in successful Blue Tits

VARIATION IN LIFETIME REPRODUCTION BETWEEN PLOTS AND SEXES

In this section I will consider only birds which produced at least one fledgling in their lifetime, the maximum being 62 in a male living seven years (this male fledged another two young in 1988, its eighth breeding season). Variation in lifetime reproduction is shown for each sex and study plot in Fig. 2.2. In all groups a clear mode is apparent at 10–12 fledglings, representing birds raising one brood. A second peak is found at 18–22 young (two successful broods), and a long-tail to the right shows that only a small proportion of individuals produced more than 40 young. This tail is longer in males (3 % Plot B, 8 % Plot C) than in females (2 % Plot B, 4 % Plot C) and in Plot C than in Plot B, but only the between-plot differences are statistically significant (Table 2.7). Fifty per cent of the individuals produced 26 % (Plot C) and 30 % (Plot B) of the fledglings, and the 10 % most productive birds produced 24 % (Plot C) and 25 % (Plot B) of the fledglings. Because all distributions are highly skewed I used Kruskal–Wallis analyses of variance to compare the values between plots and sexes. Birds of both sexes were significantly more successful in Plot C than in Plot B (Table 2.7). No differences existed between the sexes. This result is confirmed when the number of recruits is compared (Fig. 2.1, Table 2.7).

THE EFFECT OF NOT BREEDING AT AGE 1

Although lifetime production differed significantly between the study plots, this was not true when comparing individuals that survived the same number of seasons (Table 2.8). When comparing birds that reached the same age, I can therefore combine the two plots.

About 25 % of the individuals breeding in the period 1979–1983 in each plot–sex group were excluded from the analyses, because they were

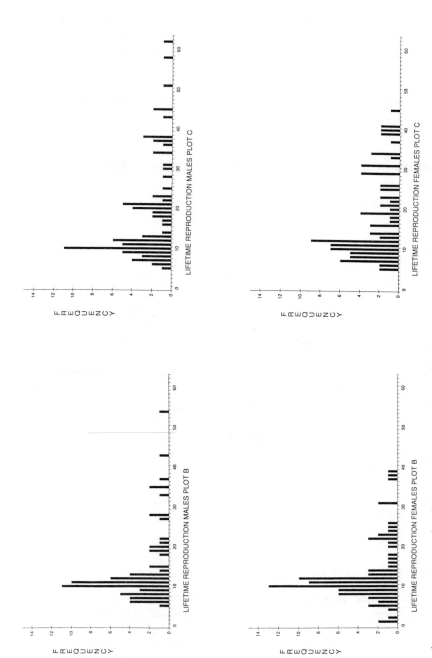

Figure 2.2 Frequency distribution of lifetime fledgling production per sex and in each plot. Note that more birds produced many offspring in Plot C than in Plot B, but that no differences between males and females exist.

Table 2.7 Results of Kruskal–Wallis analyses of variance to compare the values between plots or sexes. In both sexes lifetime fledgling production (LFP) and the number of recruits after one year (1YRSU) was higher in Plot C than in Plot B. This was caused by a higher annual survival (larger number of breeding seasons NBS) there, and not by a higher number of fledglings per year (see Table 2.8). No significant differences between the sexes were demonstrated.

Comparison of the plots per sex

Parameter/sex	df	H	P	Larger in Plot
LFP FF	1	7.735	< 0.01	C
NBS FF	1	18.91	< 0.001	C
1YRSU FF	1	5.066	< 0.025	C
LFP MM	1	4.856	< 0.05	C
NBS MM	1	7.541	< 0.01	C
1YRSU MM	1	1.867	ns	

Comparison of the sexes in one plot

Parameter/plot				
LFP C	1	0.081	ns	
LFP B	1	0.322	ns	
NBS C	1	0.352	ns	
NBS B	1	0.174	ns	
1YRSU C	1	0.005	ns	
1YRSU B	1	0.609	ns	

Table 2.8 Comparison of lifetime fledgling production in the two plots by maximum age. All birds first bred at age 1. Birds that reached the same age did not differ in lifetime fledgling production. Since annual survival in both sexes was higher in Plot C than in Plot B, lifetime fledgling production in Plot C was significantly higher than in Plot B, and this both for males and females (see Table 2.7).

Max age	Plot C			Plot B			ANOVA		
	n	mean	SE	n	mean	SE	df	F	P
Females									
1	37	9.59	0.369	60	9.33	0.373	1,95	0.221	ns
2	24	16.21	1.125	14	19.57	1.042	1,36	4.021	0.053
3	16	27.56	1.663	5	29.80	2.800	1,19	0.442	ns
4	8	36.38	2.618	2	37.50	0.500	1,8	0.042	ns
Males									
1	40	9.83	0.340	48	9.85	0.336	1,86	0.040	ns
2	15	20.07	1.076	11	18.73	1.369	1,24	0.655	ns
3	9	26.22	2.274	6	33.50	2.263	1,13	4.097	0.064
4	7	35.43	2.776	1	37		U-test:	U=4	ns

Table 2.9 Comparison of birds first found breeding as yearlings (Age 1) or later (Age >1), that bred the same maximum number of seasons (NBS).

NBS	Age 1			Age > 1			ANOVA			Combined		
	n	mean	SE	n	mean	SE	df	F	P	mean	SE	
Females												
1	97	9.43	0.269	33	8.36	0.559	1,128	3.592	0.06	9.16	0.246	
2	38	17.45	0.841	11	19.82	0.829	1,47	2.104	ns	17.98	0.682	
3	21	28.10	1.414	3	28.33	5.696	1,22	0.003	ns	28.13	1.399	
4	10	36.60	2.072	2	35.50	2.500	1,10	0.051	ns	36.42	1.823	
Males												
1	88	9.84	0.238	26	9.92	0.704	1,112	0.020	ns	9.86	0.243	
2	26	19.50	0.812	12	20.00	1.382	1,36	0.109	ns	19.66	0.705	
3	15	29.13	1.947	4	31.00	3.028	1,17	0.207	ns	29.53	1.675	
4	8	35.36	2.412	4	40.75	2.287	1,10	1.802	ns	37.33	1.799	

Table 2.10 Lifetime fledgling production of polygynous males according to the number of breeding seasons (NBS) in which they were present (both plots combined, all years). Since annual survival is not influenced by polygyny, polygynous males have a higher lifetime fledgling production than monogamous males.

NBS	Polygynous males			Monogamous males			ANOVA		
	n	mean	SD	n	mean	SD	df	F	P
1	8	14.88	2.748	214	9.71	2.407	1,220	35.11	< 0.001
2	7	25.71	4.821	56	19.25	3.564	1,61	18.93	< 0.001
3	4	32.00	6.683	23	26.13	8.308	1,25	1.77	ns
4	4	46.25	7.500	12	37.42	6.473	1,14	5.20	< 0.05

first found breeding when older than one year (see above). Comparing their breeding performance to that of birds breeding first as yearlings shows that no differences existed between the two groups of birds, if they bred an equal number of seasons (Table 2.9). Birds which delayed breeding did thus not compensate for this by producing more offspring per attempt later in life. Since their annual survival was similar to that of birds first breeding as yearlings (see above), delayed breeding resulted in a reduced lifetime fledgling production.

Although the exclusion of the birds, which did not first breed as yearlings did not bias the results, one exceptional male excluded in Plot B must be mentioned separately. It raised 61 fledglings over six breeding seasons, both values being larger than any of the included birds in the same plot.

The reproductive success of polygynous males

A small proportion of the males was polygynous. To evaluate the effect of this on lifetime reproduction, and have a reasonable sample size, I have included all males that were known to be polygynous at least once in their lifetime. I present the data (Table 2.10) in relation to the number of breeding seasons in which they were present, so that I can include all the birds, even those which bred for the first time after 1983, or for the first time when older than one year old. The data clearly show that polygynous males produced more young than comparable monogamous males, breeding an equal number of seasons. Since polygyny did not reduce male annual survival (Dhondt 1987a), males increased their lifetime fledgling production by becoming polygynous. Dhondt (1987a) mentioned a male from Plot T, not included in this analysis, who was bigamous in two different years and who fledged 53 young in three seasons.

Cohort effects on reproduction

Part of the variation in lifetime reproduction is caused by the year in which an individual was born, mainly as a result of reduced adult survival over cold winters. A reduced adult survival resulted in a reduced lifetime fledgling production. Juvenile survival was also lower over cold winters. In order to test the hypothesis that between-cohort differences existed, I calculated the partial correlation coefficient on the lifetime number of recruits, both sexes combined, for each plot, after allowing for the effect of lifetime fledgling production. In both plots this partial correlation coefficient was $r = 0.19$ ($P = 0.02$). An effect of body size in this analysis was not apparent. The effects on adult survival and on juvenile survival through cold winters had a multiplicative effect on lifetime fledgling production, and certainly on lifetime recruitment, and thus on LRS.

Conclusions

Blue Tits in northern Belgium produced between 0 and 62 fledglings in their lifetime. An estimated 86 % of the fledglings never bred successfully themselves and about 35 % of successfully reproducing parents never had grandchildren. The main factor which contributed to variation in lifetime reproduction between individuals was the number of seasons an individual bred, regardless of the age at first breeding. Delayed breeding was not compensated for by raising more fledglings per attempt nor by an increase in annual survival. No differences were found between the sexes within a plot, but the differences were very large between plots. This was mainly caused by differences in adult survival, because birds which survived the same number of years did not produce different numbers of fledglings. Although an estimated 64–97 % of the fledglings dispersed from the plot in which they were born (Table 2.5), lifetime fledgling production was significantly correlated with lifetime recruitment.

In order to explain inter-individual and inter-plot differences in LRS, one must therefore understand the causes of variation in survival. These were not studied here, but preliminary analyses suggest that selection pressures differ between plots, years and phenotype, with strong phenotype–environment interactions. The effect of polygyny was clearly beneficial to males, because polygynous males did not survive less well than monogamous males (Dhondt 1987a), but produced more fledglings than monogamous males breeding the same number of seasons.

The effect of Great Tits on Blue Tits operates mainly through the exclusion of Blue Tits from good roosting sites in winter (Dhondt & Eyckerman 1980b, Dhondt unpublished). The higher survival of Blue Tits in Plot C is most probably, at least in part, caused by the higher proportion of Blue Tits that used nestboxes for roosting in winter, because of reduced interspecific competition for roosting sites there. Since lifetime fledgling production and LRS in the Blue Tit are mainly determined by the number of seasons a bird can breed, and hence by adult survival, these will be higher when competition from Great Tits is reduced.

References

Dhondt, A.A. 1987a. Reproduction and survival of polygynous and monogamous Blue Tits. *Ibis* **129**: 327–34.

Dhondt, A.A. 1987b. Blue Tits are polygynous but Great Tits monogamous: does the polygyny threshold model hold? *Amer. Nat.* **129**: 213–20.

Dhondt, A.A. 1988. The necessity of population genetics for understanding evolution: an ecologist's view. In *Population Genetics and Evolution*, ed. G. de Jong, pp. 14–18. Berlin: Springer.

Dhondt, A.A. 1989. The effect of old age on the reproduction of Great Tits *Parus major* and Blue Tits *P. Caenuleus*. *Ibis* **131**: 268–80.

Dhondt, A.A. & Eyckerman, R. 1980a. Competition and the regulation of numbers in Great and Blue Tit. *Ardea* **68**: 121–32.

Dhondt, A.A. & Eyckerman, R. 1980b. Competition between Great Tit and Blue Tit outside the breeding season in field experiments. *Ecology* **61**: 1291–6.

Dhondt, A.A. & Schillemans, J. 1983. Reproductive success of the Great Tit in relation to its territorial status. *Animal Behaviour* **31**: 902–12.

Dhondt, A.A., Schillemans, J. & De Laet, J. 1982. Blue Tit territories at different density levels. *Ardea* **70**: 185–8.

Newton, I. 1985. Lifetime reproductive output of female Sparrowhawks. *J. Anim. Ecol.* **54**: 241–53.

3. Great Tit

R.H. McCLEERY & C.M. PERRINS

The Great Tit *Parus major* is a small passerine bird which breeds in
woodland and forest throughout much of Europe and Asia. The two
sexes are basically similar in appearance though males are slightly more
brightly coloured than their mates; in particular, males have a noticeably
wider black stripe on their underparts. In size, males (20 g) are slightly
larger than females (18 g). The species is relatively short-lived, with
roughly half the adults breeding in one year dying before the next, and
extremely few surviving to seven or eight years.

The Great Tit is primarily insectivorous, but also takes many tree seeds
in winter. It is abundant, nesting at densities of around one pair per
hectare in broad-leaved deciduous woodland. It has been much studied,
as it is a hole-nester and readily accepts nestboxes; in most cases the
whole population will use boxes, making the bird both easy to census
and to study (Perrins 1979). The species also has a large clutch, the
annual average having varied from 7.8 to 12.3 in our study. The large
variations in clutch size, both between individuals within years, and in
population means between years, have resulted in extensive study of the
relationship between fecundity and other population parameters (Perrins
1965, Lack 1966, Boyce & Perrins 1987). In our study area the birds do
not normally have a second brood.

In Britain, the species is largely resident. Individuals start to breed at
age 1 (see below) and normally attempt to breed every year thereafter
until they die. The adult birds move very little between their nesting
places in successive years (Harvey *et al.* 1979). They are sufficiently
faithful to their nesting territory that a bird may be assumed dead when
it is no longer found breeding in the vicinity of its previous nest. Indeed
in this paper we accept that birds which have not been seen for two

LIFETIME REPRODUCTION IN BIRDS
ISBN 0-12-517370-9

successive summers have died. In juveniles, dispersal is greater, the modal distance between birth-place and first nesting place in males being 558 m and in females 879 m (Greenwood et al. 1979). This is of importance in that some juveniles raised in our study area settle to breed outside it, while others may evade capture in some years; hence we underestimate the actual recruitment rates. We do not know what proportion emigrates nor whether this proportion varies between years. Some statistical approaches to estimating these quantities are explored in Clobert et al. (1988).

The study has been carried out in Wytham woods, a 230 ha area of mixed deciduous woodland near Oxford (see Elton 1966 for a general description). Work was started in 1947 by Drs D. Lack and J.A. Gibb, but the data on which this paper are based are from the years 1964–1985, since during the late 1950s and early 1960s the nestbox study area was expanded greatly until it covered more or less the whole wood; about 1,000 nestboxes are currently present (see Minot & Perrins 1986 for details) and some 120 to 350 pairs of Great Tits breed in these boxes each year.

The numbers of Great Tits breeding in Wytham varied markedly between one year and the next, sometimes doubling or halving between successive years. However, over the 40 years of study, the numbers of pairs have fluctuated around the same level of abundance with no sign of any long-term trend. One of the factors associated with the major changes in numbers is the seed (= mast) crop of beech, *Fagus sylvatica*, with tit numbers increasing from one year to the next when there was a beech crop in the intervening winter and decreasing when there was none (Perrins 1966). Crops usually occur at intervals of 1–4 years but heavy crops never occur in two successive years.

Annual adult survival is about 50 %, but varies between years, the extremes being 29 % (1984) and 73 % (1968). In the early years of this study, many birds fell prey to Weasels *Mustela nivalis*. In some years up to 50 % of the nests were lost, and in 20 % of these cases the females were also taken. In the mid-1970s we switched the type of nestbox used to one that Weasels cannot easily enter and the birds now have a higher nesting success. A few breeding adults are taken each year by Sparrowhawks *Accipiter nisus*, causing nests to fail. A very few other nests are deserted, some probably as a result of our disturbance. Over the period 1960–1987, an annual average of 67 % of the nesting pairs raised young and the mean number of chicks raised per pair was 5.0 (including those pairs which failed).

Adult survival does not vary from year to year nearly as much as the recruitment of juveniles. During the study period, as few as 3 % and as many as 21 % of the juveniles fledging in one year were found breeding in the wood the next. Not surprisingly, this variation had a major effect on the changes in numbers, the population increasing when juvenile recruitment was high and decreasing when it was low. In fact the recruitment of juveniles was the "key factor" affecting population change. The underlying cause of these changes in numbers has not been discovered, but happens soon after the young have left their nests and well before the seeds of beech are available. However, subsequent winter mortality is density dependent.

Methods

Nestboxes were checked weekly (more often if required), enabling us to obtain: the number of breeding pairs, the laying date of each pair and their clutch size, the numbers of young which hatched and the number which survived to leave the nest. On the 15th day after hatching, the young were weighed and ringed. During the nestling period the adults were caught while bringing food to their young (i.e. most failed breeders were not caught). This enabled us to obtain measures of adult survival and measures of recruitment of young from the previous year into the breeding population. We believe that all these measures are fairly accurate and that errors are negligible, except that, as mentioned, some of the young left the wood to breed elsewhere; hence recruitment was underestimated, but we see no reason why the figures which we give in this paper should be unrepresentative of any of the particular groups of birds discussed.

Results

In simple terms, two main factors might be expected to affect the number of young which a bird raises in its lifetime, namely the number of times that the bird breeds and the number of young which it raises on each occasion. In the following sections the effects of each of these aspects are dealt with separately and then combined.

NUMBER OF BREEDING ATTEMPTS

Most Great Tits are known to breed when they are one year old. We have used our long run of data to see whether there is any evidence for

Table 3.1 Number of Great Tits breeding at age 2 in relation to whether or not they had been caught breeding at age 1.

Year x	No. caught yr. x	Proportion caught yr. (x−1)	Estimated number yr. x	Actual number yr. x	% Difference (actual–estimated)
Males					
1965	10	0.255	39.2	28	−39.9
1966	7	0.446	15.7	17	7.8
1967	11	0.672	16.4	14	−16.9
1968	9	0.464	19.4	10	−93.8
1969	6	0.406	14.8	16	7.7
1970	9	0.481	18.7	19	1.4
1971	11	0.471	23.3	24	2.7
1972	10	0.523	19.1	17	−12.5
1973	17	0.646	26.3	24	−9.6
1974	11	0.764	14.4	16	10.0
1975	10	0.600	16.7	21	20.6
1976	10	0.652	15.3	17	9.9
1977	22	0.740	29.7	33	10.1
1978	19	0.842	22.6	25	9.7
1979	22	0.783	28.1	24	−17.1
1980	40	0.843	47.4	45	−5.4
1981	28	0.718	39.0	37	−5.5
1982	38	0.813	46.7	44	−6.2
1983	29	0.810	35.8	41	12.8
1984	27	0.738	36.6	45	18.7
1985	15	0.840	17.9	16	−11.6
1986	40	0.860	46.5	52	10.5

Year x	No. caught yr. x	Proportion caught yr. (x−1)	Estimated number yr. x	Actual number yr. x	% Difference (actual–estimated)
Females					
1965	23	0.833	27.6	28	1.3
1966	25	0.842	29.7	31	4.3
1967	19	0.969	19.6	22	10.9
1968	36	0.845	42.6	45	5.4
1969	14	0.877	16.0	19	16.0
1970	25	0.845	29.6	29	−2.0
1971	36	0.742	48.5	46	−5.4
1972	21	0.770	27.3	24	−13.6
1973	33	0.831	39.7	40	0.7
1974	12	0.843	14.2	16	11.1
1975	16	0.760	21.1	19	−10.8
1976	20	0.818	24.4	23	−6.3
1977	35	0.833	42.0	48	12.5

Table 3.1 Continued.

Year x	No. caught yr. x	Proportion caught yr. (x−1)	Estimated number yr. x	Actual number yr. x	% Difference (actual–estimated)
Females					
1978	21	0.901	23.3	26	10.3
1979	33	0.813	40.6	35	−15.9
1980	49	0.907	54.0	54	−0.1
1981	41	0.802	51.1	46	−11.1
1982	45	0.866	52.0	46	−13.0
1983	44	0.856	51.4	47	−9.4
1984	37	0.837	44.2	46	3.9
1985	29	0.873	33.2	30	−10.7
1986	67	0.902	74.3	73	−1.8

The calculation is made as follows. Column 2 shows the number of two-year old birds caught in year x which had also been caught breeding at age 1 (in year x−1). Column 3 shows the proportion of all breeding birds of that sex which were caught in year x−1. If all breeders had been caught in year x−1 then in year x we should have caught (col2 × 1/col3) two-year olds which had bred in year x−1; this number is given in column 4. The actual number of two-year olds caught in year x is given in column 5. If this number was consistently greater than the estimate in column 4, it would indicate that there were a number of birds which were not breeding at age 1. In fact the mean difference between actual and estimated numbers of birds present is −4.4 % for males (+1.9 % if the two very high figures for 1965 and 1968 are excluded) and −1.1 % for females. The difference between −4.4 % and −1.1 % is not significant.

non-breeding at age 1. The two-year old birds breeding in the population were divided into two groups, those that were found breeding at age 1 and those that were not. Since we do not catch every breeding bird, we would expect to encounter a number of birds breeding at age 2 that had been missed at age 1. However, since in each year we know both the number of breeding birds and the number of each sex and age class which we catch, we can estimate the number not caught. If all birds bred at age 1, then in any year the number of two-year olds which had not been found breeding the previous year should not be greater than the number of one-year old birds which were not caught in the previous year. The results of this analysis are presented in Table 3.1. The main assumption in such a calculation is that the age composition of the caught and non-caught sections of the population is the same. Because we did not catch the failed breeders, there might be some scope for error here. However, in recent years we have usually caught a high proportion of the population, so any such error is likely to be small.

Although there is considerable annual variation in these estimates, they provide no evidence that there is, overall, an important proportion of birds (at least of those which survive to breed at age 2) which do not breed at age 1. In some years this may not be the case, but we cannot tell how reliable any individual figure is. As there is little difference between the two sexes, it seems unlikely that many birds fail to breed at age one simply because of a bias in the sex ratio. This result seems to conflict with that of Bulmer & Perrins (1973), who concluded that up to one-third of the one-year old males might be unable to find a mate in this monogamous species. The discrepancy may be due to heavy Weasel predation in the earlier years of the study which, because females were more vulnerable, may have produced a more biased sex ratio than in recent years, but this is not apparent in Table 3.1.

As mentioned, the Great Tit is short-lived, with an annual adult survival rate of somewhat less than 50 %. Fig. 3.1 shows the proportion of males and females of known age which bred between 1960 and 1985 and which

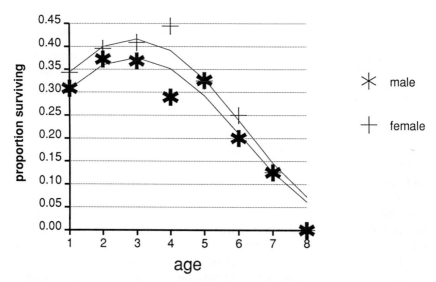

Figure 3.1 Proportion of male and female Great Tits which survived to breed in a subsequent season. The fitted lines are logistic regressions in age and age squared. Both parameters are highly significantly different from zero. The female curve is significantly higher than the male curve because females are more likely to be trapped. For both sexes the proportion re-trapped is somewhat lower than the true survival rate, because not all adults were caught in any one season.

were seen breeding in a subsequent breeding season. The true survivorship is somewhat higher because of failure to trap all the birds, but the relationship with age is probably correct. Both males and females are most likely to survive when aged more than 1 and less than about 4. The curvature of the logistic regressions is highly significant. The significantly higher return rate of females is due to fewer males having been caught in the earlier years of the study.

The data on which Fig. 3.2 is based are drawn from the birds caught breeding in nestboxes. The fact that there might be old birds which are alive, but not breeding (as shown by Dhondt (1985) for Belgium) does not affect the calculations of lifetime reproductive success.

BREEDING PRODUCTIVITY

As has often been reported before (e.g. Kluijver 1951), female Great Tits increase in fecundity after their first breeding season, on average laying about 0.5 eggs more per clutch in subsequent seasons. The increase

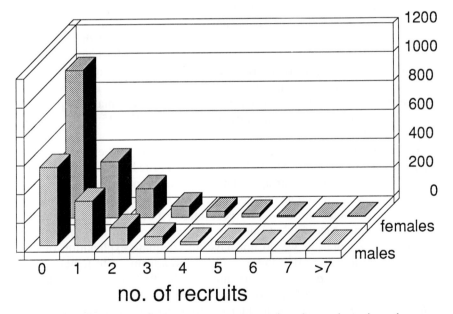

Figure 3.2 Lifetime reproductive success, measured as the total number of young which recruited into the Wytham breeding population, shown separately for males and females. Only birds which were recorded breeding in their first summer, and had no gaps in their breeding records, were included. Most individuals produced no surviving young.

occurs regardless of the age of their mate and is in addition to the fact that pairs breeding together for the second time also lay slightly more eggs on average than newly formed pairs (Perrins & McCleery 1985). The difference in clutch size between first year and older females might be an artefact: if better quality females laid larger clutches and also had better chances of surviving to the following season, then the birds for which a value could be determined at older ages would be of higher average quality than those scored at age 1. However, an analysis (Table 3.2a) of clutch size related to the age of females, using for the one-year age class only females which were also recorded breeding in a subsequent season, shows that the effect of female age is highly significant (F = 21.8 with 1,1232 df, $P < 0.001$), and that the difference in mean clutch size is about 0.4 eggs. There is a slight ($0.01 < P < 0.05$) tendency for the mates of older males to lay more eggs than mates of one-year olds in this subset of the data; but there is no interaction between male age and female age.

Table 3.2b repeats our previous analysis (Perrins & McCleery 1985), using a slightly expanded data set, and confirms the earlier result; the estimate for the difference in clutch size between one year and older females is slightly, but not significantly, larger than that estimated in Table 3.2a, and there is no sign of an effect of the age of the male.

Table 3.2c repeats our earlier analysis (Perrins & McCleery 1985) on recruitment of young in relation to parental age, again using more data, as we can now use information on birds breeding up to 1985 inclusive. As we previously reported, there is a tendency for older males to produce more recruited young independently of female age ($F = 15.6$, df. 1, 1784, $P < 0.01$).

We can conclude that older female Great Tits are no better than one-year olds at raising young which themselves enter the breeding population, even though there is a clear age-related difference in the number of eggs laid. Conversely, at any given age the mates of older males probably lay no more eggs than those of first year males, but older males may be marginally more successful at raising young which recruit into the nesting population. At present we do not have an adequate explanation of this paradox, but it follows that there would be no advantage to females, and little to males, in postponing breeding in the first season in order to improve their chance of surviving to breed when older, on the grounds that by waiting to breed when older the total reproductive success would be greater. Although we cannot test directly whether birds that did not breed at age 1 survived better than those that did, there is currently no evidence that reproductive effort in one season affects an adult's survival to the next (Boyce & Perrins 1987, and see below). Consequently it is

Table 3.2 Effects of parental age on clutch-size and recruitment of Great Tits.

(a) Effects of male and female age on relative clutch size based on 1,234 observations of pairs in the years 1964–1987. In this analysis only females which were recorded breeding beyond their first summer were included in the class of females aged 1.

Parameter	Estimate	SE	F	df	P
—	0.00	0.69	—	—	—
Female age	0.42	0.11	21.81	1,1232	0.001
Male age	0.23	0.14	3.91	1,1231	$0.01 < P < 0.05$
Interaction	−0.07	0.20	0.13	1,1230	ns

(b) Effects of male and female age on relative clutch size using all females from 2,092 pairs in the years 1964–1987.

Parameter	Estimate	SE	F	df	P
—	0.00	0.05	—	—	—
Female age	0.48	0.07	37.59	1,2090	0.001
Male age	0.15	0.10	1.87	1,2089	ns
Interaction	−0.10	0.15	0.05	1,2088	ns

(c) Effects of male and female age (analysed separately) on relative number of recruited young. Based on 1,787 observations of pairs in the years 1964–1984, with recruited survivors recorded up to 1987.

Parameter	Estimate	SE	F	df	P
—	−0.09	0.03	—	—	—
Female age	0.02	0.06	3.5	1,1785	ns
Male age	0.17	0.06	15.6	1,1784	< 0.01
Interaction	0.01	0.09	0.01	1,1783	ns

no surprise to find that, so far as we can tell, the vast majority of Wytham Great Tits start breeding at age 1.

LIFETIME REPRODUCTIVE OUTPUT

Elsewhere (McCleery & Perrins 1988) we have analysed the sources of variance between individual Great Tits in lifetime reproductive success (LRS). We took as our measure of reproductive success the number of offspring which recruited into the Wytham population as breeding birds. This ignores the problem of emigration but, provided that all offspring are equally likely to emigrate, the number recovered breeding will be well correlated with the total reproductive output including emigrants. LRS was taken to be reproductive success summed over the number of attempts to breed, assuming no more than one successful attempt per season in this population. We found that the proportion of fledged individuals surviving to breed was the biggest source of variance in LRS between adults, followed by the longevity of the individual. It is possible moreover that much of the variation in recruitment rate of fledged individuals depends on external factors such as the food supply, outside the control of individual parents. We will examine this possibility further (see below).

From the genetic point of view the success of an individual is well predicted by its success within a single season, provided that there is no relationship between reproductive effort in one season and the probability of survival to the next. In Great Tit nests at Wytham there is, if anything, a positive relationship between the reproductive success in one season and the chance of an individual surviving to breed in a subsequent year. Table 3.3a shows the proportions of females surviving to breed in a subsequent season classified by whether they raised a young which recruited into the breeding population in a subsequent season. Logistic multiple regression (Table 3.3b) shows that successful females are significantly more likely to survive than unsuccessful females, even when the (highly significant) effect of year is taken into account. A successful female is 1.46 times more likely to survive compared with an unsuccessful female; the 95 % confidence interval of this estimate is 1.26–1.71. The lack of a significant interaction term in the model indicates that the difference in survival probability between successful and unsuccessful females is the same in each season, even though the survival rates differ greatly between seasons.

The problem with using correlations between survival and fecundity in natural broods to test for trade-offs between these life history parameters is that some birds have access to the best territories or food supply which might itself have a positive influence on survival and reproduction. Within birds of particular qualities there may indeed be a trade-off between reproductive effort and survival, but because each bird chooses the best combination for itself the overall picture is that the best birds are best both at reproduction and survival (cf. Bell 1984). Boyce & Perrins (1987) analysed all the available data from Wytham on broods which had been manipulated by adding or removing young to see if birds which raised a brood bigger than they "intended" showed a lower survival to the next season. No consistent negative relationship between reproductive effort in one season and survival to the next was found, and hence they

Table 3.3 Survival of females in relation to previous breeding success.

(a) Proportion of breeding females known to have survived to the following breeding season in relation to whether they bred successfully or failed to raise young.

Year	Unsuccessful	Successful
1964	0.38	0.45
1965	0.30	0.41
1966	0.31	0.52
1967	0.50	0.39
1968	0.44	0.63
1969	0.24	0.24
1970	0.31	0.45
1971	0.19	0.40
1972	0.38	0.48
1973	0.33	0.51
1974	0.38	0.49
1975	0.26	0.34
1976	0.37	0.49
1977	0.30	0.22
1978	0.28	0.30
1979	0.35	0.38
1980	0.36	0.30
1981	0.30	0.31
1982	0.40	0.60
1983	0.21	0.22
1984	0.28	0.52

(b) Logistic Multiple Regression (using GLIM) on proportion of females which survived to breed in a subsequent season depending on (a) whether or not they raised one or more young which survived to breed, and (b) the year in which they bred.

Parameter	Deviance	df	Change in deviance	Change in df	P
—	153.8	41	—	—	—
Success	129.31	40	−24.5	−1	<0.01
Year	24.1	20	−105.2	−20	<0.01
Interaction	0	0	−24.1	−20	ns

concluded that there was no evidence of any cost of reproductive effort (see also Pettifor *et al.* 1988).

INDIVIDUAL VARIATION IN LIFETIME REPRODUCTIVE SUCCESS

From 1960 to 1985, a total of 1,773 successfully breeding females produced a total of 1,422 young which survived to breed in the wood. Thus the mean reproductive success of such females was 0.80 young per female. The comparable figures for males were 1,071 individuals producing 958 young, giving a mean success of 0.89. However, 1,023 females and 541 males produced no surviving young, so the 1,422 recruited young were produced by only 750 (42 %) of females and the 958 offspring were produced by only 530 (49 %) of males. The frequency distribution of number of surviving young is shown in Fig. 3.2. Only adults for which we had complete breeding histories, that is those for which we had a record of breeding at age 1, where there were no missing seasons, and where no breeding attempt was subject to a manipulation experiment, were used to calculate these figures. From 1960 to 1985 the population has apparently been stable so we would expect, on average, each bird to produce two viable offspring. The above figures suggest, therefore, that about half the recruited young emigrate from Wytham to breed elsewhere, being replaced by birds recruited from outside the wood. This interpretation is consistent with the fact that in any year about half the breeding birds are individuals not known to have been raised in the wood. The number of these immigrants is positively correlated with that of residents in any year, suggesting that numbers of potential recruits fluctuate in synchrony inside and outside the wood. On the other hand, recoveries of Wytham ringed birds breeding outside the wood are remarkably few and far between in most seasons and, since Wytham is supposed to be optimum habitat, it might be expected that it would be a net exporter of recruits to the surroundings rather than being in balance with them. Further work is needed on this aspect of the population dynamics.

Although only giving a relative measure of success, there are large differences between individuals in the number of young recruited into the Wytham population. We have already shown (above) that the number of recruits produced in a season is independent of age and there is no survivorship penalty for successful individuals, so it follows that longevity should be an important determinant of the total recruited young. This is apparent from Table 3.4. Because the distribution of recruited young per individual parent is so strongly skewed (Fig 3.2), we felt it desirable to fit a statistical model with Poisson rather than normally distributed errors.

Table 3.4 Effect of breeding lifespan and year of first breeding on number of young recruited to local breeding population.

The model: log recruited young/no. of parents = log (breeding season + year) was fitted to reproductive success data for 1964–1985. The change in deviance is distributed as chi-squared.

Parameter	Deviance	df	Change in deviance	df	P
Breeding lifespan	958.4	118	−118.1	1	< 0.001
Year first breed	364.7	93	593.6	25	< 0.001
Interaction	229.3	68	135.4	25	< 0.01

As expected, the number of seasons in which an individual bred had a highly significant effect on the number of recruited young per parent. Fitting a factor for the year of first breeding produced a highly significant improvement in fit of the model, indicating that the number of recruited young per parent differed, depending on which year the individual parent started as a breeder. The interaction term between year of first breeding and number of breeding attempts (i.e. lifespan) is also highly significant, showing that the relationship between number of breeding attempts and numbers of recruited young is different for adults starting breeding in different years.

Another question of interest is whether some birds are consistently successful at raising young. If they are, it could indicate either that those birds were in some way of better "quality" than others, or that they had settled in places which were particularly good for raising young (since most birds remain in the same place throughout their lives, settling in a good place would result in a tendency for such birds to have high nesting success each year). Table 3.5 shows an analysis of the repeatability of success for individual birds, in other words will a bird which raises young which recruit into the breeding population in one year be more likely than average to do so again the following year? Using contingency analysis we found no such relationship, suggesting that differences between years in factors unrelated to the individual birds are having a major effect on success in any particular year.

One such factor which obviously differs between years is presence or absence of beech mast. This effect is likely to be indirect, insofar as the main mortality of young, which determines the overall recruitment from each cohort, occurs in the summer, before the beech crop is available to the birds. However, it turned out that the presence or absence of a beech crop in the first year of breeding did have a marked effect on the total

Table 3.5 The probability that individuals will show consistent breeding success (i.e. young recruited) in different years. Based on chi-square averages of tendency to repeat reproductive success in birds which bred in at least two years.

		Recruits first year				
		0	1	2	3	4
Recruits	0	804	306	151	65	18
second	1	331	141	55	25	11
year	2	112	62	22	10	5
	3	49	26	9	6	0
	4	24	10	9	2	0

Chi-square $= 15.67$, 16 df, $P = > 0.05$.

number of young that were recruited to the breeding population by each parent. Birds that bred for the first time in a beech year had a larger LRS than those that did not, independently of lifespan (Fig. 3.3, Table 3.6). Again the interaction term is significant; the slope of the relationship between number of attempts and total surviving young per parent is slightly but significantly shallower when breeding started in a beech year than when it did not. This is probably explained by the fact that the year after a beech crop was always a non-beech year whereas a non-beech year was often (though not always) followed by a beech year. Thus for birds starting to breed in a beech year, most recruits were produced in the first season. Individuals surviving to breed a second time probably produced few recruits in their second season, so the difference in LRS for birds with two attempts and those with only one would be small for this group. For birds starting to breed in a non-beech year, none would produce many recruits in their first season, but those surviving to the next (often beech) year produced a good number of recruits, so the difference in LRS score between birds breeding in two seasons and those breeding in only one would be larger in this group than for the corresponding group starting in a beech year; hence the significant interaction term in the statistical model. An additional complication is that adult survival was also better in winters with a beech crop (Perrins & McCleery, in prep).

Determinants of Success

We wondered whether the body size of adults might affect their breeding success, particularly since both Jones (1973) and Garnett (1981) have

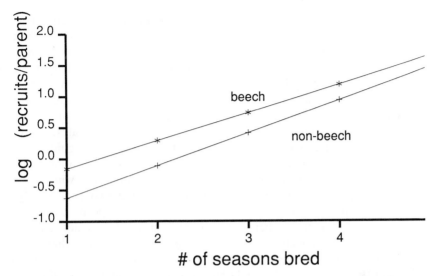

# of seasons bred

Figure 3.3 Effect of lifespan on number of recruits produced per parent for birds whose first season was either a beech or a non-beech year. Birds starting to breed in a beech year (top line) produced more young, on average, than birds starting in a non-beech year (bottom line), but for birds starting in a non-beech year the effect of each extra breeding season was more marked: the slopes of the two lines differ significantly from one another (Table 3.4).

Table 3.6 Effect on breeding lifespan and presence of a beech mast crop on the number of young recruited to the local breeding population.

(a) The model: log recruits/parents = log breeding season + beech score was fitted to reproductive success data for 1964–1985.

Parameter	Deviance	df	Change in deviance	Change in df	P
Breeding lifespan	958.4	118	−118.1	1	< 0.001
Beech in first season	829.2	117	−129.2	1	< 0.001
Interaction	822.1	116	−7.02	1	< 0.01

(b) Coefficients for beech and lifespan for model fitted as Table 3.6a. A coefficient differs significantly from zero if it is more than twice its standard error.

Parameter	Estimate	SE
	−1.155	0.039
Beech in first season	0.548	0.066
Lifespan (no beech in first year)	0.521	0.016
Lifespan (beech in first year)	0.447	0.023

shown a tendency for lighter females to nest earlier, and earlier nests produced most young. In recent years, most of the individual adults caught while breeding have had their wing length measured as an index of body size. We looked for a correlation between wing length and breeding success. Wing length increased slightly but significantly between the first and subsequent breeding seasons (Table 3.7a). However, in birds for which we have measurements in both the first and subsequent seasons, there is a high correlation between the values (0.9 for females, 0.7 for males), so the first year wing measurement is a reasonable guide to an individual's size. We compared "successful" (at least one recruited young) and "unsuccessful" (no recruited young) birds for first year wing length (Table 3.7b). In neither males nor females do the two groups differ statistically, though curiously the mean values go in opposite directions, with successful females being larger and successful males smaller. Consequently, we conclude that there is no evidence that either larger or smaller birds are noticeably more successful; if the results in Table 3.7b had been significant, it would have raised some interesting difficulties in interpreting the sexual dimorphism of the species.

Table 3.7 Wing length, age and breeding success in Great Tits.

(a) Difference in wing length (mm) of Great Tits between first and subsequent breeding seasons.

Sex	Age 1			Age > 1			t	P
	Mean	SD	n	Mean	SD	n		
Male	75.80	2.01	331	76.87	1.91	327	7.00	< 0.001
Female	71.96	3.19	398	72.99	3.11	304	−4.30	< 0.001

(b) Difference in wing length between successful and unsuccessful first year Great Tits.

Sex	Unsuccessful			Successful			t	P
	Mean	SD	n	Mean	SD	n		
Male	75.93	1.85	164	75.79	1.82	322	0.65	ns
Female	72.51	2.74	180	72.75	1.54	344	1.64	ns

Conclusions

The Great Tit is short-lived, with rather fewer than half the birds that survive to breed once surviving to breed a second time. Therefore one might expect little selection for complicated patterns of variation in reproductive output in relation to age. There is, however, some evidence that, once Great Tits have reached five years of age, their reproductive output diminishes (Perrins & Moss 1974, Dhondt 1989).

From the analysis presented above it appears that much of the variation in individual lifetime success is due to events which are outside the control of the individual and is consequently outside the scope of natural selection. Thus the random event of which year the individual starts to breed has much the largest effect on LRS, and seems to be explained, at least to some extent, by unknown factors associated with the beech crop. Probably because of the magnitude of these effects, inherent individual variations in reproductive success and survivorship are difficult to detect, if they exist. From the evolutionary point of view, the important thing is the relative success of individuals exposed to the same environmental conditions. An individual which does well when other individuals do badly obtains a big increase in relative fitness.

The largest single effect on lifetime reproduction is the number of years the individual lives. We can find no characteristics of the individuals which have a clear effect on either their reproductive output or their chances of survival. Rather it seems that the number of recruits is, in large part, a chance affair even though clutch size is an inherited feature (van Noordwijk 1980) and doubtless reproductive output is selected for. Similarly, we know no way of predicting at the outset which birds have the best chance of surviving for several years. However, this is an important part of the picture, since the longer a bird lives the more breeding attempts it has and the greater its chances of producing young which recruit into the breeding population.

Finally, it is perhaps worth speculating on how general these conclusions might be for Great Tits: is our study area or the techniques we use likely to affect the generality of our conclusions? The most obvious way in which we have altered the natural situation is by the provision of nestboxes. These certainly increase the number of nesting sites available and the readiness with which Great Tits accept them implies that there is a great shortage of natural sites as good as the boxes. However, British woodlands now have far fewer dead trees with holes than would have been found in primaeval Britain. The few reports available suggest that semi-natural woodland may have large numbers of sites available (though to what extent they are of comparable quality we cannot say). An

experiment in one area of Wytham, in which the nestboxes were removed
(East & Perrins 1988), showed that present-day Wytham had few suitable
natural nesting sites for Great Tits; hence one must conclude that we
have greatly increased the number of pairs which nest. The other factor
which affects the reproductive output of the birds is their nesting success.
As mentioned, in some years of the early part of the study, many birds
were taken by Weasels, a situation which probably exists in natural sites
where these predators can often gain easy entry. The current boxes
preclude this, thereby increasing nesting success and probably female
survival. However the number of breeding pairs has not increased as a
result of this improvement in nesting success, so presumably the density
dependent mortality which takes place in winter is acting more severely
than it did in the past. Since we have no evidence that adult survival
rates have changed and since the population has remained stable, we
presume that post-fledging mortality of juveniles has increased.

On parts of the Continent, for reasons unknown, many more of the
Great Tits raise second broods. As a result of this, densities must increase
more between spring and summer than in England and hence even more
birds (per pair or unit area) must die there between one nesting season
and the next. Since adult survival rates are comparable and the populations
are stable, juvenile mortality must be higher.

We conclude, therefore, that the structure of the breeding population
and in particular, the survival rates of adults and the numbers of recruited
young each adult produces, may not differ much between populations in
spite of the large variations in nesting success.

References

Bell, G. 1984. Measuring the cost of reproduction. I. The correlation structure
    of the life table of a plankton rotifer. *Evolution* **38**: 300–13.
Boyce, M.S. & Perrins, C.M. 1987. Optimizing Great Tit clutch-size in a
    fluctuating environment. *Ecology* **68**: 142–53.
Bulmer, M.G. & Perrins, C.M. 1973. Mortality in the Great Tit *Parus major*.
    Ibis **115**: 277–81.
Clobert, J., Perrins, C.M., McCleery, R.H. & Gosler, A.C. 1988. Survival rate
    in the Great Tit *Parus major* in relation to sex, age and immigration status.
    J. Anim. Ecol. **57**: 287–306.
Dhondt, A.A. 1989. The effect of old age on the reproduction of Great Tits
    Parus major and Blue Tits *P. caeruleus*. *Ibis* **131**: 268–280.
Dhondt, A. 1985. Do old Great Tits forego breeding? *Auk* **102**: 870–2.
East M. L. & Perrins, C.M. 1988. The effect of nestboxes on breeding populations
    of birds in broadleaved temperate woodland. *Ibis* **130**: 393–401.
Elton, C.S. 1966. *The Pattern of Animal Communities*. London: Methuen.

Garnett, M.C. 1981. Body size, its heritability, and influence on juvenile survival among Great Tits *Parus major*. *Ibis* **123**: 31–41.

Greenwood, P.J., Harvey, P.H. & Perrins, C.M. 1979. The role of dispersal in the Great Tit (*Parus major*): the causes, consequences and heritability of natal dispersal. *J. Anim. Ecol.* **48**: 123–42.

Harvey, P.H., Greenwood, P.J. & Perrins, C.M. 1979. Breeding area fidelity of Great Tits (*Parus major*). *J. Anim. Ecol.* **48**: 305–13.

Jones, P.J. 1973. Some aspects of the feeding ecology of the Great Tit *Parus major* L.D. Phil. thesis, Oxford.

Kluijver, H.N. 1951. The population ecology of the Great Tit, *Parus m. major* L. *Ardea* **39**: 1–135.

Lack, D. 1966. *Population Studies of Birds*. Oxford: Clarendon Press.

McCleery, R.H. & Perrins, C.M. 1988. Life-time reproductive success of the Great Tit. In *Reproductive Success*, ed. T.C. Clutton-Brock, pp. 136–53. Chicago: University Press.

Minot, E.O. & Perrins, C.M. 1986. Interspecific interference competition nest-sites for Blue and Great Tits. *J. Anim. Ecol.* **55**: 331–50.

Noordwijk, A.J. van, Balen, J.H. van & Scharloo, W. 1980. Heritability of ecologically important traits in the Great Tit. *Ardea* **68**: 193–203.

Perrins, C.M. 1965. Population fluctuations and clutch size in the Great Tit *Parus major*. *J. Anim. Ecol.* **34**: 601–647.

Perrins, C.M. 1966. The effect of beech crops on Great Tit populations and movements. *Brit. Birds* **59**: 419–432.

Perrins, C.M. 1979. *British Tits*. London: Collins.

Perrins, C.M. & McCleery, R.H. 1985. The effect of age and pair bond on the breeding success of Great Tits *Parus major*. *Ibis* **127**: 305–15.

Perrins, C.M. & Moss, D. 1974. Survival of young Great Tits in relation to age of female parent. *Ibis* **116**: 220–4.

Pettifor, R.A., Perrins, C.M. & McCleery, R.H. 1988. Individual optimization of clutch size. *Nature* **336**: 160–162.

4. Pied Flycatcher

HELMUT STERNBERG

This study of Pied Flycatchers *Ficedula hypoleuca* breeding in nestboxes
was made in the eastern part of Lower Saxony near Wolfsburg (52°,31′
N, 10°,54′ E) in the Federal Republic of Germany, from 1964 to 1986.
The analysis is based on 2,251 known-age breeding birds ringed as
nestlings before 1979 but retrapped up to 1986, with males and females
treated separately. As the birds live up to eight years, all young fledged
after 1978 were excluded so as not to bias the sample in favour of short-
lived individuals. Data on egg-laying dates, clutch sizes, numbers of
hatched and fledged young, numbers of offspring breeding, numbers of
nesting attempts, ages of first breeding and longevity are presented.

The data were obtained from eight study areas, totalling about 70 ha,
and all lying within an area of 5 × 5 km. Most of the woodland was
broad-leaved deciduous, but some was coniferous or mixed. More than
1,000 nestboxes were provided, as few natural tree holes were available.
Breeding density in areas with nestboxes, at 3–10 pairs per ha, was more
than 10 times greater, on average, than that in areas without boxes.

The Pied Flycatcher is sexually dichromatic, as many adult males are
a striking black-and-white, while females are brown-and-white. Male
colour varies, however, and some individuals are as dull as females. The
mean colour of the male population varies geographically, with mainly
grey or brown males in central Europe, and darker ones in England,
Fenno-scandia and Switzerland. On average, young males are more
female-like than older ones. The species is relatively small, as individuals
of both sexes weigh around 12 g.

Pied Flycatchers are specialized insect feeders, and raise their young
on both adults and larvae. They are migratory, breeding in woodland
over much of the western Palearctic and wintering in Africa south of the
Sahara, from Upper Guinea to the Congo region. Birds return to their
breeding places in central Europe from late April, males a few days

LIFETIME REPRODUCTION IN BIRDS
ISBN 0-12-517370-9

earlier than females, and adults earlier than yearlings. After arriving, the males search for suitable breeding holes, defend them against rivals and advertise their holes to potential mates. The pair bond lasts for only one breeding season at a time. Females build the nests, usually lay 5–7 eggs and incubate alone for about 13 days. First-year females lay smaller clutches than older ones. Clutch size is influenced by the date of laying; late clutches are small and produce few fledglings. Both parents feed the young in the nestling period of about 15–17 days, and the family keeps contact for about a further week in which they leave the nesting area. Birds depart at the end of June for their winter quarters. Some males mate with two females in a year (successive polygyny), while some first-year and older males and females fail to breed. Breeding females raise no more than one brood in a season, but repeat clutches are normal (Berndt & Sternberg 1972). In an average year more than 80 % of nests produce fledglings.

About half of all the one-year old males and females, whose return was proved, bred at less than 1 km from their respective birthplaces (Berndt & Sternberg 1966, 1969b). Shortage of available breeding sites probably excludes a proportion of birds, especially first-years, from breeding. But the relatively greater proportions of first-year birds in newly occupied than in previously colonized areas show that both sexes can breed when one-year old if sufficient nest holes exist (Sternberg 1972). Older males show much more marked site fidelity than older females, about 10 % of which move as far between successive breeding sites as young birds move between birthplace and first breeding place (Berndt & Sternberg 1965, 1968).

Each year in our study areas, all breeding females were trapped during incubation and most of the males when they were feeding nestlings. Polygyny was recorded directly in 3–5 % of the breeding males each year, involving a roughly similar percentage of females. Such females were classed as primary or secondary depending on which was first to lay.

I define breeding lifespan as the number of breeding seasons in which an individual (or its mate) produced at least one clutch of eggs. Individuals which failed to produce eggs (females) or young (males) are considered not to have become breeders.

While breeding, female flycatchers in our study areas were occasionally preyed upon by Pine Martens *Martes martes*, young were occasionally taken by Great-spotted Woodpeckers *Dendrocopus major*, and clutches were destroyed by Fieldmice *Sylvaemus flavicollis*, which occupied the nestboxes. Losses from predation differed between years and between study areas, but were generally small.

Results

AGE DISTRIBUTION OF THE POPULATION

The analysis in Table 4.1 is based on the age distribution of 953 males and 1,298 females, which were found breeding for the first and subsequent times in the study areas. Birds caught in more than one year figured repeatedly in the records, so that survival rates could be calculated for breeders of all age-classes.

The number of birds first found breeding in their first, second, third, fourth or fifth year respectively was known (underlined in the table), as was the number found breeding in each succeeding year. The number of pre-breeders (non-breeders up to first breeding or death) in the preceding years was estimated by extrapolating from these figures backwards (after Newton 1985), assuming that the survival rate (S) was similar for breeding and pre-breeding birds, and that pre-breeding males and females were as faithful to the study areas as were breeders.

The 363 males which first bred as one-year olds produced in their lifetimes a total of 673 broods containing 3,304 fledglings. These birds produced 39 % of all fledglings. The 408 males which first bred at two-years old (728 attempts with 3,740 fledglings) produced 44 % of all fledglings, while the 182 males which first bred at 3–5 years produced 280 broods, containing only 17 % of all fledglings.

On average, females started breeding at an earlier age than males (Tables 4.1 and 4.2). The 777 individuals that started in their first year produced 1,536 broods containing 6,547 fledglings, or 59 % of the total. The 408 females which first bred at two years had 796 attempts producing 3,692 fledglings, or 34 % of the total; while the 113 females that first bred at 3–5 years old produced only 174 broods, containing 7 % of all fledglings. Only 40 % of all males and 60 % of all females, which started breeding in their first year, produced any fledglings in their lifetimes.

The number of birds recruited into a stable breeding population per unit time will exactly equal the number of adults which are lost. On this basis, the mean annual mortality (M) of adults can be computed at $2,113/4,160 = 0.508$ for males and at $2,070/4,184 = 0.495$ for females. In other words, in a stable population about half of adult flycatchers die each year.

However, annual survival was not constant through life, but declined progressively in each successive age class from about 53 % between ages 1 and 2 to nil after age 8, with no difference between the sexes (Table 4.1). A Gompertz equation gave the best fit to the data, which helped

Table 4.1 Numbers of Pied Flycatchers alive in the following years after birth (breeders and pre-breeders), numbers of fledglings per cohort and annual survival for each age-class. Underlined figures show birds breeding for the first time, and other figures show numbers of these same individuals breeding in later years. The numbers of pre-breeders were estimated (see text).

Year of first breeding	Numbers alive in the following years after birth — Age in years								Totals	Fledglings produced (% of total)
	1	2	3	4	5	6	7	8		
Males										
1	_363_	158	83	45	19	5	–	–	673	3304 (39)
2		_408_	183	84	35	11	6	1	728	3740 (44)
3			_132_	54	20	4	–	–	210	1062 (12)
4				_42_	11	5	1	–	59	345 (4)
5					_8_	3	–	–	11	52 (1)
Nos breeding	363	566	398	225	93	28	7	1	1,681 = 40.4 %	
Nos pre-breeding	1,750	571	137	21	–	–	–	–	2,479 = 59.6 %	
Nos total	2,113	1,137	535	246	93	28	7	1	4,160 = 100.0 %	
Survival rate (S)	0.538	0.471	0.460	0.378	0.301	0.250	0.143	0.000		
Females										
1	_777_	390	204	100	42	19	3	1	1,536	6547 (59)
2		_408_	212	105	51	14	5	1	796	3692 (34)
3			_84_	31	12	6	1	–	134	619 (6)
4				_25_	5	3	2	–	35	134 (1)
5					_4_	1	–	–	5	26 (0)
Nos breeding	777	798	500	261	114	43	11	2	2,506 = 59.9 %	
Nos pre-breeding	1,293	302	73	10	–	–	–	–	1,678 = 40.1 %	
Nos total	2,070	1,100	573	271	114	43	11	2	4,184 = 100.0 %	
Survival rate (S)	0.531	0.521	0.473	0.421	0.377	0.256	0.182	0.000		

Note: Estimated mean annual mortality for all age groups together was 0.508 for males and 0.495 for females.

Table 4.2 Age at first breeding of 2,251 Pied Flycatches.

	Age in years					Mean age (±SD)
	1	2	3	4	5	
Males (n = 953)						
Number	363	408	132	42	8	1.87 ± 0.87
Proportion of first breeders	0.381	0.428	0.139	0.044	0.008	
Females (n = 1,298)						
Number	777	408	84	25	4	1.51 ± 0.73
Proportion of first breeders	0.506	0.363	0.096	0.030	0.005	

The age distribution differed between males/females: $\chi^2_4 = 116.52$, $P < 0.001$.

to confirm that survival after first breeding was not independent of age, but declined progressively through life (Table 4.3).

AGE OF PRE-BREEDERS

As Pied Flycatchers first breed at 1–5 years old (see above) pre-breeders in both sexes may be found up to four years; however, the age distribution of the sexes is different, and males tend to have a longer pre-breeding period than females (Table 4.4). On my calculations, in any one year, about 60 % of all males and 40 % of all females in the population are pre-breeders. Some 71 % of all males and 77 % of all females are one year old, but 83 % of the one-year old males and 62 % of the one year old females are pre-breeders.

LIFETIME PRODUCTION

For 953 males and 1,298 females I was able to calculate the total number of young reared to fledging during their lives, on the assumption that all breeding attempts occurred within the study areas (cf. p. 56). The number of young produced by these birds varied greatly. Some 2 % of males which attempted to breed produced no young during their lives, and, of the 98 % that did produce young, the number varied between one and 37 in different individuals. The corresponding figures for females were 8 % and 92 %, while the number of young varied between one and 36. The peak at 5–6 fledglings for males and females is explained partly by the fact that these were the most common brood sizes and most breeding

Table 4.3 Expected frequency versus Gompertz equation of total population (breeding and pre-breeding part).

Age (years)	1	2	3	4	5	6	7	8	Total 1-8	χ^2/1-8 df 7
Expected frequency										
by table (males)	2,113	1,137	535	246	93	28	7	1	4,160	0.59
by Gombertz (males)	2,113	1,137	560	246	92	28	6	1	4,183	
by table (females)	2,070	1,100	573	271	114	43	11	2	4,184	0.12
by Gombertz (females)	2,070	1,100	563	271	117	43	11	2	4,177	

The age distribution of males/females (comparisons were made using two row χ^2-test), by table males v females: $\chi^2_7 = 10.02$, $P < 0.001$; by Gombertz equation, males v females: $\chi^2_7 = 10.23$, $P < 0.001$.

Table 4.4 Age distribution of 3,043 pre-breeding Pied Flycatchers.

	Age in years				Mean age
	1	*2*	*3*	*4*	
Males (*n* = 2,479) Number	1,750	571	137	21	1.37 ± 0.63
Proportion of pre-breeders	0.706	0.230	0.055	0.009	–
Proportion of corresponding age-class	0.828	0.502	0.256	0.085	–
Females (*n* = 1,678) Number	1,293	302	73	10	1.28 ± 0.57
Proportion of pre-breeders	0.770	0.180	0.044	0.006	–
Proportion of corresponding age-class	0.625	0.275	0.127	0.037	–

The age distribution of males/females (comparison was made by two row χ^2-test): $\chi_3^2 = 21.38$, $P < 0.001$.

birds bred only once in their lifetimes. Half of all males and females produced 4–7 young. Relatively few males and females produced only 1–3 young, because such small broods were rare.

A very small percentage of males and females were responsible for most of the fledgling production: 26.7 % of the most productive breeding males and 24.3 % of the most productive females accounted for more than 50 % of the fledglings in the population. The arithmetic mean lifetime productions for males ($x = 8.92$, SD = 5.71) and females ($x = 8.49$, SD = 5.74) were about the same, and the median for both sexes was six fledglings.

ROLE OF LIFESPAN

The main source of individual variation in lifetime production was longevity (Fig. 4.1). Lifetime productions of up to 10 young (with two females) were recorded for males which survived for only one breeding season (eight young for females), productions up to 22 young were recorded for males which survived for two seasons (with two females), (15 for females); all productions greater than 22 young (15 for females) were from birds which survived more than two years. Productions greater

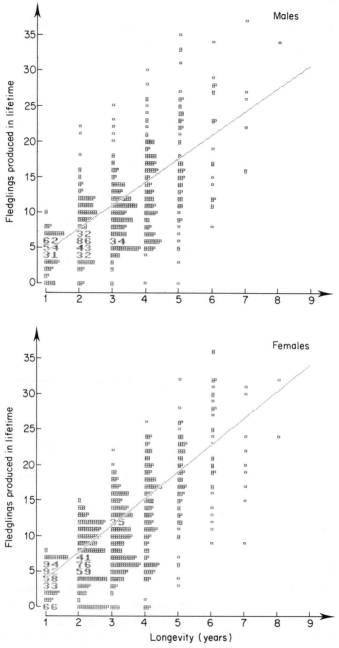

Figure 4.1 Lifetime fledgling productions of males and females in relation to longevity. Each point represents a different individual, except for large numbers where the number is given. Statistics: males: $n = 953$, $r = 0.68$, $\hat{t} = 28.34$, $P < 0.001$, $y = 2.96x + 1.12$; females; $n = 1{,}298$, $r = 0.75$, $\hat{t} = 40.21$, $P < 0.001$, $y = 3.38x + 0.21$.

than 25 young were from birds older than three years. The maximum (37) was raised by a male which lived for seven years, while 36 young were produced by two females which lived for six years. On the other hand, one male which survived for six years produced only eight young and one female which survived for seven years produced only nine young. Within birds of any one age group, the variation in productivity was great, and some individuals up to six years old raised no more young than did other individuals which lived only one year.

As Pied Flycatcher females raised up to eight young in a brood, individuals which lived for eight years could in theory produce 64 young in their lifetime. Only certain females which lived for one year produced the maximum possible for their age, and with increasing age of female, the difference between observed maximum and potential production widened (Table 4.5). Conversely, the chances of a breeding bird producing no young during its lifetime declined with increasing age (Fig. 4.1). But not until the sixth year of life had all males (fifth year for females) in the sample produced young.

Table 4.5 Potential lifetime production versus actual maximum production in relation to longevity for females.

	Longevity in years							
	1	*2*	*3*	*4*	*5*	*6*	*7*	*8*
Numbers of:								
Potential fledglings	8	16	24	32	40	48	56	64
Maximum observed fledglings	8	15	22	26	32	36	31	32
Proportions	1.00	0.94	0.92	0.81	0.80	0.75	0.55	0.50

ROLE OF AGE OF FIRST BREEDING

After longevity, the most important demographic factor influencing lifetime fledgling production was age of first breeding. In both sexes the number of fledglings produced at the first attempt was lower for birds that first bred at one-year old than for birds that first bred at later ages (Table 4.6). However, the later in life a bird started to breed, the lower its subsequent breeding lifespan, and the lower the number of fledglings and recruits it produced. In addition, significant differences between males and females were found for all age-classes in the number of fledglings produced at the first attempt ($P < 0.001$), and in the number of breeding offspring produced per lifetime in the one-year group

Table 4.6 Mean lifespan as a breeder and mean lifetime offspring production among 2,251 Pied Flycatchers of known age at first breeding (± SD).

	Ages at first breeding						
	1 year		2 years		3–5 years	
Males						
Number	363		408		182	
Lifespan as breeder (years)	1.35 ± 1.23	ns	1.28 ± 1.11	**	1.04 ± 0.80	
Fledglings at first attempt	4.99 ± 1.44	***	5.48 ± 1.70	ns	5.62 ± 1.86	
Fledglings in lifetime	9.10 ± 6.36	ns	9.17 ± 5.68	**	8.02 ± 4.04	
Breeding offspring (males)	0.43 ± 0.95	ns	0.38 ± 0.80	ns	0.31 ± 0.72	
Breeding offspring (females)	0.49 ± 0.96	ns	0.43 ± 0.82	ns	0.40 ± 0.68	
Proportion with at least one offspring	(0.48)		(0.47)		(0.46)	
Females						
Number	777		408		113	
Lifespan as breeder (years)	1.48 ± 1.28	ns	1.45 ± 1.18	***	1.04 ± 0.92	
Fledglings at first attempt	4.12 ± 1.44	***	4.85 ± 1.34	ns	4.75 ± 1.40	
Fledglings in lifetime	8.43 ± 6.01	ns	9.05 ± 5.56	***	6.89 ± 3.98	
Breeding offspring (males)	0.34 ± 0.79	ns	0.40 ± 0.83	ns	0.27 ± 0.60	
Breeding offspring (females)	0.39 ± 0.85	ns	0.45 ± 0.84	**	0.20 ± 0.51	
Proportion with at least one offspring	(0.41)		(0.47)		(0.33)	

Note: Significance values for differences between means of 1, 2 and 3–5 year olds:
** $P < 0.01$; *** $P < 0.001$; ns = not significant.

($P < 0.01$) and 3–5 years group ($P < 0.01$). A sex difference in breeding lifespan was evident in birds which started at two years old ($P < 0.05$).

ROLE OF POLYGYNY

Of 953 males studied throughout their lives, 60 (6.3 %) were recorded as polygynous in only one year and another three (0.3 %) as polygynous in two years. Polygyny was slightly more frequent among older males than among young ones (Table 4.7), but no male was recorded with more than two females in a year. Males which practiced polygyny produced more fledglings, both in annual attempts and in their lifetimes than did monogamous males (Table 4.7). In fact, the lifetime mean fledgling production was about twice as great for polygynous males (at 18.25 ± 7.63 SD) as for monogamous males (at 8.26 ± 5.10 SD). Among the females, those known to be mated to polygynous males were slightly younger, on average, than other females (Table 4.8). The annual production of primary females was slightly higher, in general, than that of secondary females, which received less support from the male.

OTHER FACTORS INFLUENCING LIFETIME PRODUCTION

Birds which had been raised as nestlings in the study areas, and later retrapped there as breeders, were classed as "residents". Breeders which were ringed as nestlings outside and first recaptured in the study areas, were classed as "immigrants". Relatively more of the resident females than the immigrants produced young. The difference in productivity was statistically significant in females which bred more than once. No differences were apparent between females which bred only once or between both groups of males.

For 1,214 females hatched in the study areas, I knew the date of hatching within the season. No correlation was apparent between birth date of female and subsequent lifetime production, nor between birth date and subsequent longevity. Only the mean lifetime production of those females which were born later than 9 June was significantly reduced compared to those born before this date.

Although the individual young in small broods survived better than those in large broods, the breeding adults which produced the largest broods also produced the largest number of subsequent breeders.

INDIVIDUAL CONTRIBUTIONS TO FUTURE GENERATIONS

The following calculations are made on the assumptions of an equal sex ratio and a stable population, using the method of Newton (1985). From

Table 4.7 Breeding performance of monogamous and polygynous males.

(a) Monogamy and polygyny in relation to age.

	Age in years								Mean age (± SD)
	1	2	3	4	5	6	7	8	
Monogamous	355	537	388	210	89	28	7	1	2.54 ± 1.26
Polygynous (%)	8(2.2)	29(5.1)	10(2.5)	15(6.7)	4(4.3)	0(0)	0(0)	0(0)	2.67 ± 1.13

In 60 males polygyny was recorded in only one year and in three other males in two years. The age distribution of monogamous and polygynous males did not differ significantly: $\chi_7^2 = 13.10$, $P > 0.05$.

(b) Annual fledgling production by monogamous and polygynous males.

	Number of males which produced the following numbers of fledglings																Mean number of fledglings (± SD)
	0	1	2	3	4	5	6	7	8	9	10	11	12	13	14	15	
Monogamous	138	10	37	95	202	391	524	185	31	2	0	0	0	0	0	0	4.85 ± 1.94
Polygynous	0	0	1	1	0	1	0	2	8	10	12	14	12	1	3	1	10.09 ± 2.29

Annual productions of monogamous and polygynous males differed significantly: $\chi_{15}^2 = 1{,}337.29$, $P < 0.001$.

(c) Lifetime fledgling production by monogamous and polygynous males.

	Number of males which produced the following numbers of fledglings					Mean number of fledglings ($\pm$ SD)
	0–10	11–20	21–30	31+		
Monogamous	653	208	27	2		8.26 $\pm$ 5.10
Polygynous (%)	12(1.8)	27(11.5)	19(41.3)	5(71.4)		18.25 $\pm$ 7.63

Lifetime productions of monogamous and polygynous males differed significantly: $\chi^2_{34} = 268.66$, $P<0.001$.

Table 4.8 Breeding performance of females of different mating status. Females of "unknown status" were mostly monogamous.

(a) Status in relation to age

	Age in years								Mean age (± SD)
	1	2	3	4	5	6	7	8	
Unknown status	748	772	487	257	112	42	10	2	2.34 ± 1.28
Primary female	14	14	9	2	1	0	0	0	2.05 ± 1.01
Secondary female	15	12	4	2	1	1	1	0	2.14 ± 1.48

The age distribution of primary and secondary females did not differ significantly: $\chi^2_7 = 3.91$, $P > 0.7$.

(b) Annual fledgling production

	Number of females which produced the following numbers of fledglings									Mean number of fledglings (± SD)
	0	1	2	3	4	5	6	7	8	
Unknown status	387	37	67	129	272	530	641	278	33	4.36 ± 2.32
Primary female	0	0	1	2	3	15	28	14	3	5.83 ± 1.14
Secondary female	2	0	6	10	15	22	8	3	0	4.26 ± 1.48

Annual productions of primary and secondary females differed significantly: $\chi^2_7 = 41.46$, $P < 0.001$.

(c) Lifetime fledgling production

	Number of females which produced the following numbers of fledglings				Mean number of fledglings (± SD)
	0–10	11–20	21–30	31+	
Unknown status	854	296	63	9	8.41 ± 6.30
Primary female	23	17	0	0	9.68 ± 4.44
Secondary female	19	17	0	0	9.83 ± 5.65

Lifetime productions of unknown (mainly monogamous) and primary females did not differ significantly: $\chi^2_{36} = 30.97$, $P > 0.5$; and nor did productions of primary and secondary females: $\chi^2_{20} = 20.25$, $P > 0.3$.

the data on survival and age of first breeding (Table 4.1), an estimated 77.6 % of all males and 76.4 % of all females which left the nest died before they could start breeding in their first to fifth year of life. At least another 0.4 % of males and 2 % of females attempted to breed, but produced no young in their lifetimes. From that batch of male fledglings which became first-year breeders, 8.3 % produced young in their lifetime. The corresponding figures for males which subsequently bred in their second, third, fourth or fifth years were 9.5 %, 3.1 %, 1.0 % and 0.1 %, respectively giving an overall figure of 22.0 % for males. The corresponding figures for females which first bred in their first, second, third, fourth or fifth years, were 12.9 %, 6.9 %, 1.4 %, 0.4 % and 0 % respectively, giving an overall figure of 21.6 %. In other words, less than one fourth of fledglings in any one Pied Flycatcher generation contributed fledglings to the next generation. These breeders produced enough young to replace not only themselves, but all the non-contributing individuals too.

The males which attempted to breed (22.0 % of those fledged) produced an average of 8.9 young, and the females which attempted to breed (21.6 % of those fledged) produced an average of 8.5 young. If those birds which died before they could breed were included, the mean number of young produced to first breeding per bird was (8.9 × 22.4 % =) 1.99 for males, and (8.5 × 23.6 % =) 2.01 for females, figures very close to the two young per pair needed to maintain population stability in the long term.

RECRUITMENT TO BREEDING POPULATION

The number of young raised to fledging was taken as a measure of reproductive output. In order to confirm the validity of this measure, the relationship between the number of fledglings and the number of subsequent recruits to the local breeding population was examined (Fig. 4.2).

Of the 8,503 fledglings produced by 953 males in the sample, only 793 (369 males and 424 females) were later found breeding in the study areas. Of the 11,018 fledglings produced by 1,298 females in the sample, only 971 (461 males and 510 females) were later found breeding in the study areas. These figures imply that 9.3 % of fledgling males and 8.8 % of fledgling females subsequently became breeders in the study areas (with others settling elsewhere). However, the numbers of local recruits derived from individual parents were significantly correlated with the number of fledglings they had produced (including birds which produced no young to fledging: $r = 0.38$ for males $n = 953$, $P < 0.001$; $r = 0.48$ for females $n = 1,298$, $P < 0.001$). The relationship was linear (Fig. 4.2), indicating

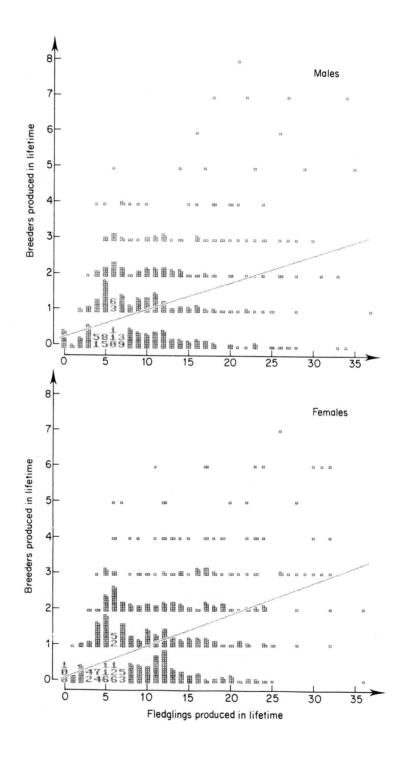

that the number of fledglings produced was a reasonable predictor of contribution to future breeding populations.

Discussion

The idea that mortality rates among adult birds in nature are fairly constant and independent of age, first introduced by Nice (1937), has been almost universally accepted (Lack 1943a,b, 1966, von Haartman 1971, Ricklefs 1973), although alternative models of age-dependent mortality were discussed by Botkin & Miller (1974). Average annual mortality rates reported for many species result in exceedingly long potential lifespans; and it seems more reasonable to assume that avian mortality is age-dependent, increasing rapidly in old age (see Chapter 26). The first report of age-dependent mortality in female Pied Flycatchers (Berndt & Sternberg 1963) was discussed by Lack (1966), von Haartman (1971) and Ricklefs (1973); while data for several other species show that the notion of age-independent mortality in birds is no longer universally true (Coulson & Wooller 1976, Hildén 1978, Perrins 1979, Curio 1983, Rattiste & Lilleleht 1987, and several chapters in this book).

The large proportion of pre-breeding individuals in the Pied Flycatcher, which may also be widespread in other hole-nesters, results mainly from shortage of nest sites. Adults return earlier than yearlings to their breeding areas, so have a better chance to find an empty hole. In a Finnish population of Pied Flycatchers, von Haartman (1982) estimated the proportion of unmated males on different methods at about 60 % (the same as my estimate) and mentioned a considerable proportion of non-breeding females (40 % on my estimate). Once they have bred, Pied Flycatchers normally breed annually until their death, but a few birds of both sexes missed a year. This occurred at all ages, but the possibility of the observer missing a breeding attempt, especially if it failed early, was high.

Matings between adults and adults and between yearlings and yearlings were more than expected by chance, while matings between yearlings and adults were less frequent (Berndt & Sternberg 1971). Such selective mating might have occurred "passively", through adults taking up nest

---

Figure 4.2 Numbers of offspring recruited to the breeding population in relation to lifetime fledgling productions of male (father) and female (mother). Each point represents a different male or female parent or the number of individuals is given. Statistics: males: $n = 953$, $r = 0.38$, $t = 12.71$, $P < 0.001$, $y = 0.08x + 0.15$; females: $n = 1,298$, $r = 0.48$, $t = 13.65$, $P < 0.001$, $y = 0.09x + 0.01$.

holes before yearlings and thus tending to pair with other adults, and through yearlings often occupying nest holes later and hence pairing with other yearlings.

It was not only survival and breeding opportunity which changed with age. Comparing different age groups, laying dates (of the females) and clutch sizes improved during the first two years of life, while the mean number of young raised per clutch continued to increase, at least to the third year (Sternberg 1980). Beyond this age, production deteriorated (4–8 year-old birds). Hence, in this species senility shows both in decreased survival and in decreased breeding success, but as few individuals lived longer than three years, the data were sparse.

Acknowledgements

I am grateful to Dr R. Berndt, P. Dancker, H. Duchrow, U. Hennig, M. Henss, H.-H. Männich, U.-I. Rahne, E. Schramm, R. Sternberg, E. Strauch, H.-J. Ulrich, H.-J. Wisniewski and B. Zeitzschel for their help during the fieldwork, and to H. Hötker, Dr I. Newton and Dr J. Wittenberg for helpful comments on the manuscript.

References

Berndt, R. & Sternberg, H. 1963. Ist die Mortalitätsrate adulter *Ficedula hypoleuca* wirklich unabhängig vom Lebensalter? *Proc. Int. Orn. Congr.* **13**: 675–84.

Berndt, R & Sternberg, H. 1965. Schematische Darstellung der Ansiedlungs-Formen bei weiblichen Trauerschnäppern *Ficedula hypoleuca*. *J. Orn.* **106**: 285–94.

Berndt, R. & Sternberg, H. 1966. Der Brutort der einjährigen weiblichen Trauerschnäpper *Ficedula hypoleuca* in seiner Lage zum Geburtsort. *J. Orn.* **107**: 292–309.

Berndt, R. & Sternberg, H. 1968. Terms, studies and experiments on the problems of bird dispersion. *Ibis* **110**: 256–69.

Berndt, R. & Sternberg H. 1969a. Uber Begriffe, Ursachen und Auswirkungen der Dispersion bei Vögeln. *Vogelwelt* **90**: 41–53.

Berndt, R. & Sternberg, H. 1969b. Alters- und Geschlechtsunterschiede in der Dispersion des Trauerschnäppers *Ficedula hypoleuca*. *J. Orn.* **110**: 22–6.

Berndt, R. & Sternberg, H. 1971. Paarbildung und Partneralter beim Trauer-schnäpper. *Vogelwarte* **26**: 136–42.

Berndt, R. & Sternberg, H. 1972. Über Ort, Zeit und Grösse von Ersatzbruten beim Trauerschnäpper, *Ficedula hypoleuca*. *Beitr. Vogelkd.*, **18**: 3–18.

Botkin, D.B. & Miller, R.S. 1974. Mortality rates and survival of birds. *Amer. Nat.* **108**: 181–192.

Bulmer, M.G. & Perrins, C.M. 1973. Mortality in the Great Tit *Parus major. Ibis* **115**: 277–81.

Cody, M.D. 1971. Ecological aspects of reproduction. In *Avian Biology*, Vol. I, ed. D.S. Farner & J.R. King, pp. 461–512. London: Academic Press.

Coulson, J.C. 1984. The population dynamics of the Eider Duck *Somateria mollissima* and evidence of extensive non-breeding by adult ducks. *Ibis* **126**: 525–43.

Coulson, J.C. & Wooller, R.D. 1976. Differential survival rates among breeding Kittiwake Gulls *Rissa tridactyla* (L.). *J. Anim. Ecol.* **45**: 205–13.

Curio, E. 1983. Why do young birds produce less well? *Ibis* **125**: 400–4.

Curio, E. & Regelmann, K. 1982. Fortpflanzungswert und 'Brutwert' der Kohlmeise. *J. Orn.* **123**: 237–57.

Fisher, R.A. 1958. *The Genetical Theory of Natural Selection.* New York: Dover Publications.

Fitzpatrick, J.W. & Woolfenden, G.E. 1988. Components of lifetime reproductive success in the Florida Scrub Jay. In *Reproductive Success*, ed. T.H. Clutton-Brock, pp. 305–20. Chicago: University Press.

Haartman, L. von 1971. Population dynamics. In *Avian Biology*, Vol. I, ed. D.S. Farner & J.R. King, pp. 391–459. London: Academic Press.

Haartman, L. von 1982. The biological significance of the nuptial plumage of the male Pied Flycatcher. *Proc. Int. Orn. Congr.* **18**: 34–60.

Haldane, J.B.S. 1955. The calculation of mortality rates from ringing data. *Proc. Int. Orn. Congr.* **11**: 454–8.

Hildén, O. 1978. Population dynamics in Temminck's Stint *Calidris temminckii. Oikos* **30**: 17–28.

Hötker, H. 1988. Lifetime reproductive output of male and female Meadow Pipits *Anthus pratensis. J. Anim. Ecol.* **57**: 109–17.

Lack, D. 1943a. The age of the Blackbird. *Brit. Birds* **36**: 166–72.

Lack, D. 1943b. The age of some more British birds. *Brit. Birds* **36**: 193–7, 214–21.

Lack, D. 1966. *Population Studies of Birds.* Oxford: Clarendon Press.

Newton, I. 1985. Lifetime reproductive output of female Sparrowhawks. *J. Anim. Ecol.* **54**: 241–253.

Newton, I. 1986. *The Sparrowhawk.* Calton: Poyser.

Nice, M.M. 1937. Studies in the life history of the song sparrow. *Trans. Linn. Soc. New York* **4**: 1–247.

Nice, M.M. 1957. Nesting success in altricial birds. *Auk* **74**: 305–21.

Perrins, C.M. 1979. *British Tits.* London: Collins.

Rattiste, K. & Lilleleht, V. 1987. Population ecology of the Common Gull *Larus canus* in Estonia. *Ornis Fenn.* **64**: 25–6.

Ricklefs, R.E. 1973. Fecundity, mortality, and avian demography. In *Breeding Biology of Birds*, ed. D.S. Farner, pp. 366–435. Washington: Nat. Acad. Sci.

Stearns, S.C. 1976. Life-history tactics: a review of the ideas. *Quart. Rev. Biol.* **51**, 3–47.

Sternberg, H. 1964. Untersuchungen über die Farbenzugehörigkeit der männlichen Trauerschnäpper, *Ficedula hypoleuca*, im Schweizerischen Mittelland. *Orn. Beob.* **61**: 90–4.

Sternberg, H. 1972: The origin and age composition of newly formed populations of Pied Flycatchers *Ficedula hypoleuca. Proc. Int. Orn. Congr.* **15**: 690–1.
Sternberg, H. 1980. Influence of age of mates on reproduction in the Pied Flycatcher *Ficedula hypoleuca. Proc. Int. Orn. Congr.* **17**: 1403. Berlin.
Sternberg, H. 1986. Age-dependent mortality rate and survival in the Pied Flycatcher. *Int. Ornith. Congr.* **19** (Abstracts): 456.
Williams, G.C. 1966. Natural selection, the costs of reproduction and a refinement of Lack's principle. *Amer. Nat.* **100**: 687–90.
Wooller, R.D. & Coulson, J.C. 1977. Factors affecting the age of first breeding of the Kittiwake *Rissa tridactyla. Ibis* **119**: 339–49.

5. Collared Flycatcher

LARS GUSTAFSSON

In this chapter I describe the effects of specific life-history traits on the lifetime reproductive success (LRS) of individual Collared Flycatchers *Ficedula albicollis*. I also discuss ultimate and proximate determinants of LRS, as well as some differences in LRS between generations. Finally, I discuss the merits of lifetime measures over annual measures of survival and reproduction, taking account of trade-offs between different fitness components. Throughout, LRS is taken as the number of recruits which enter the breeding population of the study area.

The Collared Flycatcher is an Old World flycatcher, which breeds in Europe and winters in Africa. It is considered to be an allospecies of the Pied Flycatcher *F. hypoleuca* (Alerstam *et al.* 1978). The breeding ranges of the two species overlap in central and eastern Europe and on the Baltic islands of Öland and Gotland (Wallin 1986). Sympatric populations hybridize to some extent (Alerstam *et al.* 1978, Alatalo *et al.* 1982). On the island of Gotland, where this study was made, the total populations of Collared and Pied Flycatchers have recently been estimated at 4,000 and 500 pairs respectively. The nearest Collared Flycatcher population is found about 600 km southeast of Gotland, while the Pied Flycatcher breeds abundantly on the mainland of northern, central and eastern Europe. Collared Flycatchers thus predominate in this peripheral isolated area within the range of the Pied Flycatcher.

Collared Flycatchers are sexually dimorphic in plumage, the males being black on the crown and mantle with contrasting white on forehead, neck and underparts, while the females have brownish-grey upperparts, sometimes with a little greyish-white on the nape. Males show slightly longer wing length and tail length, but there is no dimorphism in tarsus or beak. Both sexes average about 13 g.

Collared Flycatchers are easy to study because they use tree holes as nest sites. By providing nestboxes, one can study almost the whole

LIFETIME REPRODUCTION IN BIRDS
ISBN 0-12-517370-9

population because these birds seldom use natural cavities where boxes are available. The Great Tit *Parus major* and Blue Tit *P. caeruleus* are the main competitors for nestboxes on Gotland (Gustafsson 1988). They also feed their young with similar food; namely insects, including a large proportion of caterpillars (Török 1986). However, the main feature rendering the Collared Flycatcher population on Gotland especially suitable for study is the high fidelity to the island shown by returning adults (Pärt & Gustafsson 1989) and juveniles (Gustafsson 1986). Accurate data on adult survival and juvenile recruitment can thus be obtained, making it possible to calculate lifetime reproductive success with a precision seldom possible in other species and areas. Perhaps because these birds breed on an isolated island, the selection pressure to return to their natal area is unusually strong, for otherwise they would end up on the mainland where only Pied Flycatchers are known to breed. The penalty for such wide dispersal would be non-breeding or hybridization.

The first males arrive on Gotland in the first week of May and the latest probably during the first week of June. The first females arrive, on average, one week after the earliest males. Nest building starts in mid-May and the first eggs are laid around 20 May so that most clutches hatch during the first half of June. Broods contain up to eight young, which normally fledge on the 14th or 15th day after hatching. Most have left the nest by 5 July and the latest around 15 July. Only one clutch is laid, except for occasional replacement clutches. There was no predation on nest contents in the study areas, but a few clutches were destroyed each year by Wrynecks *Jynx torquilla*. Young and parent flycatchers leave for their winter quarters at the end of August.

The mating system in the Collared Flycatcher is much the same as in the Pied Flycatcher. Males first acquire a female in one territory (= the primary female), then many males (about 50 %, varies with arrival time and number of empty nestboxes) occupy a second territory and attempt to attract another female (= the secondary female). Whether successful or not, the males then return to their first females. Males provide less food to the young of secondary females, and as a result, some of the young in secondary nests starve. A further complication is that extra-pair copulations seem to be rather common. Alatalo *et al.* (1984a) estimated the frequency of multiple paternity in Collared and Pied Flycatcher broods to be approximately 25 %. A new estimate, using the same method, on a larger sample of Collared Flycatchers gave a frequency of 15–20 % (Alatalo *et al.* 1989a). All estimates of LRS given below are based on the assumption that extra-pair copulations were evenly distributed in the population, so that all individuals had the same probability of experiencing fitness losses or gains due to such behaviour. In the Pied Flycatcher the opportunity for extra-pair copulations seemed not to differ between polygynous and monogamous males, and nor did the risk of becoming cuckolded (Alatalo *et al.* 1987).

Study area and methods

The study area was situated on a narrow peninsula in the south of Gotland (57°10′N, 18°20′E), where a total of 800 nestboxes were erected on a grid system in nine woodlands. Eight deciduous woods were dominated by oak *Quercus robur* and ash *Fraxinus excelsior*, with a dense understorey of hazel *Corylus avellana* and hawthorn *Crategus* spp. The one area in coniferous forest was dominated by pine *Pinus sylvestris*, with some birch *Betula pubescens*. Nestbox densities varied between 1.5 ha^{-1} and 15 ha^{-1}. Collared Flycatchers reached breeding densities of up to seven females ha^{-1} in areas with a surplus of boxes.

For each nest I recorded laying date, clutch size, hatching date, number of hatched young and fledged young. All young were ringed in the nest and weights and tarsus lengths were measured at 13 days of age. Almost all breeding adults were captured, measured and ringed, usually while feeding their nestlings. In nests where all young died at an early stage the male was not always caught. Most of the birds were of known age, having been ringed in previous years, while others were aged on criteria described by Alatalo *et al.* (1984b) for females and by Svensson (1984) for males.

The study started in 1980 and is still running (1989). Adult annual mortality was about 50 % and the oldest bird known to be alive was seven years. For calculation of LRS, I used only birds for which I have complete records, and only from the first five cohorts (1980–1985). In 1987 less than 1 % of the birds in the first four cohorts were still alive and less than 5 % from the fifth cohort. Birds could be categorized into cohorts if they were either ringed as nestlings in my area or were immigrants breeding as yearlings. Locally born individuals that were absent during their first breeding season were assumed to have had zero reproduction that year. Likewise, individuals, both local and immigrants, that were absent during a subsequent year were assumed to have zero reproduction then.

An unknown fraction of young dispersed outside the study area to breed. Thus, the values of LRS given here considered only local recruitment. If emigration equalled immigration, about 20 % of breeding birds were emigrants. Comparative tests involving values of LRS can be correct only assuming that there is no bias in the fraction of dispersing young between compared groups. Factors affecting fledging weight can be ruled out as causes of error, since fledging weight did not correlate with dispersal distance (Gustafsson 1987).

Results

LIFETIME REPRODUCTIVE SUCCESS

There was a considerable variability in LRS, as measured by recruit production (Table 5.1). About 50 % of all birds failed to produce any local recruits and the rest produced from one to seven. On average, breeding males had a higher LRS than breeding females (0.97 ± 1.22, $n = 566$ and 0.76 ± 0.99, $n = 423$, respectively, $t = 2.90$, $P < 0.01$). Relatively more males than females produced more than two recruits and relatively fewer males than females had zero LRS (Table 5.1). This difference between the sexes was significant ($G = 13.95$, df $= 7$, $P = 0.05$). Average lifespan, in years, did not differ between males (1.74 ± 0.72, $n = 573$) and females (1.73 ± 0.75, $n = 428$), and the greater frequency of high LRS among males was due to some individuals being polygynous.

The figure for males may have been biased on the high side for two reasons: (a) more males than females remained uncaptured at nests where the young died soon after hatching, (b) some males remained unmated. If these males were first year birds which died before the next breeding

Table 5.1 Lifetime reproductive success (LRS) of male and female Collared Flycatchers, as shown by the total number of offspring recruited to local breeding populations.

LRS	Males		Females	
	Number	Per cent	Number	Per cent
0	260	45.9	222	52.5
1	168	29.7	114	27.0
2	79	14.0	61	14.4
3	31	5.5	19	4.5
4	17	3.0	5	1.2
5	7	1.2	0	0
6	2	0.4	0	0
7	2	0.4	0	0

season, they would not enter the calculations and consequently the proportion of males of breeding age with zero LRS would have been underestimated.

PROXIMATE FACTORS

In order to explore the relative importance of different factors in influencing LRS, I used an unbalanced ANOVA to assess the proportion (r^2) of variance in LRS that each variable could account for (Table 5.2). Life-history traits, such as lifespan, typically had high r^2 values, while morphological features had low values.

The cohort into which an individual was born was also of some significance for individual LRS (Table 5.2). LRS varied significantly between cohorts in both sexes (Table 5.3). This was largely because reproductive success differed between years (Table 5.3), and cohorts, by definition, bred partly in different years. However, cohorts also differed in some other characteristic, probably produced during development and

Table 5.2 Proportion of LRS accounted for (r^2) by different traits in male and female Collared Flycatchers, and standardized selection differentials (S). The latter demonstrate lifetime selection on particular traits using LRS as the fitness measure.

Trait	r^2		S	
	males	*females*	*males*	*females*
Lifespan	0.25	0.16	0.69	0.52
Fledged young*	0.10	0.17	0.30	0.45
Clutch size*	0.08	0.09	0.09	0.23
Weight of young*	0.04	0.09	0.21	0.22
Time of laying*	0.02	0.05	-0.016^{ns}	0.28
Mating status*	0.06	0.09	0.22	-0.20
Cohort	0.02	0.04		
Weight*	0.03^{ns}	0.02^{ns}	0.02^{ns}	0.03^{ns}
Wing length*	0.01^{ns}	0.00^{ns}	0.02^{ns}	0.04^{ns}
Tarsus length*	0.00^{ns}	0.02^{ns}	0.00^{ns}	-0.06^{ns}

*The variable value used was the mean of annual values, except for number of breeding seasons, which is summed as the total to approximate life span, and cohort, which is absolute. All values are significant ($P < 0.05$) except where otherwise noted (ns). Sample sizes were 543 for males and 463 for females. Standardized selection differential is a measure of the difference in mean before and after selection in units of phenotypic standard deviations. In other words, if, for example, S is positive and large, individuals with a large value of that particular trait contributed more offspring to future generations.

Table 5.3 Mean LRS and mean adult survival of males and females in five cohorts classed by the year they were one year old. To demonstrate year effects on reproduction, the average number of recruits produced per breeding pair in different years is also given at the bottom.

Cohort/Year	1980	1981	1982	1983	1984	1985	1986	P
Males								
LRS	1.08	0.82	0.89	1.07	0.99			< 0.05[a]
Survival	0.46	0.47	0.43	0.39	0.46			ns
n (males)	145	140	89	67	125			
Females								
LRS	0.88	0.96	0.73	1.27	0.67			< 0.05[b]
Survival	0.46	0.43	0.40	0.43	0.42			ns
n (females)	124	91	59	41	108			
Recruits	0.58	0.41	0.19	0.64	0.45	0.55	0.33	< 0.0001[c]
n[d]	140	198	335	290	347	338	320	

[a] Kruskall–Wallis test, $\chi^2 = 10.94$, df = 4
[b] Kruskall–Wallis test, $\chi^2 = 9.50$, df = 4
[c] Analysis of variance, F = 15.21
[d] Number of breeding pairs in each year

growth, which influenced their subsequent reproductive performance. For instance, cohorts differed significantly in morphology (Gustafsson in prep.). To separate the effects of year and phenotypic cohort differences, I had to take into account the fact that reproductive success may also have been age-dependent. I therefore made analyses of covariance for cohort differences on the data in Table 5.4, with year and age as covariates. This procedure compensated for differences between years and age, when comparing differences between cohorts. For both male and female the effect of cohort on reproductive success remained, even after the effect of year and age were removed (analysis of covariance, males; F = 3.25, $P < 0.01$, females; F = 5.48, $P < 0.0001$). There were no differences in the annual survival of males or females between cohorts (Table 5.3), so the phenotypic differences established during growth may have been associated with cohort differences in reproductive success (see later).

The polygynous mating system affected LRS of both females and males. Polygynous males did better than monogamous males, while monogamous and primary females did better than secondary females (Table 5.5). A complication with polygyny in the Collared Flycatcher concerns effects which persist beyond one generation. Male offspring from secondary nests of polygynous males were less likely to become polygynous themselves

Table 5.4 Average number of recruits produced in different years by individuals from the five different cohorts for which LRS was estimated. Values for years in which the number of surviving individuals was less than five are not presented.

Cohort	Year 1980	1981	1982	1983	1984	1985	1986
Males							
1980	0.73	0.64	0.31	0.50			
1981		0.32	0.27	0.92	0.73	0.90	
1982			0.23	0.74	0.58	0.92	
1983				0.73	0.68	0.81	
1984					0.56	0.65	0.60
Females							
1980	0.61	0.38	0.13	0.64	0.50		
1981		0.44	0.28	0.52	0.50		
1982			0.14	0.73	0.71		
1983				0.75	0.79	0.87	
1984					0.37	0.63	0.17

Table 5.5 Mean LRS for males that were always monogamous and those that, at least once, were polygynous, and for females that, at least once, were secondary and those that were always monogamous or primary.

	Males Monogamous	Polygynous	Females Monogamous or Primary	Secondary
LRS	0.88	2.12	0.80	0.57
SD	1.18	1.82	0.99	0.89
n	524	42	311	107
P[a]	< 0.0001		< 0.05	

[a]Mann–Whitney U-test
There are fewer polygynous males than secondary females in the sample, because in some secondary polygynous nests the male was not caught. This means that some polygynous males were wrongly classed as monogamous males.

than were male offspring from monogamous nests (Gustafsson MS). This will to a small degree reduce the long-term fitness differential between monogamous and polygamous males. Similarly, female offspring from secondary nests run a high risk of becoming secondary themselves. Altogether this will increase the fitness difference between monogamous and secondary females over that calculated from the numbers of recruits alone.

I have data on two more factors affecting reproductive success, namely hybridization and interspecific competition. A small ($<$ 5 %) fraction of Collared Flycatchers on Gotland ended up in mixed pairs with Pied Flycatchers (Alatalo *et al.* 1982, MSa). They produced young which, until the fledging stage, survived as well as young from pure pairs. However, the hybrid offspring were often sterile or had reduced fertility (Alatalo *et al.* 1982, 1990).

The role of interspecific competition for food with Great and Blue Tits during the breeding season was tested by the experimental manipulation of densities (Table 5.6). It was impractical to run an experiment over the entire lifespan of an individual, so the fitness measure used was on the annual survival of adults (S_a) and number of recruits produced (n), that is, fitness equals $S_a + n/2$. Fitness was lower for Collared Flycatchers breeding alongside tits, both in areas with high and low flycatcher and tit densities, compared with experimental areas where tit densities were reduced. Intraspecific densities probably also influenced fitness in Collared Flycatchers, but the experiment was not designed to test for this.

Following the definitions of Fretwell (1972), in situations with unmanipulated densities, Collared Flycatchers seemed to be "despotically distrib-

Table 5.6 Individual fitness values for Collared Flycatchers breeding in experimental (E) and control (C) plots with low (L) and high (H) densities (from Gustafsson 1987).

	Fitness*			t-test (one-tailed)	
	$\bar{X}$	SD	n	t	P
1982					
E_L	0.729	0.375	24	3.012	< 0.005
C_L	0.375	0.389	24		
E_H	0.579	0.426	20	2.981	< 0.005
C_H	0.167	0.236	12		
1983					
E_L	1.100	0.519	32	4.529	< 0.005
C_L	0.547	0.439	32		
E_H	1.062	0.414	14	2.078	< 0.025
C_H	0.710	0.449	14		

*Fitness (w) is the combined genetic value of recruited young from one year and the average of the added survival to the next year of adult male and female at each nest (w=($S_{male}+S_{female}$)/2 + $n/2$). n values represent number of nests (pairs) rather than number of individuals.

uted" among nest boxes and "ideally free distributed" over habitats (but see Alatalo *et al.* 1985). Good habitats, such as dense deciduous woodland, had much higher densities than poor habitats, such as coniferous forest (Gustafsson 1988), yet breeding success was much the same in different habitat types (Table 5.7). On the other hand, within habitats there was considerable variation among individuals, as some birds acquired good territories and nestboxes, and others poor ones (cf. Askenmo 1984). Thus, owing to density regulation, habitat probably had little or no influence on LRS.

Table 5.7 Average annual reproductive success and adult survival in a deciduous wood (high density, about 5 ha⁻¹) and an adjacent coniferous forest (low density, about 1.5 ha⁻¹). Data are from 1980–1983.

	Mean	SD	n	P*
Fledged young				
Coniferous	4.79	1.56	84	
Deciduous	4.66	1.89	252	> 0.6
Recruits				
Coniferous	0.73	0.95	83	
Deciduous	0.61	0.82	251	> 0.3
Female survival				
Coniferous	0.37	0.49	81	
Deciduous	0.48	0.50	249	> 0.1
Male survival				
Coniferous	0.48	0.50	81	
Deciduous	0.46	0.50	239	> 0.8

*t-test.

ULTIMATE FACTORS

In this section I will analyse the selection pressures that promote high LRS. First, the additive effect of genetic variation had little or no impact on LRS. A new analysis of parent–offspring resemblance in LRS, with a larger sample than that used in Gustafsson (1986), gave similar results. Thus, the heritability estimate in LRS for males was $h^2 = 0.024 \pm 0.079$, $n = 167$, $P > 0.7$; and for females $h^2 = 0.041 \pm 0.098$, $n = 127$, $P > 0.6$. The environmental effects on LRS must therefore have been paramount.

How, then, do environmental pressures interact with individual life histories to shape different phenotypes with different fitnesses? Weight at fledging is mainly an environmentally determined trait in the Collared

Flycatcher ($h^2 = 0.072 \pm 0.082$, $n = 173$, $P > 0.5$, Alatalo et al. 1989b). If an individual's weight at fledging affects its LRS, parents should be selected to optimize the weight of fledglings, as well as their number. In males, weight at fledging was positively associated with the subsequent LRS (Table 5.8). However, the number of young in the brood was negatively associated with subsequent LRS. This conclusion still held when I calculated standardized selection gradients (Lande & Arnold 1983) to separate the effects of selection on correlated characters. Presumably, therefore, parents were strongly selected to have both large broods and heavy fledglings (Table 5.2).

In females the pattern was different and less clear than in males. Females showed a trend of increasing LRS with increasing clutch (and brood) size of the mother, but weight at fledging was unrelated to LRS (Table 5.8). This difference between the sexes might be of some significance, since females inherit clutch size from their mothers ($h^2 = 0.32$, Gustafsson 1986) and clutch size is under strong directional selection (Table 5.2). Females might face opposing selection pressures on clutch size, resulting from different fitness optima for female and male offspring. Female offspring from a large clutch had a fitness advantage through the

Table 5.8 Individual LRS in relation to natal brood size and weight at fledging. For females, parental clutch size is presented instead of number of siblings for reasons explained in text. Standardized selection differentials $(S)^*$ and gradients $(\beta)^{**}$ on these traits are also given. Significance levels are from partial regression coefficients in the multiple regression analysis.

LRS	0	1	2	3	4	≥ 5	S	β	P
Males									
Natal brood size	4.81	4.59	4.27	4.00	4.20	4.00	−0.55	−0.21	< 0.01
Weight	14.7	14.6	14.9	14.9	15.0	16.0	0.10	0.21	< 0.01
n	59	53	22	9	5	3			
Females									
Natal brood size	6.08	6.19	6.20	6.38	5.00	6.50	0.11	0.15	> 0.05
Weight	14.7	14.7	14.7	14.4	14.7	15.3	0.01	0.03	> 0.05
n	47	31	25	8	12				

*See Table 5.2 for explanation.
**The direct and indirect effects of selection can be partitioned with multivariate statistics if the correlated characters that are actually under selection are included in the data. The partial regression of relative fitness on a character measures the direct force of selection on that character and is known as the selection gradient.

effect of heritability of clutch size, but a disadvantage in terms of reduced growth and fledging weight, due to the negative correlation between fledging weight and clutch size ($r = -0.11$, $n = 352$, $P < 0.05$). The outcome was that brood size of mothers was not tightly correlated with LRS of female offspring.

Trade-offs in life history evolution mean that improvements in certain fitness-related traits are associated with decreases in other such traits. Increased reproductive effort may increase the number of young that survive until independence at one breeding attempt, but may reduce the number of future young or the longevity of the parents (Sibly & Calow 1986). These types of trade-offs have important consequences for studies of LRS, because without such trade-offs, we can use measurements of annual recruitment and adult survival as good equivalents of LRS.

What evidence for the existence of trade-offs can we find in the Collared Flycatcher? Most fitness components examined in this species were positively correlated, except clutch size with fledging weight (Table 5.9). This indicates that no different reproductive strategies are likely to exist in this species as a result of opposing selection on related fitness traits. All correlations with laying date were also negative; in other words, to lay late was always bad.

Moreover, there were no trade-offs between present and future reproduction. For both number of fledged young and number of recruited

Table 5.9 Pearson correlation coefficients for fitness components. A negative correlation with Laying Date means that the particular trait is positively correlated with early laying.

	LF	FL	CL	LD	FW
Males ($n = 566$)					
LRS	0.55***	0.24***	0.07	−0.12**	0.17***
Lifespan (LF)		0.09*	0.04	−0.11**	0.13***
Fledglings (FL)†			0.31***	−0.24***	0.26***
Clutch size (CL)†				−0.38***	−0.06
Laying date (LD)†					−0.19***
Fledging weight (FW)†					—
Females ($n = 423$)					
LRS	0.40***	0.35***	0.18***	−0.17***	0.14***
LF		0.08	0.09	−0.11**	0.03
FL*			0.28***	−0.30***	0.20***
CL*				−0.47***	−0.05
LD*					−0.18***

†Mean annual values.
*$P < 0.05$, **$P < 0.01$, ***$P < 0.001$.

young, there was mainly a positive but insignificant dependence of future reproduction on present reproduction in regression analysis (females (n = 360): fledglings, b = 0.07, recruits b = 0.05, males (n = 350): fledglings, b = 0.49, $P < 0.001$, recruits b = 0.00). Phenotypic correlations cannot be used as evidence for the lack of a cost of reproduction or of trade-offs between fitness components (Reznik 1985), but they do show that individuals do not use different reproductive strategies with equal fitnesses. These results hold for unmanipulated populations. However, brood manipulation experiments show several trade-offs between fitness components, between present and future reproduction and between generations (Gustafsson 1985, Gustafsson & Sutherland 1988).

Conclusions

The proximate causes of variation in LRS are very similar in the Collared Flycatcher to those in other small passerines (McCleery & Perrins 1988, Smith 1988, van Balen et al. 1987). In general, all fecundity-related traits are positively selected, since there are no trade-offs between fitness components. Also polygyny, hybridization and inter- and intraspecific competition affect reproduction and LRS. Choice of habitat seems not to affect reproduction, possibly because density is much lower in poorer habitat. Over and above the effects of year and age on reproductive success, cohort per se seemed to affect LRS. The mechanism was probably through phenotypic differences developed during growth, as cohorts differed in morphology, and at least in males, the weight at fledging affected subsequent LRS. Heritability of LRS was close to zero. Between generation effects also had consequences for the evolution of clutch size. Over the study period (1980–1987) clutch size was positively selected through LRS. However, LRS of male offspring was reduced when parental brood size was large, because weights at fledging were then reduced. On the other hand, parental clutch size did not relate to LRS of female offspring, presumably because of opposing selection pressures. Females inherit clutch size from their mother (positive effect), and clutch size is negatively correlated with female offspring fledging weights (negative effect).

   Since, in unmanipulated situations, there are no trade-offs between major fitness components and no trade-off between present and future reproduction, the combination of annual adult survival and recruitment of offspring provides a good measure of fitness in the Collared Flycatcher.

Acknowledgements

I thank all people who helped me with field work, especially G. Carle'n, M. Ekvall, R. Korona, M. Linde'n, D. Nordling and T. Pärt. T. R. Birkhead, A. Lundberg, J. Moreno, I. Newton, T. Pärt and S. Ulfstrand kindly commented on an earlier version of the manuscript. Financial support was obtained from the Swedish Natural Science Research Council.

References

Alatalo, R.V., Gustafsson, L. & Lundberg, A. 1982. Hybridization and breeding success of Collared and Pied Flycatchers on the island of Gotland. *Auk* **99**: 285–91.

Alatalo, R.V., Gustafsson, L. & Lundberg, A. 1984a. Why do young passerine birds have shorter wings than older birds? *Ibis* **126**: 410–15.

Alatalo, R.V., Gustafsson, L. & Lundberg, A. 1984b. High frequency of cuckoldry in Pied and Collared Flycatchers. *Oikos* **42**: 41–7.

Alatalo, R.V., Lundberg, A. & Ulfstrand, S. 1985. Habitat selection in the Pied Flycatcher *Ficedula hypoleuca*. In *Habitat Selection in Birds*, ed. M.L. Cody. London: Academic Press.

Alatalo, R.V., Gottlander, K. & Lundberg, A. 1987. Extra-pair copulations and mate guarding in the polyterritorial Pied Flycatcher, *Ficedula hypoleuca*. *Behaviour* **101**: 139–55.

Alatalo, R.V., Eriksson, D., Gustafsson, L. & Lundberg, A. 1990. Male secondary plumage colour and risk of hybridization in Pied and Collared Flycatchers, *Ficedula* spp. *J. Evol. Biol.* in press.

Alatalo, R.V., Gustafsson, L. and Lundberg, A. 1989a. Extra pair paternity and heritability estimates of tarsus length in Pied and Collared Flycatchers. *Oikos* in press.

Alatalo, R.V., Gustafsson, L. & Lundberg, A. 1989b. Phenotypic selection on heritable size traits—environmental deviations distort the prediction of genetic response. *Amer. Nat.* in press.

Alerstam, T., Ebenman, B., Sylven, M., Tamm, S. & Ulfstrand, S. 1978. Hybridization as an agent of competition between two bird allospecies: *Ficedula albicollis* and *F. hypoleuca* on the island of Gotland in the Baltic. *Oikos* **31**: 326–331.

Askenmo, C.E.H. 1984. Polygyny and nest site selection in the Pied Flycatcher. *Anim. Behav.* **32**: 972–80.

Balen, van J.H., Noordwijk, van A.J. & Visser, J. 1987. Lifetime reproductive success and recruitment in two Great Tit populations. *Ardea* **75**: 1–11.

Fretwell, S.D. 1972. *Populations in a Seasonal Environment*. Princeton: University Press.

Gustafsson, L. 1985. Fitness factors in the Collared Flycatcher *Ficedula albicollis* Temm. Doctoral thesis, Uppsala University, Uppsala.

Gustafsson, L. 1986. Lifetime reproductive success and heritability: empirical support for Fisher's fundamental theorem. *Amer. Nat.* **128**: 761–4.

Gustafsson, L. 1987. Interspecific competition lowers fitness in Collared Flycatchers

Ficedula albicollis: an experimental demonstration. *Ecology* **68**: 291–6.

Gustafsson, L. 1988. Inter- and intraspecific competition for nest holes in a population of the Collared Flycatcher *Ficedula albicollis*. *Ibis* **130**: 11–16.

Gustafsson, L. MS. Polygyny and between generation effects of mating status in the Collared Flycatcher: the empirical demise of the "sexy son". unpublished manuscript.

Gustafsson, L. & Sutherland, W.J. 1988. The cost of reproduction in the Collared Flycatcher. *Nature* **335**: 813–817.

Lande, R. & Arnold, S.J. 1983. The measurement of selection on correlated characters. *Evolution* **37**: 1210–26.

McCleery, R.H. & Perrins, C.M. 1988. Life-time reproductive success of the Great Tit *Parus major*. In *Reproductive Success*, ed. T.H. Clutton-Brock, pp. 136–53. Chicago: University Press.

Pärt, T. & Gustafsson, L. 1989. Breeding dispersal in the Collared Flycatcher *Ficedula albicollis*: some possible causes and reproductive consequences. *J. Anim. Ecol.* **58**: 305–320.

Reznik, D.N. 1985. Cost of reproduction: an evaluation of the empirical evidence. *Oikos* **44**: 257–67.

Sibly, R.M. & Calow, P. (1986). Physiological ecology; an evolutionary approach. Oxford: Blackwell Scientific Publications.

Smith, J.N.M. 1988. Determinants of lifetime reproductive success in the Song Sparrow. In *Reproductive Success*, ed. T.H. Clutton-Brock, pp. 154–72. Chicago: University Press.

Svensson, L. 1984. *Identification Guide to European Passerines*. Stockholm: Naturhistoriska riksmuseet.

Török, J. 1986. Food segregation in three hole nesting bird species during the breeding season. *Ardea* **74**: 129–136.

Wallin, L. 1986. Divergent character displacement in the song of two allospecies: the Pied Flycatcher *Ficedula hypoleuca* and the Collared Flycatcher *Ficedula albicollis*. *Ibis* **128**: 251–9.

6. House Martin

DAVID M. BRYANT

Introduction

The House Martin *Delichon urbica* is a monomorphic hirundine weighing around 20 g and usually social in both breeding and feeding habits. The species spreads throughout Europe during the summer, often nesting on buildings, and in winter migrates south of the Sahara, where it is widespread but inconspicuous. The sexes take a broadly similar role building the enclosed mud nest and in incubation and raising young. Throughout much of their range House Martins attempt more than one brood each year.

The species offered a number of advantages in the study of life histories, not least that its colonial habits allowed ready access to large numbers of individuals whose behaviour could be observed unobtrusively from a few vantage points. Predation of nests was absent and the birds were very tolerant of humans. As a result, success rates were unaffected in any significant way by the study techniques. Attention was focused on two aspects because previous work indicated their importance in annual breeding success: the role of the body-size variation (Bryant 1979) and the role of food supply (Bryant 1975a). While any effects of body-size were pertinent throughout the lifespan of individuals, the role of food supply was considered mainly in terms of conditions during the year of hatching.

Study area and methods

This study of the House Martin was based on observations between 1971 and 1984 at a single colony in Central Scotland. The colony was at a farmstead, lying in a mixed agricultural landscape at 170 m a.s.l. with a

LIFETIME REPRODUCTION IN BIRDS
ISBN 0-12-517370-9

small river nearby. Artificial nests were provided in excess, and could be removed to inspect contents and to capture and ring adults and nestlings. Further details are given by Bryant (1975a, 1979, 1988), Bryant & Westerterp (1983a) and Bryant & Turner (1982).

The analysis of lifetime reproductive success by individual House Martins is based on nestlings ringed between 1971 and 1981 and on yearlings ringed between 1972 and 1982. None of these individuals survived to breed in 1984. Reproductive success was judged by several inter-related criteria, namely the total of clutches tended, eggs incubated or young reared during the lifetime of individual birds. Little information was available on recruitment, because only a small proportion of offspring, especially females, returned to breed at the colony. It is assumed here that all young which reached independence had an equal probability of recruiting to the breeding population and that lifetime reproductive success was adequately described by the number of young raised to this stage. House Martins are unusual in that fledglings return to the nest to roost after their first flight and so they can be observed until fully independent. Many recruits to the breeding colony had been reared elsewhere and could be identified as such, because they were unringed, in contrast to residents. It is likely that nearly all newcomers, the majority female, were in their first summer (Bryant 1979) and so all were assigned to this age class. The fate of most birds during their first year was unknown, so analysis was restricted to individuals which reached the first year after hatching.

Three measurements were taken from individuals at each handling; body mass (g), wing length (maximum chord, mm) and keel length (length of sternum from tracheal pit to hind margin of sternum, measured only from 1977 onwards). All measurements used here are mean values derived from samples of three or more for each individual. For longer-lived and frequently handled individuals the samples were much larger. Because wing lengths and mass of males changed slightly with age (Bryant 1979), analyses were repeated using measurements taken from birds in their first year. This did not cause any detectable change in the results, so only the conclusions based on lifetime mean body sizes are reported here. Body mass data were taken during incubation, so that changes consequent on stage of the breeding cycle were minimized. All size characters were approximately normally distributed (Endler 1986); logarithmic transformation caused no change in conclusions, so the results are not reported. As the sexes show no consistent differences in body-size or plumage, they were distinguished by size of brood patch (Bryant 1975b).

Food supply was measured using an aerial insect suction trap sited 20 km from the study colony (Bryant 1975a, Taylor & Palmer 1972). Daily catches of aerial insects were measured as settled volumes in a measuring cylinder containing a fluid preservative (methanol and glycerol). Because suction trap catches reflect aerial insect abundance across a wide area, as well as locally (Taylor 1973), and recruiting House Martins were anyway likely to be drawn from the vicinity of the study colony (Rheinwald 1975, Bryant, unpubl.), the rearing conditions for all youngsters in a cohort could be assessed. Mean insect catch volumes were derived from these data for all months of the study, except June–August 1974 because of trap breakdowns. These data have been omitted from the analysis of cohort success in relation to food supply.

Results

ANNUAL BREEDING SUCCESS, SURVIVAL AND LONGEVITY

House Martin pairs raised between zero and nine young to independence each year, giving an average of about five young from 1.7 breeding attempts (Table 6.1). Older individuals of both sexes tended to be more successful than first-time breeders (Bryant 1988).

Annual survival of House Martins which reached their first summer averaged 33 % for females and 46 % for males (Bryant 1988). The

Table 6.1 Annual reproductive success of House Martins.

	Females		Males	
	Mean	*(± SD)*	*Mean*	*(± SD)*
Number of breeding attempts year[-1a]	1.70	0.46	1.69	0.48
Number of successful attempts year[-1b]	1.49	0.16	1.49	0.60
Mean size of first clutch	3.92	0.87	3.94	0.87
Mean size of second clutch	3.11	0.62	3.12	0.61
Egg survival to fledging[c]	78%		79%	
Total eggs per pair year[-1d]	6.10	1.97	6.10	2.00
Total fledglings per pair year[-1]	4.85	2.20	4.88	2.17
First egg date[e]	37.33	15.30	37.48	15.75
n	179		180	

[a] Number of clutches incubated, excludes clutches deserted before laying was completed.
[b] One or more young raised to independence.
[c] Number of independent young/number of incubated eggs × 100.
[d] Includes only incubated clutches.
[e] Numbered so 1 May = 0.

lifespan of breeders ranged from one to five years amongst females and from one to six years in males (Fig. 6.1). The majority of individuals which survived to breed did so for only one year.

PATTERNS OF LIFETIME REPRODUCTIVE SUCCESS

The median number of eggs laid during the lifetime of females, or apparently fertilized by their mates, was seven in both sexes. The median number of clutches started, young raised to independence and years of breeding was also the same for both sexes (Table 6.2). Small differences between the sexes became apparent when mean lifetime success parameters were compared, favouring males in all cases (Table 6.2). The most obvious difference, however, was the greater variance in lifetime success of males, one aspect of which was the higher frequency of very successful individuals which raised more than 15 young (Fig. 6.2). The most productive male and female, which paired with different individuals in each year, raised respectively 42 young over six seasons and 28 young over five seasons. Fifty per cent of all young raised by the 228 individuals under study were due to 20 % of males and 28 % of females. The possibility that promiscuous matings might have distorted the apparent success of males, could not be wholly discounted (see discussion).

Only one individual of each sex was a proven non-breeder. This was likely to reflect the normal pattern for females, but probably

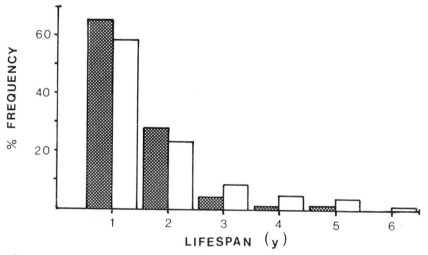

Figure 6.1 Percentage frequency of lifespans (y) for female (shaded) and male (open) House Martins which reached their first breeding year (females, $n = 125$; males, $n = 103$).

Table 6.2 Lifetime reproductive success of male and female House Martins that reached their first breeding year.

	Females (n = 125)				Males (n = 103)			
	Min	Median	Max	$\bar{x} \pm SD$	Min	Median	Max	$\bar{x} \pm SD$
Total eggs incubated (or fertilized)	0	7	38	8.5 ± 5.4	0	7	51	10.6 ± 0.9
Total young reared to independence	0	6	28	6.8 ± 4.6	0	6	42	8.5 ± 7.4
Total clutches started	0	2	9	2.4 ± 1.4	0	2	12	2.9 ± 2.2
Lifespan (years)	1	1	5	1.5 ± 0.8	1	1	6	1.8 ± 1.2
Variance in[a] egg success $(SD)^2/\bar{x}^2$				0.40				0.68
Variance in[b] young success $(SD)^2/\bar{x}^2$				0.46				0.76

[a] Variance in number of eggs laid divided by $\bar{x}^2$. (Wode & Arnold, 1980).
[b] Variance in number of young reared to independence divided by $\bar{x}^2$.

underestimated non-breeding amongst males. This was because males which failed to find a mate after a few weeks at the study colony moved away, presumably to search elsewhere, and were therefore difficult to confirm as non-breeders for a whole season. Any rise in the frequency of non-breeding would further increase the variance in male success.

Lifetime reproductive success, measured as the total young surviving to independence, can be considered as the product of three components; lifespan (i.e. number of breeding years) (L), the number of eggs laid each year (or incubated in the case of males) (F), and the mean survival rate of eggs to independence (S). The contribution of these components to variance in observed lifetime success was assessed using the method of Brown (1988). For both sexes almost all the variance in lifetime success was due to breeders (98 % ff, 99 % mm) (Table 6.3), because proven non-breeding was rare (see above). Amongst breeders, over half the overall variance explained was associated with lifespan (51 % ff, 54 % mm) (Fig. 6.3). The residual variance for females was mainly due to S (22 %) and F (17 %) (Table 6.3). About 10 % of the variance was associated with the interactive term LF. Amongst males, the interactive term LF was

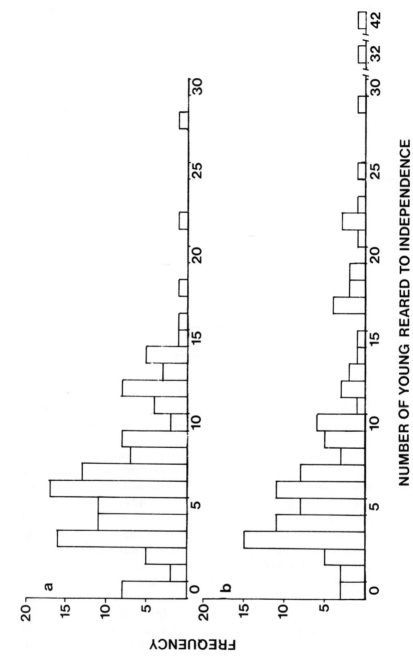

NUMBER OF YOUNG REARED TO INDEPENDENCE

Figure 6.2 The frequency of lifetime reproductive success amongst (a) female and (b) male House Martins which reached their first breeding year.

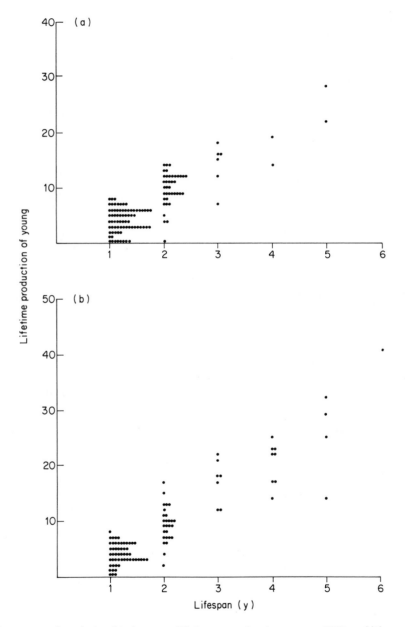

Figure 6.3 The relationship between lifetime reproductive success (LRS) and lifespan (y) in female (a) and male (b) House Martins. For females LRS = 5.00y −0.59 (r = 0.84, n = 122, P < 0.001) and for males LRS = 5.82y − 1.92 (r = 0.91, n = 101, P < 0.001).

Table 6.3 Mean, variance and percentage contribution of the components of lifetime reproductive success to its variance in House Martins.

Males

(a) Mean and variance of lifespan (*L*), number of eggs (*F*) and egg survival to independence (*S*) and their products for 102 breeding males.

	Original		Standardized variance
	Mean	Variance	
L	1.765	1.330	0.427
F	5.820	2.839	0.084
S	0.791	0.056	0.089
LF	10.800	76.405	0.724
FS	4.607	3.407	0.161
LS	1.412	1.009	0.518
LFS	8.668	54.965	0.833

(b) Percentage contributions to variation in lifetime reproductive success (V(LFS)/(LFS)^2) among breeders.

	L	*F*	*S*
L	51.29		
F	25.63	10.07	
S	0.18	−1.46	10.70
LFS	3.59		

(c) Inclusion of 1 non-breeder.

Proportion of breeders	0.99
Overall variance (O.V.)	55.16
Per cent O.V. due to non-breeders	1.32
Per cent O.V. due to breeders	98.68

(d) Percentage contributions of variation in lifetime reproductive success (V(LFS)/(LFS^2) among all males.

	L	*F*	*S*
L	50.80	9.93	
F	25.29	−1.44	10.55
S	0.18		
LFS	3.54		

Table 6.3 Continued.

Females

(a) Mean and variance of lifespan (L), number of eggs (F) and egg survival to independence (S) and their products for 124 breeding females.

	Original		Standardized variance
	Mean	Variance	
L	1.452	0.575	0.273
F	5.991	3.178	0.089
S	0.774	0.068	0.113
LF	8.794	31.234	0.413
FS	4.723	4.146	0.193
LS	1.143	0.486	0.384
LFS	7.017	23.063	0.509

(b) Percentage contributions to variation in lifetime reproductive success (V(LFS)/(LFS)^2) among breeders.

	L	F	S
L	53.63		
F	10.14	17.41	
S	0.24	−1.72	22.19
LFS	1.41		

(c) Inclusion of 1 non-breeder.

Proportion of breeders	0.99
Overall variance (O.V.)	23.27
Per cent O.V. due to non-breeders	1.69
Per cent O.V. due to breeders	98.31

(d) Percentage contributions to variation in lifetime reproductive success (V(LFS)/(LFS^2) among all females.

	L	F	S
L	53.20		
F	9.97	17.11	
S	0.23	−1.60	21.82
LFS	1.39		

ranked second after lifespan and accounted for 25 % of overall variance
(Table 6.3).

ROLE OF BODY-SIZE

Large individuals showed a generally weak tendency to be more successful
than small individuals (Table 6.4). Significant correlations between body
mass, keel length and lifetime success were found for males (Fig. 6.4),
an association which was enhanced by combining mass and keel measures
as either a multiple or sum (Table 6.4). Similar correlations amongst
females were invariably weak: only keel was significantly correlated and
then only with lifetime egg production. Combining size measures, as for
males, increased the significance of associations between body-size,
number of eggs incubated and lifespan (Table 6.4). Overall, the absence

Table 6.4 Associations between lifetime reproductive success and related parameters in the House Martin. Figures show Pearson product-moment correlation
coefficients.

	Females (n = 116)[a]			Males (n = 99)[a]		
	Total young reared	Total eggs incubated	Lifespan (years)	Total young reared	Total eggs incubated	Lifespan (years)
Total young reared	—	0.94 ***	0.81 ***	—	0.97 ***	0.91 ***
Total eggs incubated	0.94 ***	—	0.86 ***	0.97 ***	—	0.94 ***
Lifespan (years)	0.81 ***	0.86 ***	—	0.91 ***	0.94 ***	—
Body mass (g)	0.06	0.11	0.09	0.26 *	0.27 **	0.24 *
Keel (mm)	0.13	0.23 *	0.17	0.25 *	0.28 *	0.29 *
Wing (mm)	0.03	0.02	0.00	0.06	0.12	0.11
Body-size (1)[b]	0.19	0.31	0.32	0.34	0.38	0.38
Body-size (2)[b]	0.19	0.30 **	0.31 **	0.33 **	0.37 **	0.37 **

[a] For keel, body-size (1) and (2), females n = 59, males n = 54.
[b] Body-size (1) is bodymass × keel; body-size (2) is standardized mass +
standardized keel.
* $P < 0.05$, ** $P < 0.01$, *** $P < 0.001$.

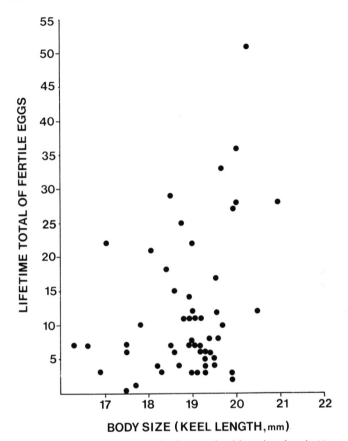

Figure 6.4 The relationship between body-size (keel length) of male House Martins and their lifetime reproductive success ($r = 0.28$, $P < 0.05$).

of significant correlations between lifetime young raised with any measure of female size, and their presence for males, suggests large size confers no selective advantage on females but does so for males, albeit in an inconsistent way. This inconsistency is illustrated by the rather low success evident for some of the larger males (Fig. 6.4).

ROLE OF FOOD SUPPLY IN ANNUAL AND LIFETIME SUCCESS

Variation in aerial insect abundance had many consequences in House Martins, both for overall breeding success and for timing of the breeding cycle (Bryant 1975a, 1979). Aerial insects tended to be scarce when temperatures were low and during prolonged rain and strong winds (Bryant 1975a). Over the period of this study annual reproductive success was closely associated with aerial insect abundance. Accordingly, the mean number of eggs laid per pair each year, mean number of young raised per pair, and mean number of broods were all significantly correlated with suction trap catches in summer (Table 6.5). Further, the timing of first clutches, itself an important correlate of seasonal breeding success (Bryant 1979) was earlier in years when food was abundant (Table 6.5). It follows that because the modal number of seasons of breeding was one, then chance must have played a major role in lifetime success. Equally, longer-lived individuals will reflect to a degree the favourability of the seasons experienced.

A less evident effect of food supply might be manifest via conditions during development, such that individuals raised in favourable years could show a greater success in later life. To examine this point, the lifetime success of each cohort (year class) was investigated in relation to food supply in the hatch year. Lifetime number of eggs incubated and young reared by females was correlated with food supply in their own year of hatching (Table 6.6). No comparable effect was detected for males, although there was a tendency for long wing-chord male cohorts to occur when food was abundant in the hatch year (Fig. 6.4). There was no similar correlation for mass or keel length, the only body-size measures which were directly associated with reproductive success (Table 6.4).

HERITABILITY

The heritability (h^2) of three body-size measures, themselves intercorrelated, was examined. A significant heritability was identified in all three measures (Table 6.6), although by comparison with some other studies, heritability was low. Heritability estimates were greatest for mid-parent

Table 6.5 Correlation coefficients for measures of annual reproductive success per pair in relation to food supply[a].

Success parameter	r	P
Total eggs	0.52	*
Young raised to fledging	0.53	*
Number of broods	0.52	*
First egg date	−0.53	*

[a] The measure of food supply used was the annual mean volume of the suction trap catch (cm^3 day^{-1}) during June and July, the months in which most first brood young were raised and second clutches were initiated.
* $P < 0.05$.

Table 6.6 Size attributes and lifetime reproductive success of eleven cohorts in relation to food supply in their year of hatching.

	Males		Females	
	r	*P*	*r*	*P*
Lifetime eggs	—	ns	0.71[b]	**
Lifetime young	—	ns	0.55[b]	*
Body mass	—	ns	—	ns
Keel length	—	ns	—	ns
Wing length	0.72[a]	**	—	ns

[a] Food supply is annual mean volume of the suction trap catch during spring (May and June, the months when most first brood young are raised). Similar correlations were evident with June–July means (r = 0.68**), or with monthly means for May–August (r ranged from 0.51 to 0.68).
[b] Food supply is annual mean volume of suction trap catch during July (log transformed). Significant correlations were also found for other measures of food supply, such as mean and minimum levels during May–August (r = 0.60 to r = 0.63).
* $P < 0.05$, ** $P < 0.01$, ns—not significant.

wing length versus offspring wing length, a result probably influenced by the lack of measurement error compared with keel length and the relative consistency in this feature compared with body mass. While a heritable component to all three-size attributes was apparent, no conclusions about the importance of this heritable component could be drawn, though body-size evidently had both genetic and phenotypic components.

Discussion

The lifetime reproductive success of male and female House Martins was largely dependent on lifespan, long-lived individuals being much more successful than the bulk of individuals which bred in only a single year. Because nearly all pairs reared some young each year, lifetime success in reproduction usually accumulated roughly in proportion to lifespan. The same pattern was shown in other species by Coulson & Thomas (1985), Newton (1985) and Hötker (1988), but as seasonal success was also positively related to age (Bryant 1979, 1988), some longer-lived birds gained a disproportionate advantage. Variation in lifetime reproductive success due to clutch size and survival rate in the nest tended to be less important. It remains possible that differences in post-independence survival of young may modify the patterns demonstrated in this study.

Table 6.7 Heritability (h^2) estimates for body-size attributes of House Martins. h^2 estimates ($\pm$ SE(n)) for mid-parent means, male and female parents, are given in relation to offspring attributes measured during their first breeding season (after H. Riley in prep).

Attribute	Mid-parent Mean	Male parent	Female parent
Body-mass	42.7 ± 15.3 (31)**	28.7 ± 11.4 (36)*	21.6 ± 12.1 (34)ns
Wing length	66.7 ± 21.2 (31)**	28.3 ± 18.8 (36)ns	41.4 ± 15.3 (34)**
Keel length	32.3 ± 17.6 (12)ns	32.2 ± 14.6 (16)*	23.5 ± 20.0 (13)ns

* $P < 0.05$, ** $P < 0.01$, ns—not significant.

However, Newton (1985) and Hötker (1988) showed that number of recruits was directly related to the number of fledged young in Sparrowhawks *Accipiter nisus* and Meadow Pipits *Anthus pratensis* respectively. In species where there is a consistent negative relationship between brood-size and fledgling quality, such a pattern would not necessarily be expected, because survival in large broods may be reduced (Perrins 1965, Nur 1984a, b). In House Martins, however, where fledgling mass and quality depends on food supply, as well as on parental quality and brood-size (Bryant 1975a, 1979, Bryant & Gardiner 1979, Bryant & Westerterp 1983a, b), yearly fluctuations in conditions also seem likely to play a substantial role in juvenile survival and recruitment. As conditions vary from year to year any negative effects on the most productive breeders each year caused by undernourishment of some of their nestlings seem unlikely to distort the relative lifetime success indicated by number of young raised.

There was no evidence of egg dumping at the study colony of House Martins, in contrast to some hirundines (Møller 1985, Brown & Brown 1988). The reproductive success of females is therefore likely to relate directly to the success parameters examined in this study. For males, however, paternity could be ascribed with less certainty (Burke & Bruford 1987). Extra-pair copulations in some bird species are sufficiently common to invalidate an assumption of paternity for all eggs in a given male's nest. That males usually guard their mates can also be taken to imply a risk to their paternity. In the House Martin, however, all matings apparently take place in the nest, so the opportunity for extra-pair associations are fewer than for most other species, and close attendance by males with fertile mates (Jones 1986) ensures that intruders are quickly ejected from nests. Hence the patterns demonstrated here for males are unlikely to be changed substantially by extra-pair fertilizations, but this needs confirmation.

Lifetime reproductive success of males was more variable than for females (ratio male: female = 1:65), but was, as expected, less variable than in species which are polygamous or which defend large, exclusive territories (Wade & Arnold 1980, Price 1984). It can be inferred that intra-sexual selection is only moderate, a view which can be easily reconciled with the lack of size or plumage dimorphism in House Martins. The fact that large males enjoyed a greater success than small males, implies directional selection for increased body-size. Yet there was no tendency for males to increase in size during the course of this study and nor has such a tendency been described for the species elsewhere. Selection may have been too weak to cause detectable changes of this type or the advantage may have been negated by counter-selection in some years. Evidence for the latter was available for females (Bryant 1988), and although lacking for males it may not have been detected because some annual samples were small. It is also possible that selection may operate against large males before they first arrive at the colony. Large Sand Martins *Riparia riparia* can be selected against during the non-breeding period (Jones 1987), so it is feasible that similar pressures operate against large male House Martins in early life, thus maintaining male size close to that of females (Bryant 1988).

Food supply, as indicated by the volume of suction trap catches during the breeding season, had predictable effects on annual breeding success, and some newly identified effects on the size (wing lengths) and lifetime success of cohorts. There was, however, no evidence that these latter effects were interdependent. Wing lengths tended to be greater in males reared when conditions were good, implying that body-size in such cohorts was larger. Yet wing length was a poor predictor of size in House Martins (Bryant & Westerterp 1982) and further, male wing length was not itself correlated with any fitness parameter. Equally, lifetime success of females was apparently independent of any aspect of body-size. The mechanism by which these effects on size and lifetime success was induced remains unclear. The size effect could have had a nutritional basis because diet is known to influence body size in other birds (Boag 1987). Alternatively, differential survival of large or potentially successful individuals in good food years might have occurred. This seems less likely, however, because good food conditions would normally slacken rather than intensify selection against less fit individuals. A third possibility was that locally favourable conditions may have attracted the largest or fittest of a cohort of prospecting juveniles, which subsequently settled to breed with the observed consequences for size and·success. A final possibility is that good years were clumped, so favourable years for nestling growth would often be followed by productive breeding years. In practice, however, this did not occur (see dates on Fig. 6.5 for example). In future the effect of poor nutrition on the survival and breeding of post-fledging juveniles may repay examination, because it could induce fitness costs like those found in other species when breeding conditions are poor or when a larger number of young than usual is reared (Askenmo 1979, Bryant 1979, Nur 1984a, Roskaft 1985, Reid, 1987).

Summary

Lifetime reproductive success is reported for 125 female and 103 male House Martins, which were studied at a colony in Scotland over 13 years. The median lifetime success of both males and females was six young reared to independence but the variance in lifetime success was 1.65 times greater in males than in females. The most successful males and females raised respectively 28 and 42 young. Most of the variation in lifetime success within sexes was due to differences in lifespan, which accounted for over half of the variance in lifetime success. Amongst females the residual variance was due to egg number and mortality in the nest, whereas in males an interactive effect of lifespan and number of eggs fertilized was important.

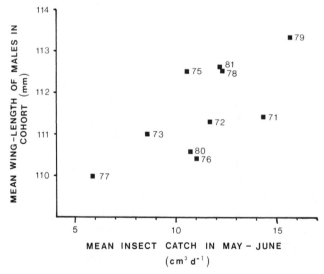

Figure 6.5 The relationship between the mean wing length of 11 male House Martin cohorts and the food supply in their year of hatching. Insect catch is mean daily volume of insects in suction trap catch (see text).

Large males tended to have a higher lifetime success than small males. Males raised under favourable conditions of food supply had longer wing-chords, but this did not apparently contribute to fitness. Females, while showing no effects of food supply on body size, were more successful when they hatched in good years. Because food supply also affects annual breeding success, its effects on fitness of females is important and yet involves a large chance element.

References

Askenmo, C. 1979. Reproductive effort and return rate of male Pied Flycatchers. *Amer. Nat.* **114**: 748–52.

Boag, P.T. 1987. Effects of nestling diet on growth and adult size of Zebra Finches *Peophila guttata*. *Auk* **194**: 155–66.

Brown, C.R. & Brown, M.B. 1988. A new form of reproductive parasitism in Cliff Swallows. *Nature* **331**: 66–8.

Brown, D. 1988. Components of lifetime reproductive success. In *Reproductive Success*, ed. T.H. Clutton-Brock, pp. 439–53. Chicago: University Press.

Bryant, D.M. 1975a. Breeding biology of House Martins in relation to aerial insect abundance. *Ibis* **117**: 180–216.

Bryant, D.M. 1975b. Changes in incubation patch and weight in the nesting House Martin. *Ring & Migr.* **1**: 33–6.

Bryant, D.M. 1979. Reproductive costs in the House Martin *Delichon urbica*. *J. Anim. Ecol.* **48**: 655–75.

Bryant, D.M. 1988. Lifetime reproductive success of House Martins. In *Reproductive Success*, ed. T.H. Clutton-Brock, pp. 173–88. Chicago: University Press.

Bryant, D.M. & Gardiner, A. 1979. Energetics of growth in House Martins *Delichon urbica*. *J. Zool. Lond.* **189**: 275–304.

Bryant, D.M. & Turner, A.K. 1982. Central place foraging by swallows (Hirundinidae): the question of load size. *Anim. Behav.* **30**: 845–56.

Bryant, D.M. & Westerterp, K.R. 1982. Evidence for individual differences in foraging efficiency amongst breeding birds: A study of the House Martin *Delichon urbica* using the doubly labelled water technique. *Ibis* **124**: 187–92.

Bryant, D.M. & Westerterp, K.R. 1983a. Short-term variability in energy turnover by breeding House Martins *Delichon urbica*: A study using doubly labelled water. *J. Anim. Ecol.* **52**: 525–43.

Bryant, D.M. & Westerterp, K.R. 1983b. Time and energy limits to brood-size in House Martins. *J. Anim. Ecol.* **52**: 905–25.

Burke, T. & Bruford, M.W. 1987. DNA fingerprinting in birds. *Nature* **327**: 149–52.

Coulson, J.C. & Thomas, C. 1985. Differences in the breeding performance of individual Kittiwake Gulls, *Rissa tridactyla*. In *Behavioural Ecology: Ecological Consequences of Adaptive Behaviour*, ed. R.M. Sibly & R.H. Smith, pp. 489–503. B.E.S. Symposium 25.

Endler, J.A. 1986. Natural selection in the wild. *Monogr. in Pop. Biol.* **21**. Princeton: Princeton University Press.

Hötker, H. 1988. Lifetime reproductive output of male and female Meadow Pipits *Anthus pratensis*. *J. Anim. Ecol.* **57**: 109–17.

Jones, G. 1986. Sexual chases in Sand Martins (*Riparia riparia*): cues for males to increase their reproductive success. *Behav. Ecol. & Sociobiol.* **19**: 179–85.

Jones, G. 1987. Selection against large size in the Sand Martin (*Riparia riparia*) during a dramatic population crash. *Ibis* **129**: 274–80.

Møller, A.P. 1985. Intraspecific nest parasitism and anti-parasite behaviour in Swallows, *Hirundo rustica*. *Anim. Behav.* **35**: 247–54.

Newton, I. 1985. Lifetime reproductive output of female Sparrowhawks. *J. Anim. Ecol.* **54**: 241–53.

Nur, N. 1984a. The consequences of brood size for breeding Blue Tits. I. Adult survival, weight change and the cost of reproduction. *J. Anim. Ecol.* **53**: 479–96.

Nur, N. 1984b. The consequences of brood size for breeding Blue Tits. II. Nestling weight, offspring survival and optimal brood-size. *J. Anim. Ecol.* **53**: 497–518.

Perrins, C.M. 1965. Population fluctuations and clutch size in the Great Tit *Parus major* L. *J. Anim. Ecol.* **34**: 601–47.

Price, T.D. 1984. Sexual selection on body size, territory and plumage variables in a population of Darwin's Finches. *Evolution* **38**: 327–41.

Reid, W.V. 1987. The cost of reproduction in the Glaucous-winged Gull. *Oecologia* **74**: 458–67.

Rheinwald, G. 1975. The pattern of settling distances in a population of House Martins *Delichon urbica*. *Ardea* **63**: 136–45.

Roskaft, E. 1985. The effect of enlarged brood-size on the future reproductive potential of the Rook. *J. Anim. Ecol.* **54**: 255–60.

Taylor, L.R. 1973. Monitoring change in the distribution and abundance of insects. *Rep. Rothamsted Exp. Stn. for 1973*, **part 2**: 202–39.

Taylor, L.R. & Palmer, J.M.P. 1972. Aerial sampling. In *Aphid Technology*, ed. H.F. van Emden, pp. 189–234. London: Academic Press.

Wade, M.J. & Arnold, S.J. 1980. The intensity of sexual selection in relation to male behaviour, female choice, and sperm precedence. *Anim. Behav.* **28**: 446–61.

7. Kingfisher

MARGRET BUNZEL & JOACHIM DRÜKE

Of the family Alcedinidae, that comprises 90 species world-wide, the Kingfisher *(Alcedo atthis)* is the only representative in Europe. Its breeding range extends from the British Isles and the European–Moroccan atlantic coast eastwards to Japan, and from the southern Baltic and Sakhalin southwards to north Morocco, Turkey and continental southeast Asia. Isolated from this main range are other breeding areas in east Indonesia and Melanesia (Bezzel 1980).

The brightly coloured, sparkling blue and orange Kingfisher weighs about 40 g. In winter the weight increases to about 46 g, or at times to more than 50 g. The sexes have similar plumage, but differ in bill colour. In the adult male the lower mandible is black, mostly with a small brownish-orange spot near the base, whereas in the adult female the lower mandible shows a large area of orange-red extending from the base almost to the tip.

Kingfishers live near various types of water, but prefer clear gently-flowing streams and still waters, with shady patches and ample lookout perches. The bird feeds mainly on small freshwater fish which it catches by diving from a perch. The breeding tunnel is dug into a vertical bank, usually at least 100 cm high, consisting of soil without too many stones or roots.

Kingfishers are essentially monogamous, but polygyny sometimes occurs in *Alcedo atthis ispida* (e.g. Heyn 1963, 1965, Probst 1982, Svensson in Cramp 1985) and probably to a greater extent in the eastern race *Alcedo atthis atthis* (e.g. Numerov & Kotyukov 1979, Podolski 1982 in Cramp 1985). Most Kingfishers start two broods per season (e.g. Hallet 1978, Svensson 1978), but up to four are possible (e.g. Guenat in Bezzel 1980, Probst 1982). The most common clutch sizes are seven or six eggs; and in our study area in central Westphalia the mean clutch size was 6.8

LIFETIME REPRODUCTION IN BIRDS
ISBN 0-12-517370-9

(n = 159). There, an egg produced a fledgling with a mean probability of 0.53 (n = 1,061 eggs of 156 clutches).

Study area and methods

The study area in central Westphalia (northwest Germany, 51° 35' N, 8° 15' E) was extended annually from 1976 to 1984–1987 to a maximum of 2,300 km². It consists of the sub-areas "Sauerland" (highland, altitude 120–550 m) and "Flachland" (lowland, altitude 60–100 m), which are separated by an area 5–12 km wide without permanently flowing streams.

During the study period the Kingfisher population in central Westphalia fluctuated considerably depending on the severity of the winters. Accordingly, between six (1987) and 46 (1984) breeding pairs were studied annually. We assume that we discovered nearly all breeding attempts in the study area.

Birds were caught with mist nets near their breeding sites, preferably when the chicks were 2–3 weeks old. Only metal rings could be used, because the Kingfishers' tarsi are too short for colour-ringing. To learn the identity of a ringed Kingfisher it was therefore necessary to catch it. Not all adults could be caught at every brood they started each year. But as changes of breeding partner within one season occurred only rarely, it was assumed that if, for example, in one year at a breeding site three broods were seen and the breeding birds were caught only while feeding the young of the second brood, all three broods belonged to the same pair. If any irregularities occured that hinted at a possible change of breeding partner (e.g. long intervals between broods), we always tried to catch the breeding birds a second time. Birds that did not breed successfully in a particular year were less likely to be caught and identified than were successful birds.

Kingfishers normally do not complete their wing moult in one go. This results in age-specific patterns of retained old and replaced new feathers. Nearly 70 % of the breeding birds were ringed before their first wing moult when age-determination presented no problems. The other birds were aged from the pattern of old and new wing feathers. In most cases this would also have given precise and reliable ages.

Results

LONGEVITY

Kingfishers are very short-lived. The oldest birds observed in population studies or in analyses of ring recoveries reached 2 years 9 months in

Czechoslovakia (Hladík & Kadlec 1964), 3 years 7 months in Westphalia (Bunzel 1987), 4 years 6 months in Great Britain (Morgan & Glue 1977), and 4 years 11 months in the Netherlands (Probst 1982). From our population study, a mean annual mortality for adult Kingfishers of 71–73 % was calculated; the mortality of young Kingfishers between leaving the breeding tunnel and the next breeding season was estimated at 78–79 % (Bunzel 1987).

In Fig. 7.1, the dark bars represent the longevity of a sample of 74 males and 51 females for which lifetime production could be estimated. Lifespan was counted from the year of birth to the year that the bird was last identified. As it never happened that a breeding bird "went missing" for a year and then re-appeared, each bird was assumed dead when it had not been found for one year. The frequency distribution of lifespans again reflects the Kingfisher's very high annual mortality. There are no significant differences between the sexes.

For about half of all adult Kingfishers caught in the breeding seasons of the study period, estimates of lifetime production could be given. The other individuals had to be excluded because of uncertainty over breeding success or lifespan. So, how representative is the sample of birds for which estimates of lifetime production are available, in terms of longevity, of the entire population? To answer this question, the lifespans of the

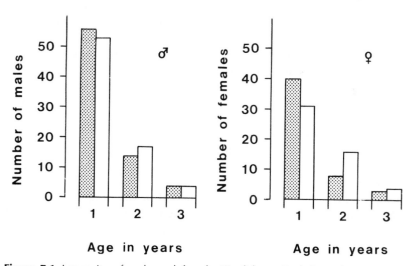

Figure 7.1 Longevity of male and female Kingfishers. Dark bars: longevity of the sample of 74 males and 51 females for which estimates of lifetime production were available. Light bars: expected longevity, calculated from the mean annual age distribution of all birds caught in the breeding seasons of 1976–1987 (n = 136 males and 111 females, including 4 male and 3 female non-breeders).

Kingfishers of the sample are compared to those calculated for all birds caught in the breeding season during the entire study (Fig. 7.1). In the males, there is no significant difference between the two groups ($\hat{\chi}_2^2 = 0.37$, ns). In the females, the sample of birds for which lifetime production is known shows a greater proportion of birds last identified in their first year than the other group. Although this difference is not significant ($\hat{\chi}_2^2 = 3.16$, ns), it hints at the possibility that the sample might be deficient in older females.

ANNUAL BREEDING SUCCESS

Most Kingfishers start breeding in their first year of life. Of 74 males, seven definitely did not breed in their first year at the breeding site they occupied in later years or in the vicinity; hence it is assumed that these birds did not breed at all in their first year. Of 51 females, three probably bred for the first time in their second year. Including these non-breeders, each male in the sample had on average 2.20 clutches per season, and each female 2.15 clutches. The mean number of successful broods per season was 1.38 for males and 1.35 for females.

Of all broods studied during 1976–1987 ($n = 477$), 58.5 % were successful. The probable cause for the complete loss of a brood could be identified in about 60 % of cases; in the other 40 % of failures clutches or nestlings were deserted or vanished. Most of the losses with identified cause (30 %) occurred during periods of bad weather with floods and turbid water, when broods were drowned or died of starvation. Another 12 % of losses were due to Fox (*Vulpes vulpes*), mustelids or other predators. Losses caused by burrowing mammals, such as voles (Microtidae), mice (Muridae) or Mole (*Talpa europaea*) and by human interference were of minor importance (8 and 7 % respectively).

The arithmetic mean of the number of fledged young per successful brood ($n = 206$) was 5.6, and the median number was 6; so for those successful broods in which chicks were not counted, 6 was chosen as an estimate of the number raised.

In Fig. 7.2 the annual reproductive success of the sample of birds for which lifetime production could be estimated is compared to the annual reproductive success of the entire population studied. For an assessment of lifetime production a bird had to be identified in each year in which it bred, but the breeding success of any bird in one season could be determined fairly accurately without knowing its identity. The sample of birds with known lifetime production shows a lower percentage of failures than the entire population. This difference is significant for males (z–test, $n = 96$, $\hat{z} = 2.03$, $P < 0.05$), but not for females (z–test, $n = 65$, $\hat{z} = 1.33$,

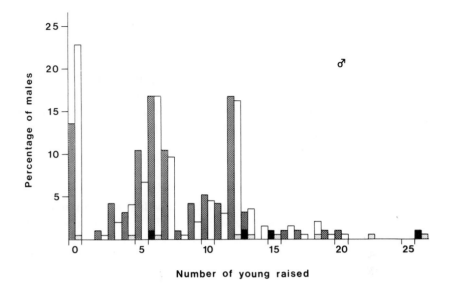

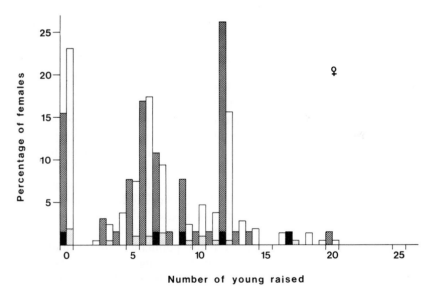

Figure 7.2 Annual production of fledglings by male and female Kingfishers ▦ sample of birds for which estimates of lifetime production were available—monogamists and non-breeders (*n* = 92 males, 60 females) ■ sample of birds for which estimates of lifetime production were available—bygamists (*n* = 4 males, 5 females). ☐ whole population—monogamists and non-breeders (*n* = 189 males, 194 females). ▦ whole population—bygamists (*n* = 9 males, 18 females).

ns). The reason for the lower failure rate of the birds in the "lifetime sample" is the lower probability with which unsuccessful breeders were identified and included in the sample. Apart from the percentage of complete failures, the annual reproductive output of the sample seemed to be fairly representative of the whole population. In general 2–7 fledged young imply one successful brood, 8–14 young imply two broods and 15–20 young three broods. Annual totals of 22 and 26 young were reached by two male Kingfishers which were each paired to two females simultaneously.

LIFETIME SUCCESS

The number of young raised to fledging by individual Kingfishers during their lifetimes varied between 0 and 32 (Fig. 7.3). The figures for the sexes were similar, the arithmetic mean for males being 9.78 (SD = 6.09) and for females 9.65 (SD = 6.19). The median number of fledglings produced per male was 10, and per female 9.

To assess whether the figures for mean lifetime production in the sample of 74 male and 51 female Kingfishers were representative of the studied population, the arithmetic means of the number of fledglings raised by the birds in the sample were compared to the mean numbers of fledglings per lifetime for all birds. The latter figures were calculated from the distribution of lifespans of all adult Kingfishers caught in the breeding season (Fig. 7.1), and the mean annual success of the whole population (Fig. 7.2). According to this, the mean lifetime reproductive success for all males in the studied population was 9.88 fledged young, and for all females 9.59 young. These figures are similar to those obtained for the sample of Kingfishers that could be followed for their whole lives.

More than 50 % of all fledged young were raised by 30 % of the 74 males studied, and by 31 % of the 51 females studied. Some 75 % of all young were raised by 51 % of both sexes.

From knowledge of breeding success, of mortality from fledging to the next breeding season (obtained from the population study) and of the percentages of adults that failed to produce young, it was calculated that 10.4 % of the eggs laid resulted in breeding birds: 5.3 % in males and 5.1 % in females.

In both sexes, lifetime production was highly correlated with lifespan (Fig. 7.4), as well as with mean annual production (males: $n = 74$, $r_s = 0.84$, $P < 0.001$; females: $n = 51$, $r_s = 0.76$, $P < 0.001$). There was no relationship between longevity and mean annual production (males: $n = 74$, $r_s = 0.13$, ns; females: $n = 51$, $r_s = 0.22$, ns).

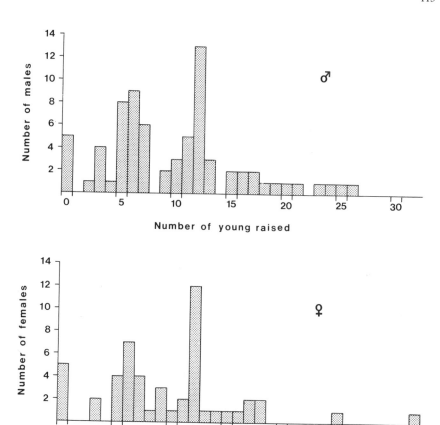

Figure 7.3 Lifetime production of 74 male and 51 female Kingfishers.

Kingfishers seemed to show preferences for particular breeding sites, and we wondered whether these preferences correlated with breeding success. We therefore calculated how often each site was chosen for breeding during the study period and whether the site was occupied in years with high or low population density (0 < preference value ⩽ 1; for details see Bunzel 1987). The question of interest was whether Kingfishers breeding at preferred sites raised more young than other Kingfishers. In Fig. 7.5 the lifetime production of male and female Kingfishers is shown in relation to breeding site preference. If birds changed breeding sites during their lifetimes, as a considerable proportion did, mean values were used. Only for the females did a correlation between preference value of

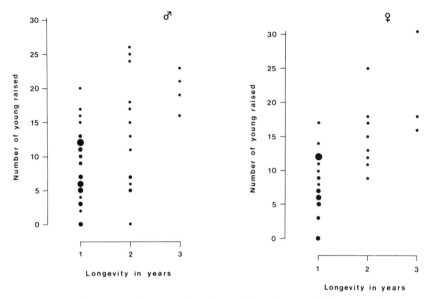

Figure 7.4 Lifetime production of male and female Kingfishers in relation to longevity. Corresponding to its size, each spot represents 1–13 different individuals. Males: $n = 74$, $r_s = 0.51$, $P < 0.001$; females: $n = 51$, $r_s = 0.67$, $P < 0.001$.

the breeding site and lifetime production emerge, but in both sexes there was a significant relationship between breeding site preference value and mean annual production (see caption to Fig. 7.5). Longevity was related to breeding site preference value only in females (males: $n = 72$ (non-breeders excluded), $r_s = 0.09$, ns; females: $n = 50$ (non-breeders excluded), $r_s = 0.31$, $P < 0.05$).

Finally we attempted to find whether lifetime production was correlated with body size as reflected in wing length and tail feather length. On samples of 59 males and 44 females, which were measured in their second calendar year before their first wing moult, no significant relationships between lifetime production and these measures of body size were found.

Discussion

Mean lifetime fledgling productions were estimated at 9.8 for the 74 male and at 9.7 for the 51 female Kingfishers that were studied. These figures were roughly as expected from the whole population, but the sample was probably slightly deficient in birds with poor success and in older birds that produced a large number of young. Only some of the

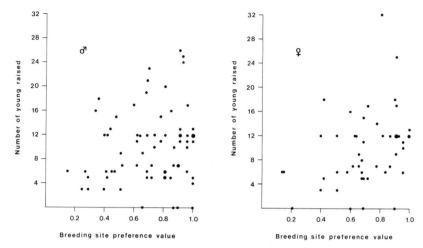

Figure 7.5 Lifetime production of male and female Kingfishers in relation to breeding site preference value. The non-breeders (2 males, 1 female) were excluded because they could not be related clearly to a certain breeding site. Corresponding to its size, each spot represents 1–3 different individuals. Males: $n = 72$, $r_s = 0.15$, ns; females: $n = 50$, $r_s = 0.38$, $P < 0.01$. Relationship between mean annual production of male and female Kingfishers and breeding site preference value (not shown): Males: $n = 72$, $r_s = 0.30$, $P < 0.05$; females: $n = 50$, $r_s = 0.41$, $P < 0.01$.

factors which might influence lifetime production could be examined. As 77 % of the individuals studied experienced only one breeding season, it was not surprising that both annual success and lifespan strongly influenced lifetime production. Breeding at preferred sites had a positive influence on annual production, but only a small proportion of the variance in mean annual production could be "explained" by the preference values of the breeding sites occupied. A significant relationship between breeding at preferred sites and lifetime production was found only in females, indicating that the breeding site was only one factor influencing annual success. Female lifespan seemed to be weakly influenced by choice of breeding site; probably longevity in both sexes depended more on survival in winter when Kingfishers could move elsewhere. Body size—at least the measures taken—apparently played no role in determining lifetime reproductive output.

Compared to other birds, a short lifetime and very high reproductive rate characterize the Kingfisher as a mainly *r*-selected species. The Kingfisher lives in a fluctuating environment, in that winter severity limits population size essentially in a density–independent manner. Population size varies considerably, and in most breeding seasons is well below the carrying capacity of the environment. As the species can breed in its first year, there should in theory be virtually no non-breeders. Intraspecific competition presumably varies in extent, but may not be absent even in years of low population, because the breeding sites differ in popularity, and gaining a preferred site increases a bird's chance of high reproductive output. A successful Kingfisher therefore cannot follow exclusively an r-strategy, but must be able to select and if necessary defend, a preferred breeding site, a feature more typical of K-selected species.

Acknowledgements

We thank all friends who helped during the field work, especially H. Bottin, J. Brackelmann, K. Dillenburger, E. Hochstein, M. Hölker, T. Jaspert, Dr H.–M. Leismann, A. Nagel and K. Schmidt. We are very grateful to Dr I. Newton for commenting and advising on an earlier draft of this chapter.

References

Bezzel, E. 1980. *Alcedo atthis*—Eisvogel. In *Handbuch der Vögel Mitteleuropas*, Vol. 9, U. Glutz v. Blotzheim & K.M. Bauer, eds., pp. 735–74. Wiesbaden: Akademische Verlagsgesellschaft.
Bunzel, M. 1987. Der Eisvogel (*Alcedo atthis*) in Mittelwestfalen. Studien zu seiner Brutbiologie, Populationsbiologie, Nahrung und Siedlungsbiologie. Inaugural-Dissertation zur Erlangung des Doktorgrades der Naturwissenschaften im Fachbereich Biologie. Münster: Westfälische Wilhelms–Universität.
Cramp, S. (ed.) 1985. *Handbook of the Birds of Europe, the Middle East and North Africa. The Birds of the Western Palearctic*, Vol. 4. Terns to woodpeckers. Oxford: University Press.
Hallet, C. 1978. Le régime alimentaire du Martin-pêcheur: étude qualitative et quantitative. Mémoire présenté pour l'obtention du grade de Licencié en Sciences zoologiques, Namur: Facultés Notre Dame de la Paix.
Heyn, D. 1963. Über die Brutbiologie des Eisvogels. *Falke* **10**: 153–8.
Heyn, D. 1965. Durch Beringung erwiesene Bigamie des Eisvogels. *Falke* **12**: 186–7.
Hladík, B. & Kadlec, O. 1964. Ergebnisse der Beringung des Eisvogels (*Alcedo atthis*) in der Tschechoslowakei. *Zoologické Listy* **13**: 1–8.
Morgan, R. & Glue, D. 1977. Breeding, mortality and movements of Kingfishers. *Bird Study* **24**: 15–24.
Numerov, A.D. & Kotyukov, Y.V. 1979. Goluboj simorodok. (The Kingfisher). *Priroda* **6**: 69–73.
Probst, J.G.A. 1982. *De Ijsvogel in Nederland. Mijn ervaringen met de Ijsvogel* Alcedo atthis ispida *en Enkele gegevens over de Nachtzwaluw* Caprimulgus europaeus. Tilburg, Gianotten.
Svensson, S. 1978. Biology and occurence of the Kingfisher *Alcedo atthis* at Klippan, province of Skåne. *Vår fågelvarld* **37**: 97–112.

Part II.   Short-lived Open Nesters

This section contains chapters on several species of short-lived, open-nesting passerines. The Meadow Pipit breeds on grassland and tundra over much of the Palearctic region, the Song Sparrow *Melospiza melodia* and Indigo Bunting *Passerina cyanea* are found in scrub habitats in North America; while the Magpie *Pica pica* is found in sparsely wooded park-type habitats in both regions. All these species are essentially monogamous, and all except the Magpie can raise up to several broods in a year, but predation on the eggs and chicks is often heavy and many nests fail. Individual breeding success is therefore highly variable.

The studies on all these species show consistent differences in LRS between localities, including individual territories. In some species this could be attributed to the quality of the territories themselves, but in the Magpie Birkhead & Goodburn discuss the problem of separating the quality of a territory from the quality of its occupants. In the Song Sparrow, population density fluctuated greatly over the study period, and emerged as a major determinant of LRS. Those individuals whose breeding lives fell mainly in a period of low population had greater LRS than others whose lives fell in a period of high population, thus providing evidence for density-dependence in LRS.

In some of these species males have higher survival rates than females, which gives rise to a surplus of males among birds of breeding age. With a longer potential lifespan the most productive males can produce more young than the most productive females, but more males than females remain unmated or become mated only late in life.

The last species in this section, the North American Red-winged Blackbird *Agelaius phoeniceus*, nests primarily in marsh vegetation, but forages on farmland. Favoured by agriculture, it has increased greatly in numbers during this century, and is now one of the commonest birds in North America, sometimes forming roosts of more than a million birds. Among passerines, it is one of the most strikingly sexually dimorphic species, in both size and colour. It is also one of the most strikingly polygynous, as territorial males can have harems of up to a dozen or more females each year. In their study of this species, Orians & Beletsky show how the reproductive rate of males increases almost linearly with the size of harem, and how extraordinarily productive the most successful males can be in their lifetime. On the other hand, in such a system many other males, but hardly any females, fail to breed at all.

8. Meadow Pipit

HERMANN HÖTKER

Lifetime reproductive outputs of male and female Meadow Pipits *Anthus pratensis* L. were studied in north-western Germany (52° 11'N, 8° 28'E; altitude 70 m) in the years 1975–1986 (Hötker 1988). In a 560 ha study plot, which held between 55 (1975) and 132 (1984) breeding territories, 864 nestlings and juveniles were individually colour-ringed. From those birds returning to the study site, 49 males and 33 females could be followed for their whole lives. None of these birds was still alive at the end of the study in 1986.

Meadow Pipits are small, mainly insectivorous migratory passerines which breed in western, northern, central and north-eastern Europe, Iceland and the north-westernmost part of Siberia. From Table 8.1, which shows some biometrical data from my study population, it may be seen that males and females differ in wing and tail lengths but not in other body measurements. There is no sexual dimorphism in colour, but during the breeding season birds may be sexed in the field by behaviour. Meadow Pipits inhabit open ground, including tundra, moorland, saltings and meadows, where they nest on the ground. The duration of the laying season varies greatly with the geographical location: about 16 weeks in the south-west of the breeding range to about five weeks in the extreme north or at high altitudes (Hötker & Sudfeldt 1982, Hötker in press). The number of broods per year varies from two to three where the season is long to only one where the season is short (Glutz von Blotzheim & Bauer 1985). In my study area the mean number of breeding attempts per year was 2.3. Many females started a third clutch even if they had already raised two broods that season. The mean clutch size of the Meadow Pipit seems to be correlated with geographical situation in the opposite way to length of breeding season. Clutches are smallest in the south, with mean clutch sizes of 4.0 eggs in north-western France (Constant & Eybert 1980) or 4.6 in my study area, and largest in the

LIFETIME REPRODUCTION IN BIRDS
ISBN 0-12-517370-9

Table 8.1 Biometrical data of adult Meadow Pipits breeding in the study area.

	males			*females*		
	n	x̄	SD	n	x̄	SD
wing length (mm)	63	82.44	1.70	65	77.28	1.50
tail length (mm)	64	56.34	2.37	60	52.58	2.04
bill length (mm)	65	11.70	0.50	65	11.56	0.59
tarsus length (mm)	64	21.66	0.67	60	21.55	0.59
weight (g)	56	18.45	1.04	54	18.13	0.87

north, for example 5.7 eggs in northern Norway (Haftorn 1971). Coulson (1956) could also show that in Great Britain clutch size increased with altitude. In my study area Meadow Pipits were not very successful in raising broods. An egg produced a fledgling with a mean probability of not more than 0.36. The main losses were due to predation, which affected 29 % of clutches and 12 % of broods (Hötker & Sudfeldt 1982). The main predators on the eggs and young were probably Carrion Crow *Corvus corone*, Stoat *Mustella erminea* and Feral Cat *Felis silvestris f. catus*. In contrast to the situation in Great Britain, the Cuckoo *Cuculus canorus* did not play any significant role in my study area. A considerable number of clutches were just abandoned. Published data indicate that breeding success increases further north in the breeding range (Glutz v. Blutzheim & Bauer 1985, Hötker in press). In summary, great differences in reproductive biology occur between the north and the south of the species' range, and my results from north-western Germany are probably not representative of the conditions in the biggest (northern) part of the range.

Coulson (1956), using British recoveries, estimated the annual mortality of adult Meadow Pipits at 57 % and that for first-year birds (after ringing in the nest) at 77 %. Seel & Walton (1979) found 54 % for adults in a colour-ringed population in Wales and I calculated 51 % for adults and 72 % for first year birds from the EURING data bank, which included recoveries from birds ringed in Belgium, Denmark, Federal Republic of Germany, Finland, France, Great Britain, Ireland, Netherlands and Switzerland. In all cases, the methods used for calculating mean mortality rates were fairly crude (Anderson *et al.* 1985).

The results of individual colour ringing of 1,078 nestlings and juveniles and 79 adults in my study showed that a higher proportion of first year males (80 individuals) than first year females (51 individuals) returned to their natal area. Similarly a higher proportion of adult males (35 %) than of adult females (32 %) returned to their territories of the previous year. Once settled in a territory, both male and female Meadow Pipits usually remain faithful to a breeding site in subsequent years. Juveniles, however, may disperse many kilometres from their birth places. Sex differences in dispersal distances (birthplace to place of first breeding and the breeding place in one year to the breeding place the next year) were not found (Hötker 1982a), so there is probably a sex difference in mortality rates. This was also suggested by a surplus of males in the breeding population, and by the fact that the maximum longevity of males (eight years) and females (three years) in my study differed considerably.

ANNUAL BREEDING SUCCESS

Male Meadow Pipits defend breeding territories and try to attract females to them. Many males, 26 % in my study, were not able to obtain a mate. Some males (one of those included in this study) were paired with two females simultaneously. Nevertheless, in my study population females were only about 80 % as numerous as males (unpublished data).

Meadow Pipits usually start to breed during their first year of life. However, some males (seven of the 49 in this study) and a few females (two of 33 in this study) were not seen in the study area until their second year. These birds were considered as not breeding in their first year. Confirmation of the existence of non-breeding and non-territorial males came from experiments in which, in 13 out of 17 territories removed territorial males were replaced within some days (unpublished data). In this study no corrections were made for the fact that more than those nine non-breeders and non-territory holders might have existed, but remained undetected because they died before settling in a territory.

Data on annual reproductive output for male and female Meadow Pipits are summarized in Fig 8.1. The peaks at four and eight fledglings are explained partly by the fact that four was the most common brood size and partly by the working assumption that four fledglings were produced in nests that were not found. In these cases the conspicuous feeding and warning behaviour of the adults (Hötker 1982b) allowed the detection of nearly all successful broods in the study area, even though the nest may have been missed. Four was chosen as an estimate of the number of fledglings in not-found but successful broods, because four was the modal and the medium number of fledglings per successful brood in the study area, and similar to the arithmetic mean of 4.07 (Hötker & Sudfeldt 1982). Females on average produced more juveniles per year ($\bar{x} = 4.26$, SD = 2.93) than males ($\bar{x} = 3.16$, SD = 3.13). The proportion

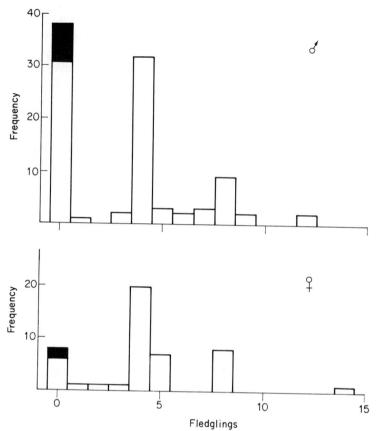

Figure 8.1 Annual production of fledglings by male and female Meadow Pipits. The black parts of the columns refer to those birds which were not seen in the study site in their first year.

of males failing in one year (40 %) was significantly higher than the equivalent percentage of females (15 %, $\hat{\chi}^2 = 7.81$, $P < 0.01$). The main reason for failing in males was their inability to attract a female.

Females showed hardly any variation of breeding performance with age (Table 8.2). Males showed an increase of mean annual production in the first three years of life and a probable (very few data available) decline thereafter. The low mean success of males in the first year was mainly due to the existence of non-breeders. Breeding first-year males were nearly as successful as their older colleagues.

Table 8.2 Numbers of fledglings raised annually by Meadow Pipits in relation to their age. Figures in brackets refer only to birds actually found breeding in that year (see text for details).

	Males								Females		
age (years)	1	2	3	4	5	6	7	8	1	2	3
number of fledglings											
0	21(14)	9	4	1	—	1	1	1	5(3)	3	—
1	—	1	—	—	—	—	—	—	1	—	—
2	—	—	—	—	—	—	—	—	1	—	—
3	—	1	1	—	—	—	—	—	1	—	—
4	19	7	4	1	1	—	—	—	11	8	1
5	—	1	1	1	—	—	—	—	6	—	1
6	1	1	—	—	—	—	—	—	—	—	—
7	2	1	—	—	—	—	—	—	—	—	—
8	5	2	—	—	1	1	—	—	8	—	—
9	1	—	1	—	—	—	—	—	—	—	—
10	—	—	—	—	—	—	—	—	—	—	—
11	—	—	—	—	—	—	—	—	—	—	—
12	—	1	1	—	—	—	—	—	—	—	—
13	—	—	—	—	—	—	—	—	—	—	—
14	—	—	—	—	—	—	—	—	—	1	—
n	49(42)	24	12	9					33(31)	14	
$\bar{x}$	2.96(3.45)	3.25	3.75	3.22					4.36(4.65)	3.92	
SD	2.92(2.87)	3.30	3.74	3.38					2.64(2.47)	3.38	

LIFETIME BREEDING SUCCESS

The lifetime reproductive output of each bird was taken as the number of young raised to fledging. This number was in turn linearly related to the number of young returning to the local breeding population (Fig. 8.2). The lifetime reproductive output of the returning juveniles did not correlate with that of their parents ($n = 20$, $r = -0.09$). Meadow Pipits raising many fledglings contributed more recruits to the next generation than did those raising few, but Meadow Pipits raised by parents with a high reproductive output did not have less chance of reproducing than did birds raised by parents with a low reproductive output. In other words, the number of fledglings raised proved to be a reasonable predictor of individual fitness.

The number of young produced by different birds varied enormously (Fig 8.3). Again, the peaks at four and eight fledglings were partly connected with the assumption of four fledglings in nests that were not found (see above). The arithmetic means for males ($\bar{x} = 6.06$, SD $=5.97$)

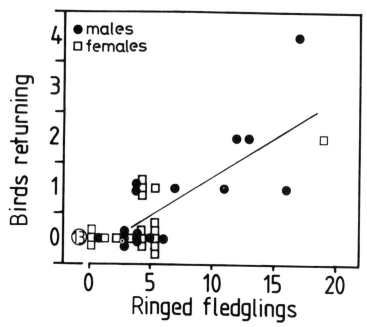

Figure 8.2 Relationship between the number of offspring returning to the breeding population and the number of ringed fledglings raised per individual parent. Each circle represents a female parent, each square a male parent. Statistics (males and females, without unsuccessful birds): $n = 32$, $r = 0.78$, $\hat{t} = 6.85$, $P < 0.001$, $y = 0.16x - 0.34$.

and females ($\bar{x} = 6.03$, SD $= 3.73$) were about the same (combined data for both sexes: $\bar{x} = 6.05$, SD $= 5.16$), but the shapes of frequency distributions were significantly different ($\hat{\chi}^2 = 9.36$, df $= 3$, $P < 0.05$). Males occurred more often at the extreme ends of the range, as many completely failed to reproduce or had extremely good success. In contrast, most females were moderately successful. The percentage of males raising either no broods or more than two broods, i.e. none or more than ten fledglings (41 % of males), was significantly higher than the equivalent percentage of females (15 %, $\hat{\chi}^2 = 6.13$, $P < 0.05$). The variance in male lifetime production was 35.6, compared with 13.9 in females. Four of the 11 unsuccessful males had no breeding success because they never acquired a mate. The one bigynous male in the study was moderately successful and raised eight fledglings; the only two females which in this study had to share their male with one another produced four fledglings each.

The productivity figures presented here should be taken as minimal estimates, because the chance of missing broods by finding neither the nest nor the alerting and food carrying parents could not be completely

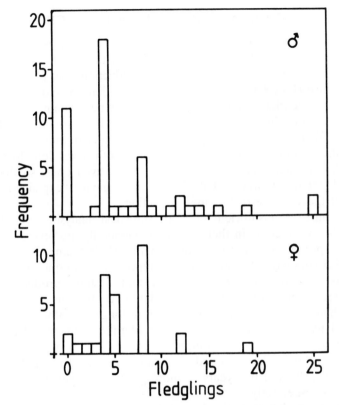

Figure 8.3 Lifetime production of fledglings by 49 male and 33 female Meadow Pipits.

excluded, and some broods may have been missed because they were raised just outside the study plot. Any error caused by overlooking broods was probably small, however,

A very small percentage of males was responsible for most of the recruitment: only 78 % of the males that reached reproductive age actually produced young; 18 % of the most productive males accounted for more than 50 % of the reproductive output of the population and 41 % for 75 %. According to the rates of breeding success and first year bird survival mentioned above, not more than 3.9 % of the eggs laid resulted in reproductive males and 0.9 % of the eggs provided half of next generation's reproducing males. The equivalent figures for the females are somewhat higher: 30 % of the females raised half the fledglings; 4.8 % of the eggs led to reproducing females; 1.5 % of the eggs yielded half the next generation's females (for details see Hötker 1988).

PROXIMATE FACTORS AFFECTING LIFETIME PRODUCTION

The lifetime production of a bird depended on its longevity and its (mean) annual productivity. In order to see which of the two factors had most influence in Meadow Pipits, lifetime reproduction was plotted against each in turn (Figs 8.4 and 8.5). Fig 8.4 shows that no birds achieved the maximum potential production of 15 fledglings per annum (three broods with five young in each, Hötker & Sudfeldt 1982); a large difference between their theoretical output and their actual output was apparent especially in older birds. Nevertheless, in both sexes long-lived birds generally produced more fledglings than short-lived ones, so that good survival enhanced fitness. Differences in longevity "explained" about 50 % of the variance of lifetime production.

Unfortunately unsufficient data for statistical analysis were obtained for birds not breeding in their first year. Generally, however, these late breeders produced more young than their early breeding conspecifics.

Fig 8.5 compares the mean annual reproductive output of each bird with its lifetime production. Annual and lifetime production were correlated significantly, but the correlation coefficients are somewhat lower than those with longevity. Therefore longevity rather than annual productivity had the greatest influence on lifetime production. Differences in mean annual reproductive output "explained" 27 % of the variance in lifetime production in males and 42 % in females. Restricting the analysis to birds which lived two years or more gave equivalent figures of 66 % and 77 % respectively.

Longevity and mean annual production of the Meadow Pipits in my study area were not related, as the small correlation coefficients show (males: $n = 49$, $r = 0.11$, $t = 0.75$, ns; females: $n = 33$, $r = 0.03$, $t = 0.15$, ns). Evidently, a high annual productivity did not reduce the lifespan.

INFLUENCE OF HABITAT ON PRODUCTION

To measure habitat quality, I covered the maps of the study site with a grid and calculated the yearly mean numbers of fledged young in each of the 140 200 × 200 m squares. I omitted those years in which any of the 49 males and 33 females involved in this study occurred, so that this measure of habitat quality was independent of the reproductive success of these particular individuals. It could not be completely excluded that "habitat quality", as defined here, was dependent on "bird quality", but in the 11 years (1975–1985) of habitat assessment, the same areas always proved to be the good ones. It is improbable that the fittest birds went

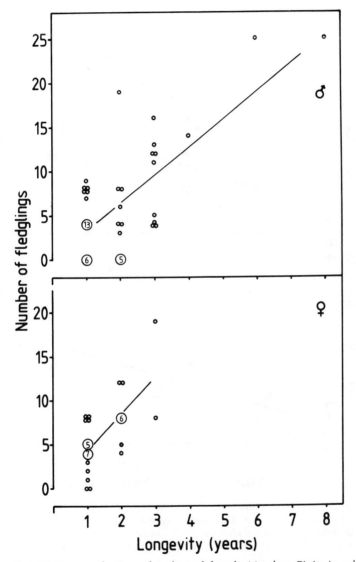

Figure 8.4 Lifetime production of male and female Meadow Pipits in relation to longevity. Each circle represents a different individual or the number of individuals given within the circle. Males: $n = 49$, $r = 0.72$, $\hat{t} = 7.11$, $P < 0.001$, $y = 3.16\,x + 0.00$; females: $n = 33$, $r = 0.69$, $\hat{t} = 5.31$, $P < 0.001$, $y = 4.19\,x + 0.06$.

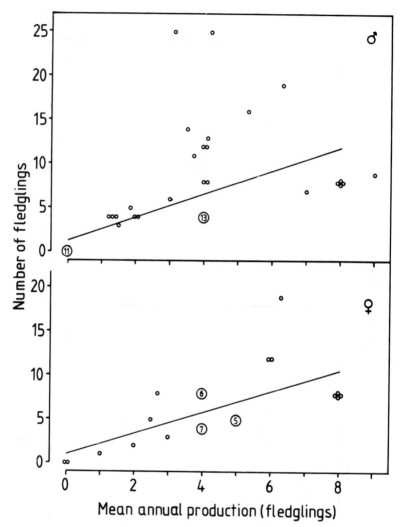

Figure 8.5 Lifetime production of male and female Meadow Pipits in relation to their mean annual production. Each circle represents a different individual or the number of individuals given within the circle. Males: $n = 49$, $r = 0.52$, $\hat{t} = 4.22$, $P < 0.001$, $y = 1.26\,x + 1.95$; females: $n = 33$, $r = 0.65$, $\hat{t} = 4.70$, $P < 0.001$, $y = 1.18\,x + 1.03$.

to the same places every year just by coincidence. The data show quite clearly that the differences in quality between the different areas were real. They also correlated with the types of vegetation found, with rough pasture being better than arable.

The lifetime reproductive output of individuals is shown in relation to the quality of their territories in Fig 8.6. Mean values were used if the birds changed territories during their lifetime. Males settling in high quality habitat usually produced more fledglings than those in poor

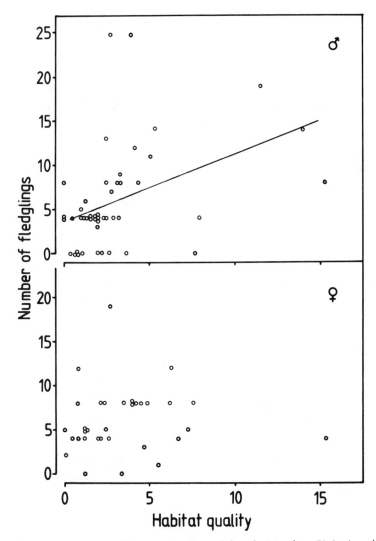

Figure 8.6 Lifetime production of male and female Meadow Pipits in relation to habitat quality. Each circle represents a different individual. Males: $n = 47$, $r = 0.49$ (Spearman), $\hat{t} = 3.79$, $P < 0.001$, $y = 0.65\,x + 3.31$; females: $n = 33$, $r = 0.19$ (Spearman), $\hat{t} = 1.05$, ns.

habitat; the correlation between lifetime reproductive success and habitat quality was significant. This did not hold true for the females which reared no more young in high quality habitat than they did in medium or even poor habitat. For the males the correlation between habitat quality and return rate was very weak ($n = 47$, $r = 0.15$, $\hat{t} = 1.02$, ns), so males in good habitat did not breed better because they survived better there, but because they had a higher mean annual production (correlation habitat quality–mean annual production: $n = 47$, $r = 0.33$, $\hat{t} = 2.37$, $P < 0.05$). This higher annual production was a result of a higher chance to find a mate and a higher breeding success in good habitat. Annual reproduction of males was significantly correlated with habitat quality ($n = 82$, $r = 0.31$, $\hat{t} = 2.92$, $P < 0.01$); annual production of females was not ($n = 44$, $r = 0.05$, $\hat{t} = 0.31$, ns).

Unfortunately the influence of biometrical body characters on reproductive success could not be measured directly because most birds included in this study were ringed as nestlings or juveniles and not retrapped later, when they were full grown. However, some adults were trapped, marked and measured and their fate could be followed for the rest of their lives. Lifetime reproductive outputs for these birds are minimal estimates because those fledglings they produced before the date of ringing could not be taken into consideration. The correlation analyses (Table 8.3) show that most body measurements were positively but very weakly correlated with the remaining lifetime reproductive success. The relationship of tarsus lengths to breeding success was significant in males (but not females). I can offer no explanation for this difference.

Discussion

The lifetime reproductive data for male and female Meadow Pipits presented here were lower than needed to maintain population stability.

Table 8.3 Correlation coefficients between biometrical characters and lifetime reproductive success (after the catching date) of adult male and female Meadow Pipits.

correlation lifetime reproductive success	males				females			
	n	*r*	*t*	*significance*	*n*	*r*	*t*	*significance*
wing length	29	0.19	1.04	n.s.	33	0.15	0.83	n.s.
bill length	29	0.09	0.48	n.s.	34	0.17	1.00	n.s.
tarsus length	29	0.42	2.43	p<0.05	31	0.05	0.25	n.s.
weight	24	0.23	1.15	n.s.	23	0.13	0.61	n.s.

With a first-year survival rate of 0.28 (from ringing results), a mean production of 6.06 fledglings (males) would yield 0.85 first-year birds and a production of 6.03 (females) would yield 0.84 first-year birds. This is less than 1.0 offspring per parent surviving to breeding age, which would be necessary to keep the population stable. As the study population did not decline during the period concerned, the first-year survival rate or the lifetime reproductive output, or both, might have been underestimated. For lifetime production a slight underestimate was likely, because of unfound broods. In addition, the population of the study area might have been maintained partly by immigration from other areas. This was indeed possible, because the biggest part of the area consisted of a mixture of pasture and arable land, which provided habitat of low quality, where few fledglings were raised.

A striking feature of lifetime reproductive data was the high frequency of failures. This was mostly due to the high annual mortality among adults and high predation on clutches and broods, which meant that, for a Meadow Pipit, the probability of surviving only one breeding season, and having all nesting attempts fail that year, was quite high. In this respect Meadow Pipits resemble most of the non-hole nesting passerines (Ricklefs 1969, Cody 1971, von Haartman 1971), so the frequency distributions shown in Fig. 8.3 may be typical for this group of species.

Only very few Meadow Pipits contributed substantially to the next generation's genepool. The effective population size (Wright 1938) is therefore very much smaller than the actual population size. In theory, this could mean that evolutionary processes, such as the spreading of favourable genotypes in the population or a reduction of the genetic variability within the population, could act very quickly, assuming the underlying traits that govern the observed patterns were inherited. This was difficult to prove, because virtually nothing was known about the causes of death in full grown birds and about those aspects of the behaviour of the parents and nestlings that enhanced nesting success. Predation, the most important cause of mortality of eggs and nestlings, apparently happened more or less randomly, so that those eggs (genes) contributing to the next generation were possibly not in the main chosen by selection but by chance.

The difference in reproductive patterns between male and female Meadow Pipits could be explained at least partly by the lower survival (return) rate of females and by the resulting relative scarcity of females in the population. Reasons for the obviously higher mortality of females are not known exactly. There are some indications that females on average migrated longer distances than males and that considerable numbers of females were killed by predators at the nest. Males do not brood and, therefore, did not suffer from this source of mortality. Scarcity of females meant that each female easily obtained a mate and bred. Failure due to lack of mate, a frequent event for males, was virtually absent in females. On the other hand, females, due to their shorter lifespan, had fewer opportunities to produce a large number of offspring. Males had to compete for females and some of them were not able to overcome this first hurdle in producing young. On average, however, males had more chance to reproduce and to raise a large number of fledglings because of their higher survival rate. Scarcity of females could also partly explain the sex difference in the relationship between habitat quality and lifetime reproduction. Males in poor territories often did not attract a female, so had a breeding success of nil, while females which settled in a poor territory had at least the chance to breed and were often successful. However, it may be that better quality males both obtained territories in the best habitat and produced more young.

Summary

During a 12-year population study in north-west Germany the number of fledglings reared in a lifetime was determined for 49 male and 33 female Meadow Pipits. Males showed a somewhat higher return (survival) rate and a somewhat lower annual production of young than females.

Males produced an average of 6.06, females an average of 6.03, fledglings during their lifetime; these figures probably being slight underestimates. The proportions of males failing completely or producing very many offspring were significantly higher than in females. The number of recruits descended from individual birds was significantly correlated with the number of (ringed) fledglings they produced, implying that lifetime fledgling output was a good measure of individual fitness. For both males and females, longevity had more influence on lifetime reproduction than mean annual production. A significant correlation between habitat quality and lifetime reproduction was found for males but not for females.

Acknowledgements

I am grateful to Christoph Sudfeldt for his help during the fieldwork and to Dr I. Newton for useful comments on the manuscript. The EURING centre in Heteren (R.D. Wassenaar) kindly provided me with recovery data from several European ringing schemes.

References

Anderson, D.R., Burnham, K.P. & White, G.C. 1985. Problems in estimating age-specific survival rates from recovery data of birds ringed as young. *J. Anim. Ecol.* **54**: 89–98.

Cody, M.L. 1971. Ecological aspects of reproduction. In *Avian Biology*, Vol. 1, ed. D.S. Farner & J.R. King, pp. 461–512, New York: Academic Press.

Constant, P. & Eybert, M.C. 1980. Donnés sur la biologie de la reproduction du Pipit farlouse dans les Landes Bretonnes. *Nos Oiseaux* **35**: 349–60.

Coulson, J.C. 1956. Mortality and egg production of the Meadow Pipit with special reference to altitude. *Bird Study* **3**: 119–32.

Glutz von Blotzheim, U.N. & Bauer, K.M. 1985. *Handbuch der Vögel Mitteleuropas*. Band 10/11. AULA, Wiesbaden.

Haastman, L. von. 1971. Population Dynamics. In Avian Biology Vol. 1 ed. D.S. Farner & J.R. King, pp. 331–459. New York: Academic Press.

Haftorn, S. 1971. *Norges Fugler*. Universitetsforlaget, Oslo, Bergen, Tromsø.

Hötker H. 1982a. Studies of Meadow Pipit dispersal. *Ring. & Migr.* **4**: 45–50.

Hötker, H. 1982b. Zum Verhalten junger Wiesenpieper (*Anthus pratensis*) nach der Nestlingszeit. *Vogelwelt* **103**: 1–16.

Hötker, H. 1988. Lifetime reproductive output of male and female Meadow Pipits *Anthus pratensis*. *J. Anim. Ecol.* **57**: 109–17.

Hötker, H. (in prep.) Der Wiesenpieper. Ziemsen, Wittenberg Lutherstadt.

Hötker, H. & Sudfeldt, C. (1982). Untersuchungen zur Brutbiologie des Wiesenpiepers. *J. für Orn.* **123**: 183–201.

Ricklefs, R.E. 1969. An analysis of nesting mortality in birds. *Smithsonian Contributions to Zoology* **9**: 1–48.

Seel, D.C. & Walton, K.C. 1979. Number of Meadow Pipits *Anthus pratensis* on mountain farm grassland in North Wales in the breeding season. *Ibis* **121**: 147–64.

Wright, S. 1938. Size of population and breeding structure in relation to evolution. *Science* **87**: 430–1.

9. Song Sparrow

WESLEY M. HOCHACHKA, JAMES N.M. SMITH &
PETER ARCESE

In this chapter, we examine some demographic and environmental influences on adult survival and lifetime reproductive success (LRS) of Song Sparrows *Melospiza melodia*. A previous analysis for Song Sparrows (Smith 1988) showed that the most important influences on lifetime success of adults are breeding lifespan, and the overwinter survival of their offspring. In Smith's (1988) analyses, variation in survival and reproductive rate among cohorts were corrected for statistically, and causes of variation among cohorts were not considered in detail. Also, the effects of environment on lifetime reproductive success were not examined. In this chapter, two main points are made. First, population density has an important influence on variation in lifetime reproductive success among cohorts, Second, some territories are predictably better than others for both adult survival and reproduction.

The Song Sparrow is a small (ca.25 g), subtly-hued, brown and grey passerine in the subfamily Emberizinae. Males are slightly larger than females, but the sexes do not differ in plumage. The species breeds over most of North America, from the Aleutian islands east to Newfoundland, and south to the highlands of central Mexico, showing wide variation in size and coloration over this range (Bent 1968, Aldrich 1984). Although migratory in some northern areas, Song. Sparrows are partially or completely sedentary in moderate climates. The preferred breeding habitat of most races is broad-leaved shrub, but a wide range of habitats is used, including riparian woodland and saltmarsh (Aldrich 1984).

The population we studied is isolated, and resides year-round on Mandarte Island, a 6 ha shrub- and grass-covered rock in the southern Haro Strait, roughly 25 km NNE of Victoria, BC, Canada. Since 1975, all birds in this population have been individually colour-banded. Numbers and reproduction have been monitored in all succeeding years, except

LIFETIME REPRODUCTION IN BIRDS
ISBN 0-12-517370-9

1980. A fuller description of the island and study methods can be found in Tompa (1964) and Smith (1981a).

This population of Song Sparrows is well suited for the study of lifetime reproductive success. We are able to study the entire population. The birds are tame, and easily observed. They live under natural conditions; their habitat is not altered by man, and the sparrows choose their own nest sites. Population density has been high but unstable, providing us with ideal material to examine relationships between density and breeding success. Previous work on this population has dealt with survival and fecundity (e.g. Smith 1981b, Nol & Smith 1987), social organization (Arcese & Smith 1985, Arcese 1987, Arcese 1988, 1989), dispersal (Arcese 1988), and morphological evolution (Smith & Dhondt 1980, Schluter & Smith 1986).

Song Sparrows are typically monogamous, with polygyny occurring at low frequency (Nice 1943, Smith et al. 1982, Arcese 1988). Males never feed their mates. The female alone builds the nest and incubates the eggs. Nests are open, and usually placed at the base of shrubs or clumps of grass. Breeding birds in the most intensively studied populations (Columbus, Ohio; San Francisco; and coastal British Columbia) typically produce more than one brood each year (Nice 1937, Johnston 1956, this study). The maximum known age reached by a Song Sparrow is eight years (Nice 1937, this study).

On Mandarte, male sparrows live for 1–8 breeding seasons (median = 2), and females for 1–5 (median = 2). All females and most males start breeding at age 1. Each year, females start from 1 to 7 nests (median = 2), usually laying a clutch of 3–4 eggs. Young are fledged from a maximum of four nests (median = 2) each year (Smith 1982). The breeding success of females varies slightly with age, with two- and three-year old females producing more independent offspring than others; for males, reproductive success increases up to four years of age (Nol & Smith 1987). Eighteen per cent of males failed to fledge any young in a given year, but some males have fledged up to 11 young in a year (median = 3); over 15 % of females fledged no young in a given year, and the most successful female fledged 10 young in a season (median = 3). Median lifetime output of fledglings is only slightly greater than annual values: over their lives males produced a median of four fledglings, and females five fledglings. However, 12 % of males and females failed to fledge any young over their lives, and the most successful males fledged 30 young and the best females fledged 24 young. The principle cause of nest failure is predation; Deer Mice (*Peromyscus maniculatus*) and Northwestern Crows (*Corvus caurinus*) are the major culprits. Brown-headed Cowbirds (*Molothrus ater*) parasitize Song Sparrow nests, but parasitism has been ignored in this chapter as it has little effect on the breeding success of Song Sparrows (Smith 1981a).

The sparrows are highly sedentary outside the breeding season. Even in winter they spend much of their time on territories. Only four sparrows born on Mandarte are known to have settled elsewhere; in general, movement of sparrows between Mandarte and nearby islands appears limited (Arcese 1988). Immigration into the population is low, with only 12 immigrants (about 3 % of breeders) entering the breeding population in 11 years. Only two of 12 successful immigrants were males, and at most two immigrants have settled in one year.

Rationale and analysis

Our aim was to explore variation in LRS both among and within cohorts. Since LRS is the product of lifespan and annual reproduction, each section of this chapter separately considers lifespan and annual breeding success before examining LRS. Some male sparrows ("floaters") never establish territories and breed, and some females settle on territories but die before nesting. These birds are excluded from consideration, because they did not contribute to the reproductive success of their parents (see below), and themselves were not members of the breeding population, not even attempting to breed. The analyses are confined to those birds that gained territories (males), or survived long enough into their first breeding season to have nested (females).

We defined breeding success as the number of young recruiting into the breeding population (offspring holding territories in subsequent breeding seasons). A more proximate measure, the number of offspring that reach independence from parental care (ca. 30 days post-fledging), was also used in previous studies of this population. The number of independent young produced in a lifetime was positively correlated with number of recruits produced ($r^2 = 0.76$). However, the slopes of the yearly relationships (with intercepts fixed at the origin) varied significantly among cohorts ($P < 0.001$), because the survival of independent young varied among years. Thus, although the number of independent young is useful for comparison of LRS within cohorts, the number of recruits is better for comparisons between cohorts.

Data on LRS presented here come from cohorts born in 1975–1983 (four more cohorts than used by Smith 1988). Twenty males and 15 females of the 1981–1983 cohorts were still alive at the time of writing, but their further reproduction can have little effect on the patterns described. Also, the population was not studied intensively in 1980, after a major population crash the previous winter. As a result we have underestimated LRS for the few surviving individuals in the 1975–1979 cohorts. Measures of annual reproductive rates and survival rates of breeding birds are based on data collected in 1975–1979 and 1981–1987. As we could not study individuals that moved away from Mandarte, we measured reproductive success as the number of offspring that remained on the island. The inclusion of data from a year of high density that preceded a dramatic population crash (see below) may have accentuated negative correlations between density and survival. Thus, these correlations are presented both with and without the year in question.

Variation among cohorts

LONGEVITY AS BREEDER

Some males lose their territories, but remain as floaters in the population for a breeding season before dying. Thus our measure of longevity is breeding lifespan of adults: the number of breeding seasons in which they occupied territories. Dramatic variation in mean breeding lifespan among cohorts was caused by substantial mortality in the winter 1979–1980, when the breeding population declined from 65 females in 1979 to nine in 1980 (Fig. 9.1). This decline was apparently not caused by extreme weather,

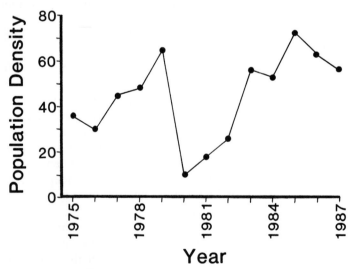

Figure 9.1 Population size (number of breeding females) of Song Sparrows over the course of the study on Mandarte Island.

nor by high density alone: numbers have since exceeded 65 breeding females without a subsequent sharp decline (Fig. 9.1).

Breeding lifespan was not consistently correlated with population density in the year of recruitment (Fig. 9.2). The latter was a good estimator of the densities experienced by sparrows throughout their lives, as life expectancy was short (median = two years) and densities were similar in succeeding pairs of years ($r = 0.43$, $n = 12$, $P < 0.20$ for all years; $r = 0.79$, $n = 11$, $P < 0.01$ when the pair of years before and after the crash were excluded). Hence, there was no good evidence that population density affected adult survival.

ANNUAL BREEDING SUCCESS

Annual breeding success declined with increased density (Fig. 9.3), and the distributions of annual breeding success were identical for males and females. Further, the range of variation in annual breeding success declined in both sexes as density increased; correlations of inter-quartile range (Fig. 9.3) against population density were: $r = 0.75$, $n = 11$, $P < 0.01$ for both males and females. The decrease in variation of breeding success with increasing density was due solely to a lowering of the upper quartile at increased densities; at all densities a large proportion of birds produced no young (lower quartile of reproductive success was zero at all but one density, Fig. 9.3).

LIFETIME REPRODUCTIVE SUCCESS

The distribution of LRS varied among cohorts (Fig. 9.4); only data for males are shown as LRS of females was similar. The cohort of breeders most affected by the population crash was that hatched in 1978, most of whose members bred only in 1979. Eighty per cent of males in this cohort produced no recruits. By contrast, only 42 % of males in the cohort with the next highest proportion of failures (1983), produced no young. This pattern was even stronger for females (not illustrated): 97 % of the 1978 cohort produced no recruits, whereas only 50 % of the next poorest cohort (1975) failed completely. The few males that survived the crash produced many recruits (see 1978 and 1979 in Fig. 9.4), but the very few adult females that survived were relatively unsuccessful.

More subtle variation in LRS occurred among cohorts, as a function of breeding density at the time of recruitment. The greater the density of the breeding population that the sparrows entered in their first year, the lower was their LRS. This was true for both females and males (Fig. 9.5). This pattern did not result from an effect of density on breeding

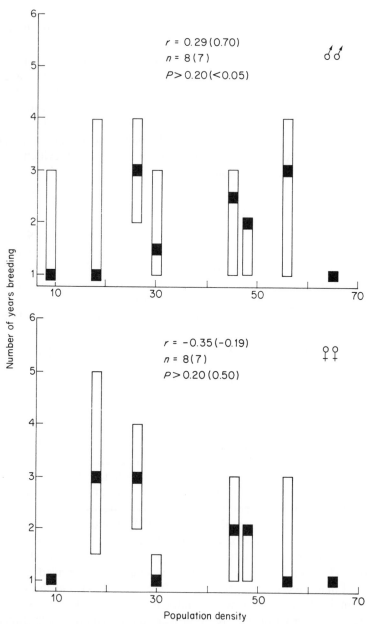

Figure 9.2 Longevity among cohorts as a function of population density in the year of recruitment. The boxes represent median, and upper and lower quartiles. Points beyond the quartiles were not plotted for clarity of presentation. Correlations of median longevity on density were performed both including and excluding (results in parentheses) the data points for those birds that recruited the year before the population crash (density = 65). None of the correlations is statistically significant.

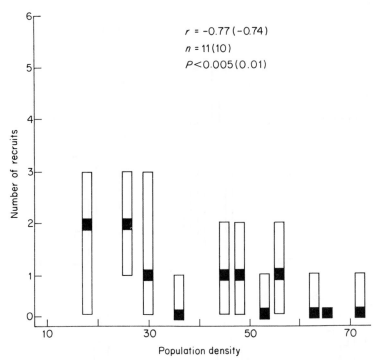

Figure 9.3 Annual reproductive success as a function of population density. The format of the plot is as described for Fig. 9.2. The distributions were identical for male and female reproductive success. The correlations between density and median reproductive success were statistically significant ($P < 0.01$), both with and without the values from the 1979 breeding season (at density = 65) that preceded the population crash.

lifespan, as these two factors were not significantly correlated for either sex (Fig. 9.2). Rather it was a function of: (a) a negative relationship between annual breeding success and density (Fig. 9.3); and (b) the fact that the median number of years that birds breed is two, and densities in succeeding pairs of years were correlated (see above).

Variation within cohorts

LONGEVITY AS BREEDER

The sexes differed in breeding lifespan. Overwinter survival between the first and second breeding seasons was higher for males than for females in six of eight cohorts examined ($P < 0.07$, sign test), and between the second and third seasons was higher in five of eight cohorts ($P > 0.20$,

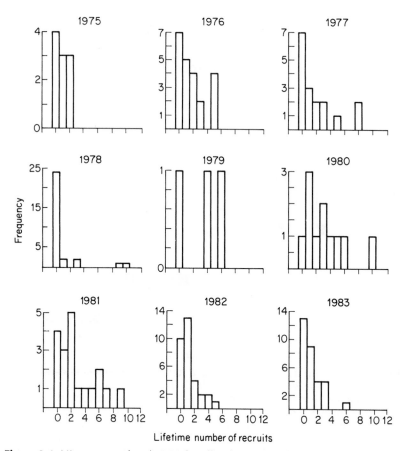

Figure 9.4 Histograms of male LRS for all cohorts considered in this study. Within any given cohort the distribution of female LRS did not differ significantly from that of the males (Kolmogorov–Smirnov two-sample tests, all $P > 0.50$), and so only data from male LRS are presented.

Figure 9.5 Lifetime reproductive success as a function of population density in the year of recruitment. The format of the plot is as described for Fig. 9.2. Correlations of median LRS on density were performed both including and excluding the cohort that recruited in 1979, the year before the population crash. All correlations were statistically significant ($P < 0.05$), except for that for females with the 1979 cohort excluded ($P = 0.07$ in that instance).

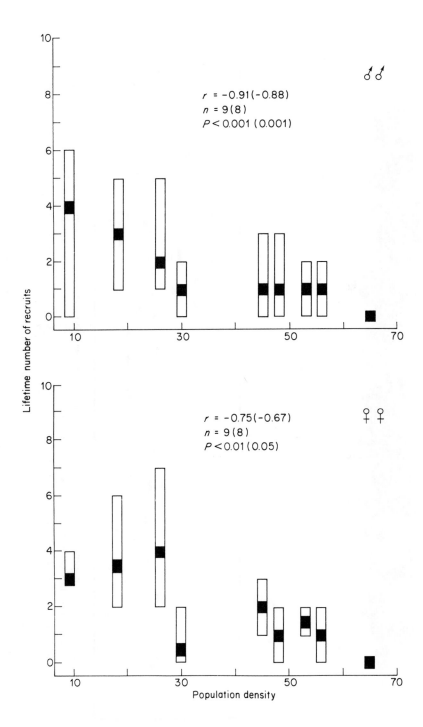

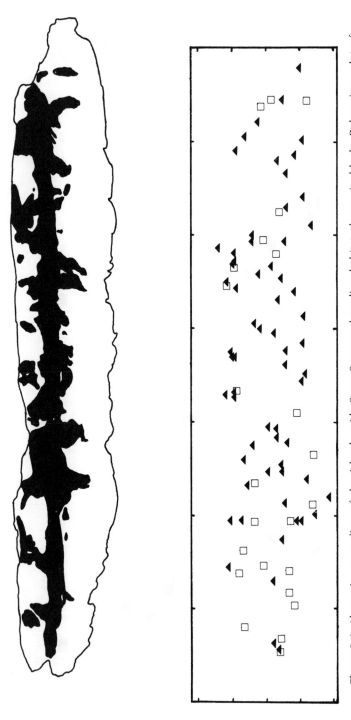

Figure 9.6 Above is an outline of the island, with Song Sparrow breeding habitat shown in black. Below is a plot of coordinates of nest sites of yearling males that died after one breeding season (open squares), or lived for more than one year (solid triangles). The upper figure and x-axis of the lower figure were drawn to the same scale, with the distance between tick-marks on the x-axis being 100 m; the y-axis of the lower graph was expanded so that the distance between tick-marks is 20 m. Analysis of nearest neighbour pairs showed that males that died after one year were more likely to be nearest neighbour pairs than would be expected by chance.

sign test). Differences in annual survival between sexes ranged from 0–60 % (median = 9 %) from ages 1 to 3. Beyond age 3 too few data were available to examine sex differences in survival. Although breeding males survived better than breeding females on average, some males ("floaters") failed to hold breeding territories, and hence to produce offspring. Most floaters were yearlings, tended to be socially subordinate, and came from nests late in the previous breeding season (Arcese & Smith 1985). Those males that floated as yearlings but gained territories at two years of age tended to lose their territories sooner than males that acquired territories as yearlings (Arcese 1988). Females always bred as yearlings, and breeding lifespan of females was not obviously affected by their social status as juveniles.

Smith (1981b) found a positive relationship between lifespan and annual breeding success. Nol & Smith (1987) showed that females that were relatively successful as yearlings were more likely to survive to age 2 than females with low success as yearlings. This pattern was not repeated for older females, or for males of any age. Positive correlations between lifespan and reproduction could be caused by variation in the local environment; birds in better habitat might both live longer and produce more offspring (van Noordwijk & de Jong 1986).

We therefore tested whether some areas of Mandarte had better survival of breeders than others, by checking whether sites occupied by surviving or dying birds were spatially aggregated. The breeding habitat on Mandarte is distributed irregularly (Fig.9.6, top panel), and on a fine scale. Hence we looked for aggregations of surviving and dying birds on a fine scale, examining clusters of only two birds. We calculated the average (x, y) coordinates of all the nests in a given year of each bird in the data set. The coordinates were used to represent the location of birds' territories. For every territory, we found the closest neighbour of the same sex within the data set, and classified each neighbour–neighbour pair as having: both neighbours surviving to breed in the next year, only one neighbour surviving, or both neighbours dying. The data come from several years, and each bird is only represented once in the data set. Thus, a bird cannot be its own nearest neighbour, and a "nearest neighbour" bird may actually be another bird occupying the same territory in a different year. Any local clustering of surviving or dying birds would indicate either that areas differed consistently in their quality over a series of years, or that birds of similar intrinsic quality consistently settled on the same parts of the island.

As survival varied between years and age classes, we performed the analyses using only data from groups of years with similar, intermediate survival rates (Table 9.1). Only data from one-year old birds were

Table 9.1 Cohorts of Song Sparrows used in nearest neighbour analyses, and ranges (%) in annual survival and reproductive success (0 or 1+ recruits produced) rates represented by the cohorts. The values from the combined samples are given in brackets.

	Cohorts Included in Analysis	Range in % Survival/Success Included
Yearling Males	1976, 77, 78, 83, 84	20–29(27)
Survival Rate		
Yearling Females	1976, 77, 84	30–36(33)
Yearling Males	1975, 76, 77, 82, 83	43–56(52)
Successful Nesting		
Yearling Females	1976, 77, 80, 82, 83	50–64(56)
Age 2 Males	1982, 83, 84	63–68(66)
Successful Nesting		
Age 2 Females	1981, 84	52–53(53)
Lifetime Males	1976, 77, 80, 81, 82, 83	48–91(65)
Successful Nesting		
Lifetime Females	1975, 77, 82	48–61(45)

analysed as too few data existed for older birds. Birds that changed territories within a year were excluded.

The frequency of nearest neighbour pairs with both birds dying was greater than expected by chance for both yearling females ($P = 0.06$) and males ($P < 0.01$) (Table 9.2, Fig. 9.6). These results suggest that some areas of the island led to lower survival of yearling sparrows than did others.

ANNUAL BREEDING SUCCESS

The local environment within the island influenced annual breeding success slightly more than did intrinsic differences among individuals. This is suggested by the group of females that moved their mean nest site more than 40 m between two breeding seasons so that their territories did not overlap between years. Each female and territory was then classified as either poor (no recruits produced) or good (1 + recruits produced) in the years before and after the move. Contingency table analysis suggests that territories were more likely to maintain their reproductive success (poor or good) with different females, than were females that moved to different territories (Table 9.3), although the results only border on statistical significance ($P < 0.06$). Too few males

Table 9.2 Observed and expected[1] frequencies of nearest neighbour pairs for yearling males and females, categorizing pairs as: both neighbours surviving more than one breeding season ("s–s"), one neighbour surviving more than one year but the other dying after one year ("s–d"), or both neighbours dying after one year ("d–d").

	Males			Females	
	frequencies			frequencies	
	obs.	exp.		obs.	exp.
s–s	44	48.0	s–s	33	35.5
s–d	26	35.5	s–d	30	34.9
d–d	20	6.5	d–d	16	8.6
	$P < 0.01$, G-test			$P = 0.06$, G-test	

[1] Expected frequencies were calculated by binomial expansion. Given a population size of n, probability of survival s, and probability of dying $1 - s = d$, expected frequencies are: ns^2 for both surviving, $2nds$ for one bird dying, and nd^2 for both dying.

Table 9.3 Contingency table, showing the frequencies with which females on new territories, and territories with new females, changed reproductive success (either producing or failing to produce recruits) between years.

		Territory with new female	Female with new territory
Reproductive status	same	27	9
	changed	18	16

$P < 0.06$, G-test

changed territories between years for us to conduct a similar analysis for them.

The nearest neighbour analysis described above (for survival) was also used to test for clustering of sites that had high or low breeding success. Clumping of good sites, or of bad sites, implies either a spatially predictable effect of the local habitat on reproduction, or that birds of similar reproductive quality settle in similar areas. Nearest neighbours were classified as: both neighbours producing recruits in a given year, only one of the neighbours producing recruits, or neither neighbour producing recruits. The analysis was performed separately for birds aged

1 and 2, and separately for each sex (Table 9.4a–d). Unsuccessful yearlings were more likely to have unsuccessful nearest neighbours than expected, but the pattern was statistically significant only for females (Table 9.4b). At age 2, an unsuccessful bird was less likely to have an unsuccessful neighbour than expected; this pattern was statistically significant for males but not for females. These results suggest that some environmental

Table 9.4 Observed and expected[1] frequencies of nearest neighbour pairs for males and females, for classes: yearling, age 2, and lifetime. Neighbour pairs are classified as: both neighbours succeeding in producing 1 + recruits ("s–s"), one neighbour succeeding in producing 1 + recruits while the other failed to produce recruits ("s–f"), or both neighbours failing to produce recruits ("f–f").

a) *Yearling males*

	frequencies obs.	exp.
s–s	26	24.3
s–f	41	44.9
f–f	23	20.7
P > 0.50, G-test		

b) *Yearling females*

	frequencies obs.	exp.
s–s	29	35.4
s–f	51	55.7
f–f	33	21.9
P < 0.05, G-test		

c) *Age 2 males*

	frequencies obs.	exp.
s–s	7	7.5
s–f	38	29.2
f–f	20	28.3
P = 0.06, G-test		

d) *Age 2 females*

	frequencies obs.	exp.
s–s	8	10.0
s–f	22	17.9
f–f	6	8.0
P > 0.10, G-test		

e) *Lifetime males*

	frequencies obs.	exp.
s–s	62	57.7
s–f	47	62.3
f–f	28	16.8
P < 0.01, G-test		

f) *Lifetime females*

	frequencies obs.	exp.
s–s	13	18.2
s–f	30	29.7
f–f	17	12.1
P > 0.20, G-test		

[1] See footnote to Table 9.2 for explanation of the calculation of expected frequencies.

factor(s) caused higher breeding success in certain areas of the island, but only for birds in their first breeding season.

Morphological characteristics of females occasionally influenced their annual breeding success (Schluter & Smith 1986). Females with relatively long tarsi and short beaks produced more independent young than other females. However, this effect was noticed only in the year of highest breeding density of the five years examined. No effect of male morphology on breeding success was found.

LIFETIME REPRODUCTIVE SUCCESS

Although more than 90 % of variance in male LRS, and 55 % of variance in female LRS, was within cohorts, previous work identified only one factor involved. Socially-dominant males are more likely to be territorial in their first year (Arcese & Smith 1985), and males that acquire territories in their first year have higher LRS than males that acquire territories only later in life (Smith & Arcese, in press). The only factor identified as affecting dominance of juveniles was their data of fledging, the earlier ones being superior.

Given that both lifespan and annual breeding success may be influenced by the local environment (Tables 9.2, 9.3), the same pattern is expected for LRS. The nearest neighbour analysis performed on annual survival and reproduction was repeated for LRS, with birds grouped as either producing 0 or 1 + recruiting offspring in their lifetimes. The frequency of unsuccessful birds being neighbours was significantly greater than expected for males, but not significant for females, although the same pattern held (Table 9.4e, f). Thus, there is some evidence that where a bird lives affects its LRS. This conclusion does not change when a different criterion is used to separate successful and failed birds (i.e. if failures produced 0–1 recruits in a lifetime, and successful birds produced 2 + recruits in a lifetime).

As lifespan and annual breeding success, which are the two determinants of LRS, and LRS itself, are all affected by local environment (Tables 9.2, 9.3, 9.4), one might expect that lifespan and annual breeding success were both poor on the same areas of Mandarte Island. This was not the case. Using the nearest neighbour analysis, we examined whether the survival of one bird would predict the likelihood of breeding success of its nearest neighbour. For neither yearling males or yearling females was this found. Thus, the effect of local environment on LRS reflects an effect of environment on survival or breeding success, but not necessarily both.

Discussion

Density-dependent annual breeding success (Fig. 9.3) has been demonstrated in other bird species (e.g Kluijver 1951, Alatalo & Lundberg 1984). However, density-dependence of lifetime reproductive success (Fig. 9.5) is a surprising and new result. For Song Sparrows, LRS could be predicted by density in the year of recruitment because birds were short-lived and densities in succeeding years were generally closely correlated. The short median lifespan of the Song Sparrows meant that most individuals experienced only a narrow range of breeding densities over their lives, causing variation in LRS among cohorts.

However, more than 90 % of variance in male LRS and over 55 % of variance in female LRS was within cohorts. LRS is expected to have very low heritability (e.g. Gustafsson 1986, and this was also true for the Song Sparrows on Mandarte Island, Smith 1988). Natural selection has been demonstrated on a morphological correlate of breeding success. Schluter & Smith (1986) noted that, although female morphology is heritable and related to annual breeding success, females with the morphological type with higher reproductive success (measured as the number of independent young) had poorer survival as juveniles. Thus, the two types of selection tended to balance each other, and no net change in morphology was expected. In fact, although the majority of variance in LRS occurs within cohorts, most of this variation appears to be of no evolutionary significance.

On Mandarte, some territories consistently had higher breeding success than others (Table 9.4). In addition, the territory a bird settled on is related to its subsequent survival (Table 9.2). Our data are consistent with van Noordwijk & de Jong's (1986) prediction that a positive relationship between lifespan and reproductive rate results from variation in the quality of the territories that birds occupy. We showed that both survival rate (Table 9.2), and annual and lifetime reproductive success (Table 9.4), were influenced by where birds settled. Surprisingly, although unsuccessful yearlings were more clumped than expected (Table 9.4a, b), unsuccessful two-year olds showed the opposite pattern, being less likely to have unsuccessful two-year old neighbours (Table 9.4c, d).

These spatial effects on survival and breeding success were detected only on a very fine scale: the patterns appeared only when birds were examined with respect to their nearest "neighbours", who were often other birds occupying the same territory in different years. These patterns held for groups of birds whose cohorts had overall similar rates of survival or breeding success; when the analyses were repeated with all available data, the patterns disappeared in three of eight analyses. Thus, spatial patterning of survival or breeding success is perhaps best explored using data from single seasons, or from single cohorts of animals.

In conclusion, in this short-lived species, breeding density had an important role in determining lifetime reproductive success on Mandarte Island. Within cohorts, differences among territories may be more important than differences among individuals in determining annual breeding success (Table 9.3). Differences in territory quality affect not only breeding success (annual and lifetime), but also the probability of survival of adults. Where density or territory quality affected annual breeding success and survival, these features also affected lifetime reproductive success. This reassures us that results from studies of annual breeding success of short-lived species are helpful in understanding lifetime reproductive success.

Acknowledgements

We thank the many people who have helped collect the data presented here. A. Breault, J. Eadie, and B. Gregory made helpful suggestions on data analysis, and C. Catterall, I. Newton, D. Schluter, and an anonymous reviewer commented on the manuscript. The work was supported financially by the Natural Sciences and Engineering Research Council of Canada. We thank the Tsawout and Tseycum Indian Bands for allowing us to work on their island.

References

Aldrich, J.W. 1984. Ecogeographical variation in size and proportions of Song Sparrows (*Melospiza melodia*). *Ornithological Monographs* **No. 35**. Washington, D.C: The American Ornithologists' Union.

Arcese, P. 1987. Age, intrusion pressure and defence against floaters by territorial male Song Sparrows. *Anim. Behav.* **35**: 773–84.

Arcese, P. 1988. The role of intra-sexual competition in dispersal and the territorial and mating system of the Song Sparrow. Unpubl. Ph.D. thesis, Univ. of British Columbia, Vancouver.

Arcese, P. 1989. Territory acquisition and loss in male Song Sparrows. *Anim. Behav.* **37**: 45–55.

Arcese, P. & Smith J.N.M. 1985. Phenotypic correlates and ecological consequences of dominance in Song Sparrows. *J. Anim. Ecol.* **54**: 817–30.

Alatalo, R.V. & Lundberg, A. 1984. Density-dependence in breeding success of the Pied Flycatcher (*Ficedula hypoleuca*). *J. Anim. Ecol.* **53**: 969–77.

Bent, A.C. 1968. Song Sparrow. In *Life Histories of North American Cardinals, Grosbeaks, Buntings, Towhees, Finches, Sparrows and Allies, Part Three*, ed. O.L. Austin, Jr. pp. 1512–1564. New York, Smithsonian Institution Press.

Gustafsson, L. 1986. Lifetime reproductive success and heritability: Empirical support for Fisher's fundamental theorem. *Amer. Nat.* **128**: 761–4.

Johnston, R.F. 1956. Population structure in salt marsh Song Sparrows part 1. Environment and annual cycle. *Condor* **58**: 24–44.

Kluijver, H.N. 1951. The population ecology of the Great Tit, *Parus m. major* L. *Ardea* **40**: 123–41.

Nice, M.M. 1937. Studies in the life history of the Song Sparrow I. A population study of the Song Sparrow. *Transactions of the Linaean Society of New York, Volume I*. New York.

Nice, M.M. 1943. Studies in the life history of the Song Sparrow II. The behavior of the Song Sparrow and other passerines. *Transactions of the Linnaean Society of New York, Volume VI*. New York.

Nol, E. & Smith, J.N.M. 1987. Effects of age and breeding experience on seasonal reproductive success in the Song Sparrow. *J. Anim. Ecol.* **56**: 301–13.

Schluter, D. & Smith, J.N.M. 1986. Natural selection on beak and body size in the Song Sparrow. *Evolution* **40**: 221–31.

Smith, J.N.M. 1981a. Cowbird parasitism, host fitness, and age of the host female in an island Song Sparrow population. *Condor* **83**: 152–61.

Smith, J.N.M. 1981b. Does high fecundity reduce survival in Song Sparrows? *Evolution* **35**: 1142–8.

Smith, J.N.M. 1982. Song Sparrow pair raise four broods in one year. *Wil. Bull.* **94**: 584–5.

Smith, J.N.M. 1988. Determinants of lifetime reproductive success in the Song Sparrow. In *Reproductive Success*, ed. T.H. Clutton-Brock. Chicago: University Press.

Smith, J.N.M. & Arcese, P. In press. How fit are floaters? Consequences of alternate territorial behavior in a non-migratory sparrow. *Amer. Nat.*

Smith, J.N.M. & Dhondt, A.A. 1980. Experimental confirmation of heritable morphological variation in a natural population of Song Sparrows. *Evolution* **34**: 1155–8.

Smith, J.N.M., Yom-Tov, Y. & Moses, R. 1982. Polygyny, male parental care, and sex ratio in Song Sparrows: An experimental study. *Auk* **99**: 555–64.

Tompa, F.S. 1964. Factors determining numbers of Song Sparrows *Melospiza melodia* (Wilson) on Mandarte Island, B.C., Canada. *Acta Zool. Fenn.* **109**: 3–73.

Van Noordwijk, A.J. & de Jong, G. 1986. Acquisition and allocation of resources: their influence on variation in life history tactics. *Amer. Nat.* **128**: 137–42.

10. Indigo Bunting

ROBERT B. PAYNE

Indigo Buntings *Passerina cyanea* are small songbirds, common in the spring and summer in temperate North America, particularly in the inland midwestern region. They feed mainly on insects, including small caterpillars and grasshoppers, and spiders, and also small grass seeds and fleshy fruits. They feed insects to their young. Buntings live in abandoned farm fields, in brushy habitats, and along the edges of cultivated lands, woods, swamps, roads and railway sidings. Male buntings are variable in plumage; the adults are bright blue and the first-year males range from blue to mixed blue and brown, sometimes more brown than the females, which usually are brown. Female buntings build the nests in low vegetation, especially in highbush blackberries *Rubus allegheniensis*, black raspberries *R. occidentalis* and gray dogwoods *Cornus racemosa*, and, though well concealed, nearly all nests can be found with sufficient field work. Buntings are long-distant migrants and winter mainly in Mexico and Central America.

Indigo Buntings have been observed in studies of song behaviour (Thompson 1970, 1972, Emlen 1971, Shiovitz 1975, Payne 1981b, 1983b, Payne *et al.* 1981, 1987, 1988), mating and parental behaviour (Carey 1982, Carey & Nolan 1979, Payne 1982, Payne & Payne 1989, Payne *et al.* 1987, 1988, Westneat 1987a,b, 1988a,b, Westneat *et al.* 1986) and population structure (Payne *et al.* 1987, Payne & Westneat 1988). In the present study, lifetime reproductive success was determined for marked buntings in two populations to determine variation in reproductive success and the proportion of individuals that contribute to the next generation. The amount of variation in breeding success is of interest because the rates of natural selection depend on the amount of variance in the population (Fisher 1958, Wade & Arnold 1980). Data on lifetime success also may be useful in a comparison of the population biology of different species (Lack 1966). In addition, observations of the breeding success of

LIFETIME REPRODUCTION IN BIRDS
ISBN 0-12-517370-9

marked individuals were used to test predictions from life history theory. The effects of age on breeding success and survival were examined to test whether success in one year influenced the survival and breeding success in later years, as is expected according to the idea of a cost of reproduction (Williams 1966). Data on the lifetime breeding success of individuals was also used to determine whether the variance of breeding success was greater in males than in females, a prediction of some hypotheses of sexual selection (Trivers 1972, Payne 1979, Wade and Arnold 1980).

Methods and study areas

THE STUDY POPULATIONS

Buntings were observed for more than 20,000 hours over 10 years by 30 observers (Payne 1981b, 1982, 1983a,b, Payne & Westneat 1988, Payne et al. 1981, 1987, 1988, Westneat 1987a,b, 1988a,b, Westneat et al. 1986). Birds were observed in two study areas in southern Michigan. The larger area was located at the E.S. George Reserve and the neighbouring Pinckney State Recreation Area (42° 27'N, 84° 00'W). The study area was enlarged from 6 km² in 1977 to 12 km² in 1983 to include more birds and to find dispersing birds. The other area was located near Niles (41° 55'N, 86° 14'W). It was enlarged from 1.4 km² in 1978 to 4 km² in 1983. Buntings on the study areas were continuous with populations over a wider region. The sizes of the study areas were determined by our efforts in the field and success in locating nearly all the buntings and their nests. All buntings in both study areas were observed through 1985, and the survival and breeding success of the older surviving birds were followed through 1987.

BREEDING BIOLOGY

Buntings nested from mid May to late August. The first egg in the earliest nest in 10 years was laid on 14 May, the latest date we found nestlings was 7 September, and a few breeding pairs were active for 14 weeks. Nest-building takes up to eight days in early spring but sometimes only two days in midsummer. Clutch size is three or four and the incubation period is 12–13 days. Young usually remain in the nest for 9–10 days, sometimes longer in cool weather. Males are sometimes polygynous with more than one nesting female at a time, and some females change males between nestings (Payne 1982, 1983b, Payne et al. 1988). First-year males do not feed the young; most older males do not feed nestlings but about half feed their offspring after they have left the nest. In pairs where the male feeds the young from an early nest, the female re-nests more quickly than in pairs where the male does not feed and the female alone cares for the fledglings (Westneat 1988b). Males may breed in their first year and many are successful (Payne *et al.* 1988), but if they are unsuccessful in attracting a female within a few weeks they may move to another territory. Females breed in every year of their life including their first year. Females alone build the nest, incubate and provide most of the food for the nestlings. The buntings often raise two broods and sometimes three broods in a season. The female often re-nests after the loss of a nest and may attempt as many as seven nests in a season. Most nests fail due to predators or to cold, rainy weather.

Male buntings ranged in weight from 12.5 to 17.5 g ($n = 1159$, mean 14.96 ± 0.74 SD g). Females were slightly smaller on the average and ranged from 11.9 to 17.5 g ($n = 398$, mean 14.33 ± 0.96 g).

FIELD METHODS AND SOURCES OF DATA ON LIFETIME SUCCESS

Males were captured in mist nets as they appeared on the study area in spring. They were measured, weighed and colour-banded, and aged by their greater primary coverts, which are partly or all brown in birds born the year before (first-year males) and blue in older males, as determined from recaptures of known-age ringed birds. Females were netted near the nest. No plumage differences were found between first-year and adult females, but we determined the breeding success of first-year females of known age from females that had been banded as nestlings in the previous year.

The rate of survival between the time at which birds fledge and the first spring following their birth is unknown because birds that disappear may have dispersed outside the study area. Fewer than 10 % of the young that fledged were seen in a later year. Most surviving older males returned to the same territory year after year. A few moved from one territory to a different one in the next year, mainly after their first year and sometimes as far as 3 km. New males arrived on the study area throughout spring and summer, perhaps after they had bred outside the study area. The latest first-year male to arrive and breed appeared on 2 August; he mated and nested later in the month. Females were more likely than males to change territories within a season and between years (Payne in prep.).

Lifetime breeding success was determined for 357 males that were banded in first-year plumage from 1979 through 1983. These comprised birds from five cohorts born in the year before they were banded, and they comprised about half of all birds that we observed in the study area. The criteria used for including a bird in the comparison of lifetime breeding success were (1) we knew its breeding success in all years, and (2) it remained on the study area for at least 28 days (the minimal time for a successful nesting) in its first season or ($n = 39$) for a shorter time but it returned to the study area in the following year. An allowance was made for (1) to include birds present in two or more years where we might have overlooked a successful nest early or late in the season (these involved 51 of the 791 cases, where a case is a male in a year, or a male-year). Excluding these birds from the estimates of success would erroneously lower the mean lifetime success for their cohort, since the more years a bird is present the more seasons its success might be overlooked, and cumulative success increases with age. One male was absent during his third season but then reappeared the next year (one out of 794 male-years); we assume he was unsuccessful in his missing year. The population sample included 26 to 51 males in each of the five year-cohorts at the Reserve (total 184 male lifetimes) and 14 to 53 males in each cohort at Niles (173 male lifetimes).

Lifetime breeding success was determined for 360 females that were banded in the same period. Because their ages were unknown unless they had been banded as nestlings, we assumed that new females captured on the parts of the study area where most females had been banded in the previous year were in their first year. Some females were missed for a year or more (27 out of 592 female-years of observation), then were seen again in the next year, usually in a different territory. In 21 cases, breeding success may have been overlooked when the female mated with a male that was not closely observed. For these 48 cases we approximated the breeding success by randomly assigning the fledging success of another female in the same cohort, year and population. A female may be overlooked in the same way in her final year, so to estimate the lifetime success for the sample group, one-year "resurrections" were allowed for the same number of female-years where a female was not seen for a year but then returned, and breeding success for these years was assigned in the same random manner as in the mid-life years. The approximations involved 12 % of the bird-year cases in Table 10.1.

A few birds in the lifetime study cohort were still alive through the end of the study period: 17 males (4.8 %) and five females (1.4 %) were observed through the 1987 breeding season. Most buntings that return in future years will increase their longevity and lifetime success, and their individual success will slightly increase the overall mean and variance in lifetime breeding success for the populations.

Table 10.1 Statistics of age-related and lifetime individual breeding success in number of fledglings in Indigo Buntings.

Population	Sex	Age (year)	No.	Min	Max	Mean	SD
					Breeding success		
Reserve	male	lifetime	184	0	22	3.33	4.21
		1	184	0	8	1.33	1.73
		2	82	0	8	2.06	2.15
		3	52	0	8	1.60	1.89
		4	31	0	7	2.03	2.06
		5	18	0	6	1.67	2.06
		6	9	0	3	1.56	1.51
		7	4	0	3	2.25	1.50
	female	lifetime	178	0	20	3.20	3.41
		1	178	0	7	2.07	1.71
		2	51	0	7	2.04	1.95
		3	19	0	6	2.70	1.87
		4	10	0	7	2.60	2.27
		5	5	0	6	2.40	2.51
		6	2	3	3	3.00	—
Niles	male	lifetime	173	0	26	4.32	4.58
		1	173	0	9	1.48	1.75
		2	111	0	10	1.94	2.09
		3	59	0	10	2.49	2.17
		4	36	0	6	1.89	1.82
		5	21	0	6	1.57	1.75
		6	11	0	6	1.73	1.90
		7	3	0	6	3.00	3.00
	female	lifetime	182	0	27	4.36	4.32
		1	182	0	8	2.23	1.74
		2	91	0	7	2.16	2.00
		3	45	0	7	2.16	2.06
		4	22	0	6	2.14	1.28
		5	11	0	6	2.64	2.38
		6	4	0	6	4.50	3.00

Results

LIFETIME PRODUCTION

Indigo Buntings varied in their lifetime breeding success, measured as the number of offspring that fledged from all nests on the territory of a male and all nests attended by a female (Fig 10.1, Table 10.1). An analysis of variance (ANOVA) of lifetime success showed no significant

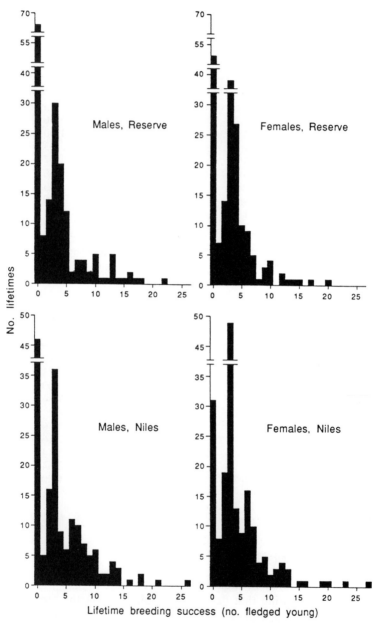

Figure 10.1 Frequency distribution of lifetime breeding success in Indigo Buntings.

variation among the cohorts from the five different years in mean lifetime success at Niles (mean $= 4.32 \pm 4.58$ SD, $F_{4,173} = 1.04$). At the Reserve, the differences among cohort means were significant (mean $= 3.33 \pm 4.21$, $F_{4,183} = 2.81$, $P < 0.05$), but the variances also differed significantly among years ($F = 2.77$, $df = 4$, $P < 0.05$) due to a few males from 1981 surviving and breeding with high success. Lifetime success varied nearly as much among the females. The mean lifetime success of females did not vary among the cohorts from the five different years at the Reserve (mean $= 3.20 \pm 3.41$, $F_{4,173} = 1.32$, ns) or at Niles (mean $= 4.36 \pm 4.32$ SD, $F_{4,177} = 0.82$, ns). The data for all years were combined into a composite cohort for each sex and area.

AGE AT FIRST BREEDING

Males establish territories in spring and summer with considerable overlap during the first month in arrival dates between first-year and older males. First-year males sing, defend territories, and attempt to breed; many attract a mate, most of the first-year males that remain as residents have at least one nest, and about 50 % succeed in producing fledglings (Payne 1982, Payne *et al.* 1988). Many first-year males disperse from one territory to another within the season; it is uncertain how many breed, but most have a nest. Nearly all older males have a mate and nest, and most succeed in fledging offspring. Females regularly breed in their first year. Fifty-one of 52 females that have been banded as nestlings in the previous year and were seen in their first summer were seen to nest.

ANNUAL SURVIVAL

The overall survival of males of all ages was 56.7 % at the Reserve, 58.2 % at Niles. The survival of females was 33.1 % at the Reserve, and 48.6 % at Niles (Table 10.1). No evidence of decreased survival with increasing age was apparent in either sex (Fig 10.2). Survival was slightly lower from the first to the second year than in later years at the Reserve, though not at Niles. The low apparent survival of females at the Reserve was probably due to emigration but may be affected also by predation. The difference in the apparent survival with age in males may involve breeding dispersal as well as actual survival between years.

DOES BREEDING SUCCESS CHANGE WITH AGE?

In neither sex was there an indication of increased or diminished breeding activity or success with increasing age of the adults, except that first-year

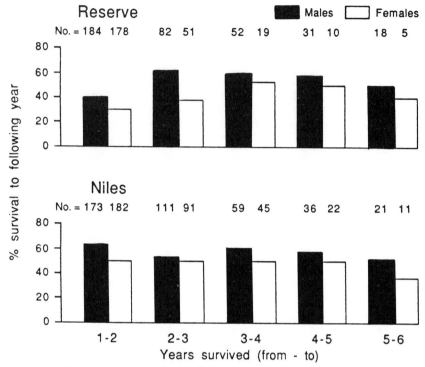

Figure 10.2 Survival of Indigo Buntings in relation to age.

males were less successful than certain older age classes (Table 10.1, Fig 10.3).

About 15 % of the males that mated had more than one female nesting on their territory in a season, at the same time or in succession (Payne 1982, Payne *et al.* 1988). Breeding males had from one to four mates and the mean number did not increase with age. One of the two males with four mates in a season was a yearling. Females sometimes mated with two or three males in a season when her first mate disappeared and she mated with the male that replaced him or when she left her mate and dispersed to another territory.

An ANOVA of the complete data on mean annual breeding success in Table 10.1 showed no significant difference with age at the Reserve (mean = 1.61 ± 1.90, F = 1.81, P > 0.05); first-year males were less productive than second-year males (pairwise Scheffé tests, P < 0.05). At Niles, breeding success varied with age (mean = 1.80 ± 1.95, F = 2.41, P<0.05) due to first-year males being less successful than third-year males. The highest observed mean success was at two years of age at the

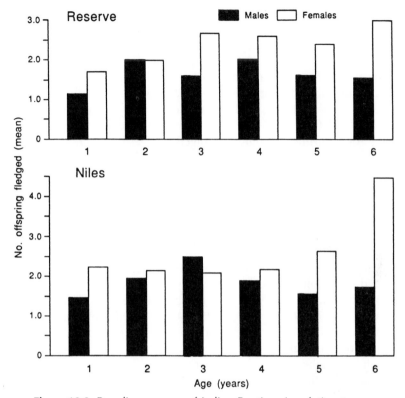

Figure 10.3 Breeding success of Indigo Buntings in relation to age.

Reserve and three years at Niles, but neither differed significantly from the other adult years. In the males two years and older, success did not vary with age in either area.

Breeding success of females did not vary with age, either in females whose age was unknown when they were banded or in females that were banded as nestlings and returned to breed in their natal area. Analysis of the data summarized in Table 10.1 showed no significant differences in mean success among ages in either area (Reserve, mean = 2.13 ± 3.34 SD, $F_{5,260}$ = 0.87, ns; Niles, mean = 2.09 ± 3.55 SD, $F_{5,349}$ = 0.62, ns). The exact age of most females was unknown. To test whether including some older females as "first-year" females affected the estimate of lifetime success, we compared first-year females that were banded as nestlings with the other females whose age was unknown when they were captured. The lifetime success of the three females banded as nestlings in the Reserve area and captured in the following year as breeding birds averaged 3.67; lifetime success of the 175 other females with complete data averaged 3.17 ± 3.43 SD. For the three females captured at the Reserve in the year after they were banded as nestlings, mean fledging success in their first year was 2.67; for the other females it was 2.08 ± 1.73 SD. For the 20 females banded at Niles as nestlings from 1979 to 1983 and recaptured in their natal area in the following year, lifetime success averaged 3.55 ± 3.61 SD, and for the other 162 females it averaged 4.14 ± 3.97 SD (Niles, $t = 0.63$, ns). Breeding success in their first year at Niles for returning nestlings averaged 2.05 ± 1.61 SD, and for the other females it averaged 2.23 ± 1.77 SD ($t = 0.44$, ns). In all respects the seasonal and lifetime breeding success of the females that were of unknown age when captured were like those of females known from their nestling year to be in their first year when they were captured.

DEMOGRAPHIC SOURCES OF VARIATION IN LIFETIME SUCCESS

The lifetime breeding success of the males indicated in Table 10.1 was correlated with their success in the first year (Reserve, $r = 0.38$, Niles, $r = 0.43$, $P < 0.01$) and with the number of years a male survived (Reserve, $r = 0.79$, Niles, $r = 0.77$, $P < 0.01$). The demographic variable that was more closely associated with individual lifetime breeding success in both areas was lifespan. The males that survived over more breeding seasons had a higher lifetime breeding success.

In females, lifetime breeding success was correlated with success in the first year (Reserve, $r = 0.55$; Niles, $r = 0.49$; $P < 0.01$ and with lifespan (Reserve, $r = 0.75$; Niles, $r = 0.71$; $P < 0.01$). In neither area was success in the first year significantly correlated with lifespan (Reserve, $r = 0.09$; Niles, $r = 0.08$, $P > 0.05$). Therefore, in both sexes the variance in lifetime success was explained in large part by lifetime itself, that is, by survival.

MORPHOLOGICAL AND BEHAVIOURAL SOURCES OF VARIATION IN LIFETIME SUCCESS

Song behaviour, plumage colour and body size varied among individuals. Male lifetime success was independent of variation among males in these traits. Neither the number of mates, seasonal breeding success, survival, or lifetime breeding success varied significantly among males with different song types (Payne et al. 1988, Payne & Westneat 1988). The proportion of blue and brown in male plumage in the first year was not associated with total lifetime breeding success (Reserve, $b = -0.13$, $r^2 = 0.06$, $F = 3.21$, ns; Niles, $b = 0.11$, $r^2 = 0.05$, $F = 2.08$, ns). In male buntings, lifetime success was not significantly associated with wing length (Reserve, $b = -0.14$, $r^2 = 0.02$, $F = 3.32$, ns; Niles, $b < 0.001$, $r^2 < 0.001$, $F = 0.003$, ns) or body weight (Reserve, $b = 0.01$, $r^2 = 0.07$, $F = 0.03$, ns; Niles, $b = 0.09$, $r^2 = 0.01$, $F = 1.20$, ns).

Females with longer wings tended to be more successful at Niles ($n = 181$, $b = 0.39$, $r^2 = 0.02$, $F = 4.07$, $P < 0.05$), but size explained little (2 %) of the overall variance, and females at the Reserve showed no similar trend ($n = 174$, $b = -0.13$, $r^2 = 0.003$, $F = 0.58$, ns). In females with nestlings, body weight was not related to lifetime breeding success (Reserve, $n = 61$, $b = -1.14$, $r^2 = 0.04$, $F = 2.17$, ns; Niles, $n = 69$, $b = 0.12$, $r^2 = 0.0004$, $F = 0.03$, ns).

ENVIRONMENTAL SOURCES OF SUCCESS OR FAILURE

Most nests were lost to predators, brood parasites or weather. No predation was directly observed, though once a Fox *Vulpes fulva* was seen 30 m from a nest site where the parents called excitedly at a nest that had young the day before. Other mammals on the study areas included Raccoons *Procyon lotor* (one was seen to lift an egg from an abandoned bunting nest), Skunks (*Mephitis mephitis*), Mink (*Mustella vison*), and small squirrels and mice. Blue Jays *Cyanocitta cristata* were suspected to be the major predators and we avoided visiting a nest when we knew they were nearby. Snakes prey on bunting nests and were common. Brood parasitism by Brown-headed Cowbirds *Molothrus ater* involved 21 % of the bunting nests; the proportion is similar in other populations in the midwest (Carey 1982). Adult female cowbirds remove some eggs from the nest of their foster species and the nestling cowbirds sometimes outcompete the foster young. Bunting nesting success was reduced in nests where a cowbird hatched, though many nests fledged both cowbird and bunting young. Egg removal by cowbirds or partial loss of the clutch by a predator was suspected in nests where we found only one deserted bunting egg. During cold, rainy periods a few broods starved. Nests were also lost to weather with the tipping of the supporting vegetation.

Territories varied in size and habitat. Most territories were less than 1 ha at Niles and 2 ha at the Reserve, but a few were as large as 10 ha in swamps. The George Reserve and neighbouring public lands were protected from cutting and development and the vegetation has grown beyond the low stage that attracts Indigo Buntings. In contrast, the Niles area is more disturbed, trees are cut and roads built and abandoned, and these activities have maintained a low bushy habitat. Also, the climate is milder and the breeding season is longer at Niles (Payne *et al.* 1988).

These differences may account for the higher lifetime success of buntings at Niles than at the Reserve.

PROPORTION OF ONE GENERATION THAT CONTRIBUTED OFFSPRING TO THE NEXT

The lifetime success of individual buntings was uneven (Fig. 10.1). Some had a high success, while many males fledged no young. At the Reserve, the most successful 10 % of the males accounted for 40.2 % of the fledglings, and 50 % of the fledglings were produced by 13.6 % of the males. At Niles, 10 % of the males accounted for 33.3 % of the fledglings, and 50 % of the fledglings were produced by 17.9 % of the males. The top 10 % of the females at the Reserve produced 34.7 % of the fledglings, and 50 % of the fledglings were produced by 18.5 % of the females. At Niles, 10 % of the females accounted for 32.8 % of the fledglings, and 50 % of the fledglings were due to 20.3 % of the females. The most successful female fledged 27 young between 1982, when she was marked as an adult, and 1987. The most successful male had 26 fledglings and for three years he was mated to the most successful female.

COMPARISON OF VARIATION IN BREEDING SUCCESS IN MALES AND FEMALES

The variance in lifetime breeding success was greater in males than in females within each population (variance $\sigma^2 = SD^2$; Table 10.1). The difference between sexes was significant at the Reserve (ANOVA, equality of variances, $F = 6.59$, $P = 0.01$) but not at Niles ($F = 0.69$, ns). The difference may be explained by the greater variation in annual survival, within-season mating success, and the first-year success of males.

RECRUITMENT OF LOCALLY BORN BIRDS TO THE BREEDING POPULATION

About 6 % of the banded nestlings that fledged were found on a breeding territory in a later year. We recovered 126 birds that had been banded as nestlings, and 124 of these were in their natal area. The proportion that returned to the natal area was 1.6 % at the Reserve and 9.7 % at Niles (Payne et al. 1987), even though more than 1,000 nestlings were banded in each area and the Reserve is four times larger than the Niles study area. Other buntings may have considerably longer natal dispersal distances. A male banded as a nestling was recovered in the following May at a distance of 350 km, and another was 52 km from its natal site.

Fewer local breeders contributed to the next local breeding generation at the Reserve than at Niles. At the Reserve, seven males (3.8 %) produced young that returned in a later year; at Niles, 31 males (20.2 %) did so. At the Reserve, seven females (4.0 %) produced young that returned, and at Niles 41 females (22.5 %) produced young that returned to their natal area.

IS THERE A DEMOGRAPHIC COST OF REPRODUCTION?

Breeding successfully may involve a cost in the future fitness of an individual (Williams 1966). The concept of a cost of reproduction suggests three predictions: (1) Birds that are successful in breeding are less likely to survive. (2) Among individuals that are successful in one year, the more successful birds are less likely to survive. (3) Among individuals that survive from year to year, high breeding success in one year is followed by low success in the next. These predictions were tested in the buntings where we were certain of breeding success in both years.

In comparing the breeding histories of individual buntings, no evidence was found for a cost of reproduction and the trends that were seen were in directions away from the predictions. (1) Successful males were more likely to return the next year than were unsuccessful males. At the Reserve, 43 % of 73 successful first-year males and 34 % of 85 unsuccessful birds returned in the next year ($\chi^2 = 1.16$, ns), and 74 % of 98 successful older males and 52 % of 67 unsuccessful older males returned ($\chi^2 = 7.87$, $P < 0.01$); at Niles, 65 % of 78 successful first-year males and 55 % of 69 unsuccessful males ($\chi^2 = 1.63$, ns), and 63 % of 126 successful older males and 54 % of 76 unsuccessful older males returned in the following year ($\chi^2 = 1.51$, ns). (2) The mean breeding success in the previous year of males that returned tended to be higher than in males that did not return (Fig. 10.4). (3) For males that produced offspring in the previous season, and whose success was certain in both years, breeding success was not significantly correlated over two successive years (Table 10.2).

Females were perhaps more likely to suffer from a cost of breeding because they provide most of the parental care to the young, but no cost of reproduction was apparent. (1) Females that fledged one or more offspring in the previous year were at least as likely to return as those that did not, and the successful females were more likely to return (Reserve, $n = 245$, $\chi^2 = 11.57$, $P < 0.001$; Niles, n = 323, $\chi^2 = 7.99$, $P < 0.01$). Survival was also compared in birds that fledged at least one young, because unsuccessful females might survive but disperse to a new area in the following year. Females that fledged four or more young were as likely to return in the following year as females with only one to three

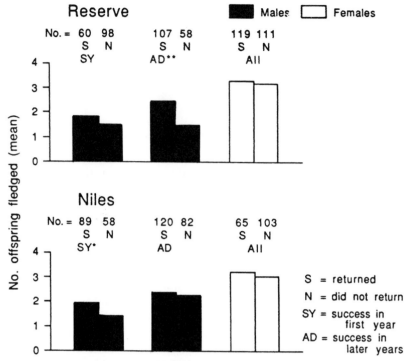

Figure 10.4 Breeding success in relation to survival, in birds that fledged at least one young in the previous year, t-tests, *P < 0.05, **P < 0.01.

Table 10.2 Correlation of breeding success across years, in birds that fledged at least one young in the previous year.

Population	Sex	Years	n	r	P
Reserve	male	1–2	30	−0.04	ns
		2–6	68	0.04	ns
	female	all	56	0.0001	ns
Niles	male	1–2	49	0.18	ns
		2–6	78	−0.16	ns
	female	all	104	0.10	ns

young (Reserve, $n = 168$, $\chi^2 = 0.002$, ns; Niles, $n = 230$, $\chi^2 = 0.12$, ns). (2) The successful females that returned had raised the same average number of young as the successful females that did not return (Fig. 10.4). The results suggest that the chance of survival from year to year was not related to the degree of breeding success. (3) For females that bred successfully in one year and returned, breeding success was not significantly correlated across years (Table 10.2). In females with the highest breeding success in one season (four or more fledglings, nearly as many as the average lifetime success), breeding success was no lower in the following year (Reserve, $n = 16$, $r = 0.03$, ns; Niles, $n = 25$, $r = 0.37$, ns). No negative effects of previous success in breeding were found between years in the older females (three years and older, returning in the following year, Niles, $n = 10$, $r = -0.32$, ns; Niles, $n = 23$, $r = -0.06$, ns), and there was no evidence of a decreasing tolerance to breeding effort with age. Although a female might be in weak condition after breeding, her breeding effort in one year was not associated with any apparent demographic cost in later years. These comparisons of individuals across breeding seasons suggest that success does not carry with it a demographic cost of survival or success in future years, at least within the normal range of breeding success in Indigo Buntings. However no experimental studies have been carried out to test the full range of response of the breeding birds.

Discussion

BREEDING SUCCESS AND POPULATION BIOLOGY

Lifetime breeding success varied in both males and females in this small migratory songbird. The variation was due in large part to the number of years that a bird survived and returned to breed. Survival and breeding success were independent of differences in size, plumage, and behaviour, and variation among years in weather and in seasonal success. The difference in the variance in lifetime success between males and females was related to differences in the first-year breeding success and in the lifespan of males and females, rather than to the mating system, as both males and females sometimes had more than one mate in a season. The sex ratio, another source of difference in the variance of males and females in breeding success in some species (Price 1984), was not important in the buntings, where males and females occur in similar numbers in the population. In addition to survival, the other main source of variation was between areas, and this points out the importance of observing birds in more than one population.

Breeding success and survival do not change with age, except for a lower breeding success in first-year than in older males. The absence of a decrease in breeding success and survival of the older birds and the lack of an increasing demographic cost of reproductive effort with age in either sex in Indigo Buntings has been noted in other small songbirds, suggesting that there is no senescence in either sex at least through the first five years of life. After that time fewer than 10 % of the birds are still alive, so any later effect of age would be negligible in population terms.

Two features of the biology of the buntings, dispersal and the parentage of the offspring, may affect the estimates of lifetime breeding success. The extent of both features has been determined in the Indigo Buntings. The extent of local dispersal can be estimated from the proportion of breeding buntings that were captured on the study area as adults and were not present locally in their first breeding season. The incidence of dispersal into the study area was considerable: 34 % of the adult males in 1983, 1984, and 1985 at the Reserve and 31 % at Niles were originally caught as adults and were not present in their first breeding season (Payne et al. 1988); they were not included in the description of survival or lifetime breeding success. In addition some birds that disappeared may have been successful elsewhere. One estimate of how well they did would be the lifetime success of birds that arrived for their first year on the study area as adults. Most males that we first saw and captured as adults were probably in their second year, since breeding dispersal within the study area involved mainly the first-year males: nearly all banded males that returned to a different territory from year to year did so between their first and second years (32 % at the Reserve and 21 % at Niles), whereas in later years only 4 % of the males at the Reserve and 2 % at Niles effected breeding dispersal between years. The males captured as adults had a mean lifetime success as great as the success of birds followed from their first year of life. Females were about twice as likely as males to change territories from year to year (Payne in prep.) The lifetime estimates of success are minimal because it is impossible to distinguish individual emigration from mortality, especially in females where the ages were not distinguished at the time of banding. The lower overall estimates of survival and lifetime success in females than in males might also involve a higher incidence of dispersal beyond the margin of the study area in females.

Secondly, the paternity of the nestling buntings is not known exactly because the fledglings on a male's territory may not have been his own offspring. Extra-pair copulations were observed (1.6–3 % of all behaviourally complete copulations), several nestlings (14.4 %) were genetically inconsistent with the resident male, and the heritability of size was more closely explained by maternal size than by size of the resident male or "father" (Payne 1983a, Westneat 1987a,b, Payne & Westneat 1988, Payne & Payne 1989). Because not all males can be distinguished by protein electrophoretic techniques, the degree to which extra-pair fertilization and multiple paternity may occur is indeterminate (Fisher 1951). The estimates from electrophoretic data suggest that 40 % of the young were not fathered by the resident male (Westneat *et al.* 1987). The difference in size of offspring in relation to size of mother and "father" also suggests that 40 % of the young were the result of an extra-pair fertilization (Payne & Payne 1989). Extra-pair mating attempts were observed during the same period of days before laying as were the within-pair copulations. If fertilization is independent of whether a male is the first or last to copulate with a female, as suggested by the timing in relation to laying of the attempted extra-pair affairs in the buntings, then the overall effect of extra-pair fertilizations on the variance among males in breeding success is probably negligible (Schwagmeyer *et al.* 1987). The extra-pair fertilizations tended to involve females that were on the territories of first-year males and were forcibly inseminated by neighbouring adults. If extra-pair fertilizations are important, the mean success for first-year males may be an overestimate and for older males an underestimate, and the population variance and the contribution of later years to overall lifetime success may be underestimated. In all cases but one, the genotype of the nestling was consistent with the adult female at the nest (Westneat 1987b), so the incidence of intraspecific brood parasitism or multiple maternity of broods is negligible.

RELATION TO LIFE HISTORY THEORY

"Life history theory" refers to the interacting effects of age, condition, survival, and natality, where one trait may affect another in the lifetime of a bird, and where one trait may have been affected in an evolutionary response to another trait (Williams 1966, Trivers 1972, Stearns 1977). Traits may vary within an individual from season to season, among individuals, among populations in different environments, and among populations and species. Life history studies involve different approaches including experimental tests, comparison of individual case histories, and demographic comparisons across ages, sexes, and breeding performances. Life history theory has also been applied to comparison of the sexes and to sexual selection (Trivers 1972, Payne 1979, 1983a, Wade & Arnold 1980).

The following predictions of life history theory and the reasoning behind the predictions were stated before the study or early in its course. (1) Breeding success is affected by a cost: buntings with higher success in one year are less likely to survive or to reproduce well in the future because of a negative effect between parental effort and residual reproductive value. (2) The cost of reproduction is greater in females than in males, because females form the eggs and provide most of the parental care. Neither prediction was supported by the observations in the Indigo Buntings. It appears desirable to test theories of population biology both with observations of individual birds throughout their lifetimes and with experiments, and not simply with experiments to alter parental effort with changes in brood size and food (Högstedt 1981, Reznick 1985, Murphy & Haukioja 1986, Zammuto 1987, Nur 1988). Field experiments may show demographic effects of reproducing at intensities higher than normal, but within the range of variation observed, no costs of reproduction were obvious in the bunting populations.

Acknowledgements

I thank the field observers for finding nests, especially Susan Doehlert Kielb, Laura L. Payne and David F. Westneat. For comments I thank F. Stephen Dobson, Geoffrey E. Hill, H. Lisle Gibbs and Ian Newton. The study was supported by National Science Foundation grants (BNS78–03178, BNS81–02404, BNS83–17810, BSR85–01075) and by the University of Michigan Faculty Research Fund.

References

Carey, M. 1982. An analysis of factors governing pair-bonding period and the onset of laying in Indigo Buntings. *J. Field Ornithol.* **53**: 240–8.
Carey, M. & Nolan, V. 1979. Population dynamics of Indigo Buntings and the evolution of avian polygyny. *Evolution* **33**: 1180–92.
Emlen, S.T. 1971. The role of song in individual recognition in the Indigo Bunting. *Z. Tierpsychol.* **28**: 241–6.
Fisher, R.A. 1951. Standard calculations for evaluating a blood group system. *Heredity* **5**: 95–102.
Fisher, R.A. 1958. *The Genetical Theory of Natural Selection*, 2nd edn. New York: Dover.
Högstedt, G. 1981. Should there be a positive or negative correlation between survival of adults in a bird population and their clutch size? *Amer. Nat.* **118**: 568–71.
Lack, D. 1966. *Population Studies of Birds*. Oxford: Clarendon Press.
Murphy, E.C. & Haukioja, E. 1986. Clutch size in nidicolous birds. *Current Ornithology* **4**: 141–80.
Nur, N. 1988. The consequences of brood size for breeding Blue Tits. III. Measuring the cost of reproduction: survival, future fecundity, and differential dispersal. *Evolution* **42**: 351–62.
Payne, R.B. 1979. Sexual selection and intersexual differences in variation of mating success. *Amer. Nat.* **114**: 447–52.

Payne, R.B. 1981a. Population structure and social behavior: models for testing the ecological significance of song dialects in birds. In *Natural Selection and Social Behavior* ed. R.D. Alexander & D.W. Tinkle, pp. 108–20. New York: Chiron Press.

Payne, R.B. 1981b. Song learning and social interaction in Indigo Buntings. *Anim. Behav.* **20**: 688–97.

Payne, R.B. 1982. Ecological consequences of song matching: breeding success and intraspecific song mimicry in Indigo Buntings. *Ecology* **63**: 401–11.

Payne, R.B. 1983a. Bird songs, sexual selection, and female mating strategies. In *Social Behavior of Female Vertebrates* ed. S.K. Wasser, pp. 55–90. New York: Academic Press.

Payne, R.B. 1983b. The social context of song mimicry: song-matching dialects in Indigo Buntings. *Anim. Behav.* **31**: 788–805.

Payne, R.B. & Payne, L.L. 1989. Heritability estimates and bahaviour observations: extra-pair matings in Indigo Buntings. *Anim. Behav.* **38**: 457–467.

Payne, R.B. & Westneat, D.F. 1988. A genetic and behavioral analysis of mate choice and song neighborhoods in Indigo Buntings. *Evolution* **42**: 935–948.

Payne, R.B., Thompson, W.L., Fiala, K.L. & Sweany, L.L. 1981. Local song traditions in Indigo Buntings: cultural transmission of behavior patterns across generations. *Behaviour* **77**: 199–221.

Payne, R.B., Payne, L.L. & Doehlert, S.M. 1987. Song, mate choice and the question of kin recognition in a migratory songbird. *Anim. Behav.* **35**: 35–47.

Payne, R.B., Payne, L.L. & Doehlert, S.M. 1988. Biological and cultural success of song memes in Indigo Buntings. *Ecology* **69**: 104–17.

Price, T.D. 1984. The evolution of sexual size dimorphism in Darwin's finches. *Amer. Nat.* **123**: 500–13.

Reznick, D. 1985. Costs of reproduction: an evaluation of the empirical evidence. *Oikos* **44**: 257–67.

Schwagmeyer, P.L., Coggins, K.A. & Lamey, T.C. 1987. The effects of sperm competition on variability in male reproductive success: some preliminary analyses. *Amer. Nat.* **130**: 485–92.

Shiovitz, K.A. 1975. The process of species-specific song recognition by the Indigo Bunting, *Passerina cyanea,* and its relationship to the organization of avian acoustical behaviour. *Behaviour* **55**: 128–79.

Stearns, S.C. 1977. The evolution of life history traits: a critique of the theory and a review of the data. *Ann. Rev. Ecol. Syst.* **8**: 145–171.

Thompson, W.L. 1970. Song variation in a population of Indigo Buntings. *Auk* **87**: 58–71.

Thompson, W.L. 1972. Singing behavior of the Indigo Bunting *Passerina cyanea.* *Z. Tierpsychol.* **31**: 39–59.

Trivers, R.L. 1972. Parental investment and sexual selection. In *Sexual Selection and the Descent of Man 1871–1971* ed. B.E. Campbell, pp. 136–79. Chicago: Aldine.

Wade, M.J. & Arnold, S.J. 1980. The intensity of sexual selection in relation to male sexual behaviour, female choice, and sperm precedence. *Anim. Behav.* **28**: 446–61.

Westneat, D.F. 1987a. Extra-pair copulations in a predominantly monogamous bird: observations of behaviour. *Anim. Behav.* **35**: 865–76.

Westneat, D.F. 1987b. Extra-pair fertilizations in a predominantly monogamous bird: genetic evidence. *Anim. Behav.* **35**: 877–86.

Westneat, D.F. 1988a. Male parental care and extra-pair copulations in the Indigo Bunting. *Auk* **105**: 149–60.

Westneat, D.F. 1988b. The relationship among polygyny, male parental care, and female breeding success in the Indigo Bunting. *Auk* **105**: 372–4.

Westneat, D.F., Payne, R.B. & Doehlert, S.M. 1986. Effects of muscle biopsy on survival and breeding success in Indigo Buntings. *Condor* **88**: 220–7.

Westneat, D.F., Frederick, P.C. & Wiley, R.H. 1987. The use of genetic markers to estimate the frequency of successful alternative reproductive tactics. *Behav. Ecol. Sociobiol.* **21**: 35–45.

Williams, G.C. 1966. Natural selection, the costs of reproduction, and a refinement of Lack's principle. *Amer. Nat.* **100**: 687–90.

Zammuto, R.M. 1987. Life histories of mammals: analyses among and within *Spermophilus columbianus* life tables. *Ecology* **68**: 1351–63.

11. Magpie

T.R. BIRKHEAD & S.F. GOODBURN

The Magpie *Pica pica* is a small to medium sized corvid, with striking black and white plumage, a long tail, and no obvious difference in appearance between the sexes. The species occurs throughout much of the Palearctic and western North America, breeding in a wide range of climatic zones, from semi-desert in North Africa, temperate parkland in Britain to boreal forest in southern Alaska. The Magpie is a ground feeder, utilizing open grassland, with a predominantly insect diet in summer and a vegetarian diet in winter: it exploits carrion as available throughout the year. Population densities vary markedly, from two breeding pairs km^{-1} in parts of North America (Reese & Kadlec 1985) to 26 pairs km^{-1} in parts of Britain (Birkhead *et al.* 1986). In our study area Magpies occupied type A territories, containing nesting and foraging areas (Hinde 1956); they were remarkably sedentary, defending their territories (mean size 4.9 ha) throughout the year, and in some cases throughout their lives (Goodburn 1987). In other areas territory boundaries break down in winter, or the birds disperse away from breeding areas (Reese & Kadlec 1985).

Magpies are monogamous, and single brooded, with a mean clutch of about six eggs. Replacement clutches (up to three) are laid if earlier breeding attempts fail at the egg stage. Breeding success is often nil, but successful pairs fledge in the order of two to three chicks (Reese & Kadlec 1985, Goodburn 1987). Breeding starts in either the first or second year, rarely later: in our study, of 41 females which survived to breed, 20 (49 %) bred in their first year, 19 (46 %) in their second, and two (5 %) in their third year. Corresponding values for males were, first year: 17 (31 %), second: 31 (56 %) and third: seven (13 %). The difference between the sexes was not significant.

Prior to breeding, Magpies live in loose flocks. Within these flocks a dominance hierarchy exists (Baeyens 1981, Eden 1987a) and it is generally

LIFETIME REPRODUCTION IN BIRDS
ISBN 0-12-517370-9

the high ranking individuals that subsequently obtain territories and breed. Territory acquisition often occurs aggressively through "ceremonial gatherings"; these are noisy aggregations of Magpies that occur as a result of a pair of non-breeding birds aggressively confronting an established pair in their territory. Gatherings sometimes result in the young birds ousting the original occupants and obtaining the territory or, more commonly the young birds obtain a small area between existing territory boundaries which they later expand into a full territory (Birkhead & Clarkson 1985).

In our study annual survival rates for breeding adults were 75 % for males, and 60 % for females (mean further expectation of life: 3.5 years and 2.0 years respectively). The maximum age of Magpies recorded in our (nine year) study was shown by one bird still alive at eight years old. The oldest known Magpie recorded in the national ringing scheme was 9.7 years old (Hickling 1983). G. Hogstedt (pers. comm.) recorded two Magpies ringed as nestlings in his study area in Sweden killed by Goshawks *Accipiter gentilis* when 11 years old. Juvenile survival in our study was relatively low: of 720 chicks ringed between 1977 and 1984, 96 (13.3 %) survived to breed in the study area (41 [43 %] females; 55 [57 %] males). Most juvenile mortality occurred in the first three months after ringing; thereafter mortality was almost constant. Clarkson (1984) demonstrated density dependence in juvenile survival and the low survival rate of young Magpies was probably linked with the high breeding density in our study area.

There have been several recent studies of the ecology and behaviour of Magpies: Hogstedt (1980, 1981a, b), Baeyens (1981), Vines (1981), Buitron (1982), Reese & Kadlec (1985) and Birkhead *et al.* (1986). Some followed individually marked, known-age birds over several seasons, but ours is the first attempt to examine lifetime reproductive success. The results presented here are derived from a study conducted in the Rivelin valley, on the outskirts of Sheffield, England (52°23'N, 01°33'W), between 1977 and 1986 inclusive. During the study a total of 37 adults and 854 chicks were individually marked. Since dispersal was extremely low (Eden 1987b), approximately 60 % of the breeding and non-breeding birds in the main part of the study area were marked and of known age and parentage. Birds could be sexed only once they started to breed (only the female incubates).

Procedure

Inevitably in a study of this type some long-lived individuals were still alive when we analysed the data on lifetime reproductive success. Since our sample sizes for long-lived birds are fairly low, rather than omit the data from these birds we performed two analyses: (a) using only birds with complete histories (i.e. now dead). These were birds that we had not seen for three years, despite extensive searching in surrounding areas. Since dispersal of both adults and juveniles is low (Eden 1987b), we assumed that these birds had died and not moved elsewhere. (b) We also used those individuals that were still alive but which were four or more years old. For the first analysis we had 24 males and 28 females, and for the second, we added a further 12 males and four females. Lifetime reproductive success was measured in three ways: (a) the number of young reaching 14 days of age (the latest age at which nests could be checked (and chicks ringed) without the risk of them "exploding"), (b) the number of young surviving for three months (i.e. until September of their first year), and (c) the number surviving to breed in the study area. In an evolutionary sense these are positively ranked in order of their importance, but in terms of sample sizes the rank order is reversed.

Results

LIFETIME REPRODUCTIVE SUCCESS

For both sets of data and all three measures of success, lifetime reproductive success was significantly and positively correlated with longevity, in both sexes (Fig. 11.1.). It is difficult to assess the relative importance of male and female effects since there were too few data to determine whether pairing was assortative with respect to age. The results in Fig. 11.1 suggest that, as in other species, lifespan was an important factor determining lifetime success.

We attempted to find features that would allow us to identify at an early age potentially long-lived individuals. We performed two analyses using data combined over all years of the study; in the first we divided birds ringed as chicks into four survival categories: (a) those not seen after ringing (i.e. at 14 days old), those surviving for (b) three months, (c) 12 months, and (d) > 12 months. There was a small cohort effect in terms of the proportion of birds surviving to breed (range: 3.2 to 10.7 %; $\chi^2 = 12.74$, 5 df $P < 0.05$), but no significant effects were found for any other variable (Table 11.1).

In the second analysis we divided birds that survived to breed into "short-lived" individuals (surviving three or less years) and "long-lived" individuals (surviving more than three years) and compared several aspects of their history and reproductive biology. Short lived birds tended to be

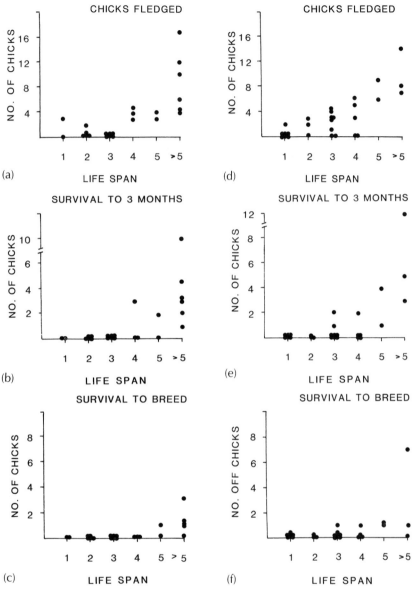

Figure 11.1 Relationship between lifespan and three measures of breeding success in Magpies; number of chicks fledged (top), number of chicks surviving for three months (middle) and the number of chicks surviving to breed (lower) for both females (left) and males (right). Data are for dead birds only and are for birds that had bred at least once. All relationships are positive and significant. (a) $r_s = 0.792$, $P < 0.002$, (b) $r_s = 0.672$, $P < 0.002$, (c) $r_s = 0.588$, $P < 0.002$ ($n = 28$), (d) $r_s = 0.805$, $P < 0.002$, (e) $r_s = 0.789$, $P < 0.002$ (f) $r_s = 0.569$, $P < 0.002$ ($n = 32$). Results using data that included birds that were four or more years old but still alive were similar: (a) $r_s = 0.716$, $P < 0.002$, (b) $r_s = 0.698$, $P < 0.002$, (c) $r_s = 0.514$, $P < 0.002$, (d) $r_s = 0.734$, $P < 0.002$, (e) $r_s = 0.734$, $P < 0.002$, (f) $r_s = 0.522$, $P < 0.002$, ($n = 36$).

Table 11.1 Factors affecting the survival of young Magpies.

Variable	Not seen			Survival								
				3 months			12 months			> 12 months		
	x̄	SD	(n)	x̄	SD	(n)	x̄	SD	(n)	x̄	SD	(n)
Median laying date[1]	5.60	± 13.9	(181)	2.50	± 12.4	(208)	2.12	± 12.5	(167)	4.30	± 12.1	(45)
Territory quality[2]	51.1 %	± 29.0	(200)	51.0 %	± 26.5	(223)	51.1 %	± 26.7	(173)	50.6 %	± 25.9	(50)
Clutch size[3]	6.03	± 1.13	(176)	6.14	± 1.09	(206)	6.18	± 1.11	(163)	5.93	± 0.96	(43)
Brood size[4]	3.45	± 1.40	(201)	3.56	± 1.34	(224)	3.57	± 1.33	(174)	3.52	± 1.40	(50)
Mean chick weight[5]	151.6	± 31.4	(261)	157.41	± 26.8	(221)	158.1	± 27.2	(175)	154.4	± 25.6	(46)

[1] Laying date (deviations from median) of parent for clutch in which chick was reared (Kruskall H = 6.019 $3df$ ns)
[2] Quality of territory (see Table 11.2) of parents for clutch in which chick reared (Kruskall H = 0.042 $3df$ ns)
[3] Size of clutch in which chick reared (F = 0.97 df = 3, 584, ns)
[4] Size of brood in which chick reared (F = 1.12 df = 3, 645, ns)
[5] Mean weight (g) of chicks at 14 days of age (Kruskall H = 6.048 $3df$ ns)

NOTE (n) = Number of broods.

those which had been hatched relatively late in the breeding season, although this effect was not significant. The other comparisons made were for breeding parameters in the bird's first breeding attempt. Short-lived birds had significantly poorer territories than long-lived birds. Laying date at the first attempt did not differ significantly between short-lived and long-lived birds, but short-lived individuals produced significantly smaller clutches and were less likely to be successful in raising young than were long-lived birds (Table 11.2). Both these results are probably related to the later age of first breeding among long-lived individuals, since both clutch size and breeding success increased with age up to age three or four years (see below). It is difficult to evaluate the importance of territory quality in longevity: by obtaining a good territory early in life a bird may increase its survival chances and improve its breeding performance. On the other hand, it may be that good quality birds obtain good territories initially; other information (see below) indicates that this is the most likely explanation for the territory effect in Table 11.2.

Table 11.2 Comparison of characteristics of short and long lived Magpies.

Variable	Short-lived			Long-lived		
	$\bar{x}$	SD	(n)	$\bar{x}$	SD	(n)
Hatch date[1]	5.91 ±	7.57	(32)	3.84 ±	5.93	(25)
Territory quality[2]	41.28 ±	31.24	(24)	59.82 ±	25.39	(29)
Lay date[3]	7.35 ±	9.79	(31)	5.00 ±	10.9	(32)
Clutch size[4]	4.90 ±	1.49	(31)	5.74 ±	1.21	(31)
Fledging success[5]	20 % (7/34)		(34)	42 % (14/33)		(33)

[1] Hatch date of short- and long-lived birds, expressed as deviations from the annual median; the difference is not significant (Mann–Whitney U test, $P = 0.34$).

[2] Territory quality: this is the percent grazing land in the birds' first breeding territory; the difference is significant ($t = 2.31$, $5df$, $P < 0.05$; analysis on arc-sin transformed data).

[3] Lay date of the first breeding attempt, expressed as deviation from the annual median laying date: the difference is not significant (Mann–Whitney U test, $P = 0.44$).

[4] Clutch size in the first breeding attempt ($t = 2.43$, $60df$, $P < 0.02$).

[5] Per cent of pairs fledging at least one chick ($\chi^2 = 3.71$, $1df$, $P < 0.1$).

NOTE: Data are for dead and live birds combined (see text).

THE EFFECT OF AGE ON BREEDING PERFORMANCE

In common with many other bird species, several breeding parameters changed with age. These are summarized for females in Fig. 11.2. Relative to the annual median laying date, older birds laid significantly earlier in the season than young birds (Fig. 11.2a), while clutch size, egg size, hatching success and fledging success increased with age.

Discussion

Longevity is an important factor influencing lifetime reproductive success in Magpies. Although there were some differences in parental characteristics between survival categories (e.g. greater survival among early hatched young; Table 11.1), and in the first breeding attempts of short- and long-lived Magpies, several of these were inter-related and we were not able to predict at an early age which individuals were likely to be long-lived. Results from other parts of this study indicate that longevity may not be simply due to chance, as several breeding parameters were determined by bird quality (Goodburn 1987). By using repeatability analysis (Falconer 1961, Lessells & Boag 1986), Goodburn was able to determine the separate effects of bird and territory quality on breeding parameters. Bird effects were generally much greater than territory effects. Among females both clutch size and egg size were determined mainly by bird effects. Among males two measures of breeding success (hatching success and juvenile survival to one year) were determined by bird quality. These results contrast with those obtained by Hogstedt (1980), who found in a different area that clutch size in Magpies was determined largely by territory quality, and that bird effects were minimal.

The bird quality effects demonstrated by Goodburn (1987) suggest that genetic differences between individuals might account for some of the variation in lifetime reproductive success (at least in males, in which two measures of success were highly repeatable). However, we had insufficient data on parents and offspring to examine this directly.

Acknowledgements

Many people have helped us in our Magpie studies in a variety of ways. We are grateful to the Birch family, the Revitt family, the Guites, Mr H. Thompson and H. White for allowing us to work on their land. The assistance of Jayne Pellatt and Pete Jackson was invaluable. Drs K.

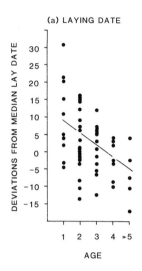

(a) LAYING DATE

DEVIATIONS FROM MEDIAN LAY DATE

AGE

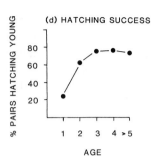

(d) HATCHING SUCCESS

% PAIRS HATCHING YOUNG

AGE

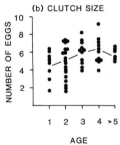

(b) CLUTCH SIZE

NUMBER OF EGGS

AGE

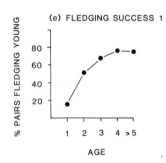

(e) FLEDGING SUCCESS 1

% PAIRS FLEDGING YOUNG

AGE

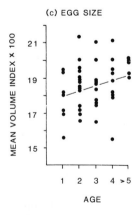

(c) EGG SIZE

MEAN VOLUME INDEX × 100

AGE

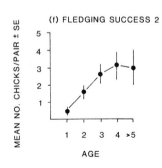

(f) FLEDGING SUCCESS 2

MEAN NO. CHICKS/PAIR ± SE

AGE

Clarkson and S.F. Eden helped to collect data upon which this study is based: we are very grateful to them.

References

Baeyens, G. 1981. The role of the sexes in territory defence in the Magpie, *Pica pica*. *Ardea* **69**: 44–5

Birkhead, T.R. & Clarkson, K. 1985. Ceremonial gatherings of the Magpie, *Pica pica*: territory probing and acquisition. *Behaviour* **94**: 324–32.

Birkhead, T.R., Eden, S.F., Clarkson, K., Goodburn, S.F. & Pellatt, J. 1986. Social organisation of a population of Magpies, *Pica pica*. *Ardea* **74**: 59–68.

Buitron, D. 1982. Behaviour of Black-billed Magpies during the breeding season. Unpublished PhD Thesis: University of Minnesota, Minneapolis.

Clarkson, K. 1984. The breeding and feeding ecology of the Magpie *Pica pica*. Unpublished PhD Thesis: University of Sheffield.

Eden, S.F. 1987a. Dispersal and competitive ability in the magpie: an experimental study. *Anim. Behav.* **35**: 764–72.

Eden, S.F. 1987b. Natal philopatry of the Magpie *Pica pica*. *Ibis* **129**: 477–90.

Falconer, D.S. 1981. *Introduction to Quantitative Genetics*, 2nd Edn. London and New York: Longman.

Goodburn, S.F. 1987. Factors affecting breeding success in the Magpie *Pica pica*. Unpublished PhD Thesis, University of Sheffield.

Hinde, R.A. 1956. The biological significance of the territories of birds. *Ibis* **98**: 340–69.

Hickling, R.A.O. 1984. *Enjoying Ornithology*. Poyser, Calton.

Hogstedt, G. 1980. Evolution of clutch-size in birds: adaptive variation in relation to territory quality. *Science, N.Y.* **210**: 1148–50.

Hogstedt, G. 1981a. Should there be a positive or negative correlation between survival of adults in a bird population and their clutch size? *Amer. Nat.* **118**: 568–71.

Hogstedt, G. 1981b. The effect of additional food on reproductive success in the Magpie *Pica pica*. *J. Anim. Ecol.* **50**: 219–29.

Figure 11.2 Relationship between female age and breeding variables. (a) Laying date. Relative to the annual median, older birds laid earlier than younger birds, $r = -0.45$, $49df$, $P < 0.002$. (b) Clutch size: $r = 0.311$ $55df$, $P < 0.025$. The data suggest that clutch size increases with age between one and four years and then decreases; however, a second order polynomial did not improve the fit. (c) Egg size: Older females tended to produce larger eggs. Egg size is expressed as the volume index; length $\times$ breadth^2. Each point is the mean volume index for a clutch, $r = 0.30$, $45df$, $P < 0.05$. (d) Hatching success: the proportion of pairs which hatched at least one chick; Linear trend on proportion, $Z = 2.524$, $P = 0.011$. (e) Fledgling success: the proportion of pairs fledging at least one chick; $\chi^2 = 10.74$, $4df$, $P < 0.05$. Data are for all breeding pairs. (f) Fledging success: the mean number of chicks fledged per pair ($+/-$ SE), Kruskall Wallis $H = 12.64$, $4df$, $P < 0.025$.

Lessells, C.M. & Boag, P.T. 1987. Unrepeatable repeatabilities: a common mistake. *Auk* **104**: 116–21.

Reese, K.P. & Kadlec, J.A. 1985. Influence of high density and parental age on habitat selection and reproduction of Black-billed Magpies. *Condor* **87**: 96–105.

Vines, G. 1981. A socio-ecology of Magpies (*Pica pica*). *Ibis* **123**: 190–202.

12. Red-winged Blackbird

GORDON H. ORIANS & LES D. BELETSKY

The Red-winged Blackbird *Agelaius phoeniceus*, one of the most abundant and widespread birds in North America, breeds from east-central Alaska and the Yukon south to northern Costa Rica, and from the Atlantic to the Pacific. Birds from the northern breeding areas migrate south for the winter, but many remain as far north as southern Canada. However, the largest wintering concentrations are found in the Gulf States and California, where some roosts exceed one million individuals (Meanly 1965, Dolbeer *et al.* 1978).

Over much of its range, the Redwing breeds primarily in marshes, where it is often the most abundant passerine. During recent decades, especially in eastern North America, Redwings have become common breeders in upland pastures and crops. The birds also use marshes for roosting in winter, but feed there only in spring and summer. During the remainder of the year they forage in open habitats and eat primarily seeds, sometimes becoming crop pests. During the breeding season, males are territorial, each defending a section of marsh or upland, within which up to a dozen or more females construct their nests. Redwings also defend their territories against Yellow-headed Blackbirds (*Xanthocephalus xanthocephalus*), but they are often displaced from better quality areas by the larger Yellowheads (Orians & Willson 1964). Redwings are strongly polygynous throughout their range: harems tend to be larger in the more productive marshes of western North America than elsewhere (Orians 1980). Male Redwings take no part in nest-building or incubation, and most do not feed nestlings. Feeding of nestlings is more prevalent among older males, and in eastern than in western North America (Searcy & Yasukawa 1983, Muldal *et al.* 1986). No strong bonds are formed between males and females, and even during the breeding season, mated individuals move around independently of one another.

LIFETIME REPRODUCTION IN BIRDS
ISBN 0-12-517370-9

Female Redwings are often aggressive to other females near their nests (Yasukawa & Searcy 1982, Hurley & Robertson 1984, Searcy 1986). Early in the breeding season some females apparently defend territories within the larger territories of males, but this behaviour weakens as the season progresses. Timing and spacing of nests give little evidence that defence of space by females has an important influence on the settling patterns of other females (Yasukawa & Searcy 1981).

The modal clutch size during our study was four, but clutches of three or five eggs were common. Successful nests often fledged their full complement of three or four young, but rarely five. Starvation was not a major source of nestling loss in our study. Many females built several nests each year (up to five) and laid replacement clutches in response to failures, but usually less than 10 % fledged two broods in a year ($\bar{x} = 5.9 \pm 4.6$ % of breeding females each year, 1977 to 1986).

As is typical of polygynous species, Redwings show considerable sexual dimorphism in plumage and size. The jet black males with red epaulets weigh 65 – 85 g in eastern Washington, the location of our studies, whereas the streaked brown females weigh 40 – 55 g. Breeding male mortality was approximately 40 % annually (the average return rate to territories was about 53 %, but the figure increased to about 60 % when males that returned but were no longer territorial were included; Beletsky & Orians 1987a). Female return rates to the core study area between years averaged 52 % from 1977 through 1986. However, females apparently moved further and more frequently between breeding seasons than did males (see below), and thus we must have failed to record a fraction of our marked females that left the study area. Overall, the annual mortality for males and females was probably similar, about 40 %.

Redwings obtain much of their food from their territories, and the extent to which they forage outside the nesting marshes varies between regions. In our study area most food delivered to nestlings consisted of insects, such as dragonflies, damselflies, caddisflies and midges, both aquatic larval stages and terrestrial adult stages (Orians 1980). During the afternoons, when few aquatic insects emerged, the birds exploited upland foods, such as grasshoppers, caterpillars and spiders. Breeding adults also ate mainly arthropods, but took seeds too, especially in early spring when insects were scarce.

The birds moult near their breeding areas. Many males begin while they are still defending territories. When they have finished, birds gather in large flocks, and from that time until the start of breeding the following spring, they feed primarily in fields and other open habitats.

Study area and methods

We studied Redwings in the Columbia National Wildlife Refuge and adjacent areas, in the Columbia Basin desert of eastern Washington State. Water levels in the marshes were relatively constant because they were regulated by weirs. Uplands were dominated by sagebrush *Artemisia tridentata* and other shrubs (see Orians 1980 for a complete description of the study area and Redwing ecology).

Our estimates of lifetime reproductive success are based upon a 10-year study of a core area containing 70 – 80 territories and several times that number of breeding females each year. All breeding males were leg-banded with unique colour combinations. Not all females were identified, partly because they were more difficult to trap than males, and partly because many nests were destroyed by predators before we were able to identify the females associated with them.

Each year nearly all of the several hundred nestlings that fledged on the core study area were banded. However, return rates of banded nestlings of both sexes were very low (about 8 % for males and 3 % for females). Most surviving nestlings clearly bred outside the core study area. In addition, most of the new males found breeding on our core study area were unbanded or, if banded, they were first captured on the study area when they were at least one year old.

Subadult (yearling) male Redwings are readily distinguished from older birds by their brown primaries and orange epaulets with black flecks. Yearling females cannot be uniquely recognized; apparently they always lack red epaulets, but some older females lack this plumage marker too. As a result, although we were able to age all males captured on the study area when they were one year old, we have accurate ages for only those few females banded as nestlings that returned to the area to breed.

Some determinants of annual reproductive success

HAREM SIZE

Males secured nesting success by maintaining a territory of sufficient quality to attract females. The average harem size for the population was 4.1 ± 2.4 females (range = 1 to 14, $n = 729$ male breeding years, Fig. 12.1). Harems were partially sequential; that is, females arrived on, and departed from, male territories throughout the breeding season. However,

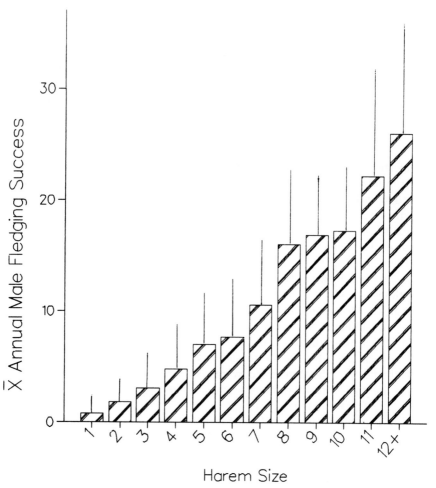

Figure 12.1 Annual number of young fledged per male as a function of harem size. Vertical lines are standard deviations. Total $n = 729$ male breeding-years: Harem size = 1, 100 male years; harem size = 2, 101 male years; 3, 134 male years; 4, 118 male years; 5, 100 male years; 6, 71 male years; 7, 37 male years; 8, 25 male years; 9, 21 male years; 10, 11 male years; 11, 7 male years; 12, 1 male year; 13, 2 male years; 14, 1 male year.

simultaneous harems of five or more females were common. Harems differed significantly in size among eight marshes in the study area (average range 3.2 – 5.2; ANOVA, $F = 7.18$, $P < 0.001$) and among 10 different years (average range 3.5 (1986) – 5.5 (1980); ANOVA, $F = 3.69$, $P < 0.001$). Harems were, on average, larger in years with a higher return rate of males, suggesting that overwinter survival was correlated for both sexes. Harem size fluctuated with number of available females because territorial behaviour prevents increases in the number of breeding males. Harem sizes did not increase significantly for males between their first and second breeding years (average first year harem size = 4.0 ± 2.4; average second year harem size = 4.3 ± 2.1; $n = 148$, $t = -1.27$, one-tailed $P = 0.10$). There was a strong positive correlation between harem size and annual male reproductive success (Beletsky & Orians 1987a) and, in fact, the data displayed in Fig. 12.1 indicate that annual male success rose almost linearly with average harem size, at least up to 12.

PREDATION AND BROOD PARASITISM

Predation on eggs and young was the major source of nest failure. Both sexes defended nests, but their efforts were often futile. The most destructive predator was the Black-billed Magpie (*Pica pica*), which sometimes emptied 25 or more nests in a locality in a single day. This depredation depended largely on the proximity of a marsh to an active magpie nest and on the availability of alternative foods for the magpies. Nest losses to other predators, such as Mink *Mustela vison*, Racoon *Procyon lotor*, Harvest Mouse *Reithrodontomys megalotis*, Marsh Wren *Cistothorus palustris* and snakes (*Pituophis melanoleucus*, *Crotalus viridis*), were more sporadic.

The Redwing is the most common host in the study area for the parasitic Brown-headed Cowbird (*Molothrus ater*). The percentage of nests parasitized increased from less than 10 % in early May (the peak of Redwing breeding) to 60 % or more by mid-June, when the final nests of the season were begun. Parasitized nests fledged, on average, about one Redwing nestling less than unparasitized nests (Røskaft *et al.* in press).

Nesting success varied considerably from year to year. The average number of young fledged per adult male varied from 2.4 ± 4.0 to 10.9 ± 9.1 in different years ($\bar{x} = 5.0$, $n = 729$ territorial male years), with 1978 and 1979 being good years, and 1977, 1981, 1982, and 1986 being poor years (ANOVA, $F = 10.18$, $P < 0.001$, $n = 729$ male years). Success also varied spatially. The number of young fledged per breeding male varied from 3.1 ± 3.9 to 8.7 ± 7.8 in different marshes (ANOVA, $F = 10.49$, $P < 0.001$, $n = 729$ male years). The prime causes of this variability in nesting success were the heavy but patchy predation by Black-billed Magpies, and differences in mean harem sizes.

AN UNUSUAL EVENT

Only one male hatched in the core study area in 1980 (0.4 %) has been recorded breeding, far fewer than from any other year ($\bar{x}$ = 2.7 % for 1977 – 1984). The eruption of Mount St Helens covered the area with about 5 cm of ash on 18 May 1980, destroying most nests. Many broods fledged prior to the eruption and others fledged later from nests started after the eruption, but our failure to observe individuals hatched that year in subsequent years suggests that survival of fledglings was very poor in 1980.

Lifetime reproductive success

AGE AT FIRST BREEDING

The ages at which males of known age were first observed to hold breeding territories were: age 1, 11 males; age 2, 84 males; age 3, 34 males; age 4, 16 males; age 5, one male; age 6, two males. Thus, only 7 % of 148 males held territories and were successful in attracting at least one female, or held a territory for a substantial part of the breeding season, when they were one year old. Most males (57 %) first held a territory when they were two years old, but 23 % did not do so until they were three years old and 13 % until they were four or older. Because we have a complete census of territorial males each year, none was missed. A few males may have bred elsewhere before nesting on our study area, but that number was certainly very small. About 70 % of surviving males returned to the same territory in subsequent years, and of those that changed territories, most moved less than 200 m. Due to these facts, and the geographical distribution of breeding marshes in the study area, we are confident that we detected almost all territorial moves (Beletsky & Orians 1987a). Presumably, therefore, our data accurately reflect the age at which males first bred.

Of the 41 females hatched on the study area for which we have breeding records, 22 first bred when they were óne year old, 12 when they were two years old, and seven when they were three or older. Nevertheless, for several reasons we believe that nearly all females attempt to breed when they are yearlings. No students of Redwings have found evidence of a floating population of non-breeding females such as is found among males. Also, in contrast to males, there are no territorial constraints on breeding by females. Experimental removal of females from their territories sometimes has resulted in an influx of more new females on experimental than on control territories (Holcomb 1974, Hurley & Robertson 1985). However, this does not demonstrate the existence of non-breeding females, because these females could have moved in from other breeding situations (see below). Moreover, because females move a great deal, both within and between breeding seasons, many of the females first observed breeding on our study area when they were more than one year old had probably bred elsewhere.

NUMBER OF YEARS BREEDING

The average number of years on territory for 264 males that initiated breeding from 1978 to 1985 was 2.1 ± 1.4 years (range = 1 to 9). Because of high annual mortality, about half of the Redwings which bred did so for only one year (Fig. 12.2). Slightly more females (55 %) than males (48 %) were recorded as breeding in only one year but, because of the

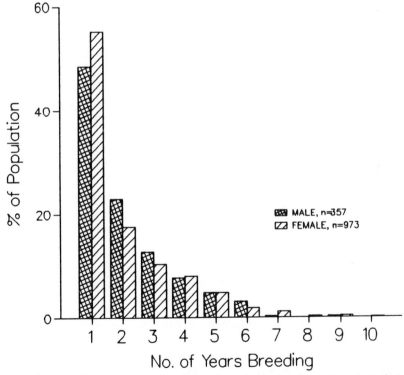

Figure 12.2 The number of years breeding for individual male and female Redwings that attempted to breed at the Columbia National Wildlife Refuge, 1977–1986.

incompleteness of the female data, this difference may not be real. Moreover, our data do not reveal the number of birds of either sex that fail to breed at all. This number is certainly much higher for males than for females, because males must survive for at least two rather than one year before breeding and because some males never gain territorial status. Between 1977 and 1986 only 14 males re-established territories after having failed to hold one for one year, indicating that, in most cases, once a male disappeared from his territory, he was either dead or had permanently lost breeding status.

Estimating the number of breeding years for females is more difficult than for males, because not all females were colour-banded and they moved around more than males. During the years of our study, 18.7 % (annual range 13.8 – 24.7 %) of nesting females changed territories within breeding seasons, and 11.4 % (n = 38) of these changed more than once (total n = 332 female switches in 1,801 opportunities between nesting attempts). An average of 7.8 % (annual range 3.3 – 11.5 %) of nesting females changed marshes within breeding seasons, and 3.4 % (n = 5) of these changed more than once (total n = 146 female moves in 1,862 opportunities between nesting attempts). Between breeding seasons, moves were much more frequent. Among those individuals with complete histories, females changed breeding marshes in 28.5 % (annual range 21.7 % – 40.4 %) of the 687 between-year opportunities. Of these females, 38.0 % (annual range 25.9 – 46.1 %) bred with a different male the subsequent year, even though their former mate was still alive and defending a territory.

Whereas females often moved to new areas for re-nesting after their nests were destroyed by predators, they apparently did not follow males. Among 30 instances of territorial moves by males, only six females changed breeding locations along with them, all to an adjacent territory. In no instance did a female accompany a male to a non-adjacent site. Moreover, females often moved to adjacent territories for subsequent nests whether or not males moved.

REPRODUCTIVE SUCCESS

To estimate lifetime reproductive success, we assumed certainty of paternity, no egg dumping by females, and correct identification of all cowbird nestlings. Of these assumptions, paternity was the least certain, because female Redwings copulate with other males, usually those from adjacent territories (Bray et al. 1975, Roberts & Kenelly 1980, Monnett et al. 1984, Davies & Orians, unpublished).

For this analaysis, a nestling was considered to have fledged if it survived at least until the age of eight days (fledging occurred at 10 – 12 days of age) and if subsequent visits to the nest revealed no signs of predation. Lifetime production of fledglings, so defined, is plotted for our population in Fig. 12.3. The curves are not monotonic because the mean number of young fledged per successful nest was 2.69 ± 0.99 (range 1–5, n = 1,364), so fewer individuals produced only one or two fledglings than three. Thirty-seven per cent of females known to have attempted to breed were not recorded to have produced any young, but this is an overestimate for the reasons already discussed. To reduce our underestimates of lifetime success for females with gaps in their breeding histories (about one-third of individuals known to have bred for at least three years), we credited such females with the average number of

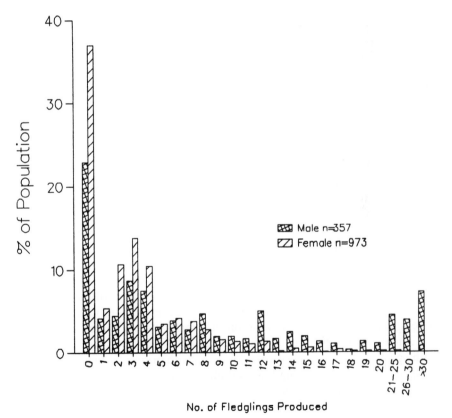

Figure 12.3 Lifetime fledgling production of male and female Redwings, that attempted to breed, 1977–1986.

fledglings produced by experienced females during the year we missed them. Data for males that held territories are generally accurate because no nests other than a very few at the end of the breeding season produced young undetected by us. Nonetheless, of the individuals that actually bred, fewer males than females produced no fledglings, because it was rare that all females in a harem failed to fledge any young for an entire breeding season, whereas many individual females failed in the 1.3 nests each attempted, on average, in a given year.

Among individuals known to have bred, there was, inevitably, greater variance in lifetime reproductive success among males than among females (195.65 vs. 13.13, 264 males vs. 795 females starting breeding between 1978 and 1985). One male who held a territory for nine years produced 159 young, much greater than the largest number (24) recorded for any female. Indeed, only if males bred fewer years on average than females, which was not the case, could there have been more variance in reproductive success among females than among males.

For both males and females, the number of young fledged rose linearly with the number of years breeding (Fig. 12.4). The correlation between number of years breeding and total number of young fledged is very strong for males ($r = 0.550$, $P < 0.001$, $n = 264$). There is little evidence, however, that the yearly reproductive success of males improves with age. For 148 males that held territories for at least two years, beginning between 1978 and 1985, slightly more nests were built per territory the second year than the first (6.3 ± 4.0 vs. 6.7 ± 4.4; $z = -1.11$, one-tailed $P = 0.13$, Wilcoxon matched pairs test). However, there was no difference in the number of young fledged during the first and second year on territory (4.6 ± 5.7 vs. 4.7 ± 5.7, $z = -0.78$, one-tailed $P = 0.22$).

Multiple regression of lifetime fledgling production on the number of years a male bred, his average harem size, and the average fledging success of his females, explained 95 % of the variance in lifetime success ($1977 - 1986$; $r^2 = 0.949$, $F = 5.42$, $P < 0.0001$). Number of years bred explained 39.8 % of the variance, harem size explained 23.6 %, and fledging success of females explained 31.5 %. Thus, more of the variation among males in lifetime success was due to differences in longevity, but harem size differences and variation in female success contributed also.

Females also appeared to improve very little in reproductive success with age and experience. However, females with prior breeding experience fledged slightly more young than did inexperienced females in every year of our study. This measure is clouded by the fact that, whereas all experienced females were known to have bred before, probably about one-third of the "inexperienced" females had prior breeding experience off our study area where they were not observed. In addition, females

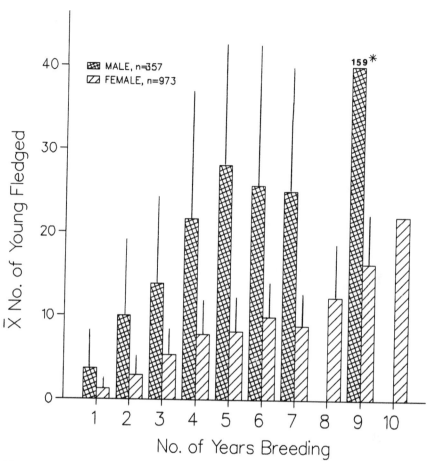

Figure 12.4 The average number of young fledged by male and female Redwings, as a function of the number of years breeding (*one male bred in nine years, producing 159 fledglings). Vertical lines are standard deviations.

that bred for more than one year had greater success in their first year than females that bred for only one year (Table 12.1). This was due primarily to greater persistence by females that bred for more than one year (1.43 vs. 1.27 nests/female during the first year of breeding). Both groups of females laid their first eggs on approximately the same date (Table 12.1). Given the very large sample size, the difference is statistically ($t = -1.86$, $P = 0.03$), but probably not biologically, significant.

Lifetime reproductive success of males was strongly correlated with success during the first year of breeding ($r = 0.582$, $P < 0.001$, $n = 148$). This was because of consistent differences in breeding success between

Table 12.1 First year reproductive success of female Redwings.

Female group	Number of nests $\bar{x} \pm SD$	Number of fledglings $\bar{x} \pm SD$	Date of first egg $\bar{x}$	n
Known to breed in only one year	1.27 ± 5.67 *	1.25 ± 1.66 *	5 May	537
Known to breed in at least two years	1.43 ± 7.09	1.70 ± 1.75	3 May	436

*t-tests, $P < 0.001$

marshes, and the strong site faithfulness of males regardless of their breeding success. Interestingly, fewer nests ($\bar{x} = 4.4 \pm 3.6$) were built on territories of males that bred only in one year than during the first year on territories of males that bred in more than one year ($\bar{x} = 6.3 \pm 4.0$, $t = -4.04$, one-tailed $P < 0.001$, $n = 264$), but there was no difference in the number of young produced ($\bar{x} = 4.4 \pm 5.7$ vs. 4.6 ± 5.7, $t = -0.26$, one-tailed $P = 0.40$).

No study of Red-winged Blackbirds has demonstrated a negative correlation between breeding success and harem size (Holm 1973, Weatherhead & Robertson 1977). Females may co-operate in defence of their nests (Picman 1980), but the increased reproductive success typically found among larger harems is probably because territories with large harems generally have better nest sites, better food supplies and, especially later in the season, lower predation rates.

Discussion

The Red-winged Blackbird is typical of small passerines in having high mortality between breeding seasons, short life expectancy, and a small average number of breeding years per individual. Males and females that bred did so, on average, about two years, although most males started breeding when they were two years old, whereas females bred as yearlings.

Delayed breeding among male Redwings is not due to a physiological constraint among first-year birds (Payne 1969). Rather, it is the result of strong territorial behaviour, combined with the presence of many more adults than available territories. In all experiments, including ours, in which territorial male Redwings were removed, unoccupied territories were rapidly taken over by floating males or expanding neighbours

(Orians 1961, Beletsky & Orians 1987b). Moreover, given the open nature of the breeding habitat and the presence of extensive foraging areas nearby, it is easy for floaters to remain close to occupied territories and to respond quickly to any opportunities that arise. Because of the presence of undefended areas rich in food, survival of non-territorial Redwing males may be no lower than that of territorial males, contrary to what is found in other bird species whose territories fill all suitable habitat (Krebs 1971). We cannot test this possibility, because of our inability to track non-breeding males as accurately as breeders, but return rates of non-breeders are suggestively high.

Given the short life spans of most males, failure to gain a territory during a particular breeding season carries a high probability of not breeding at all. Therefore, it is surprising that individuals who lack a territory do not fight more vigorously to obtain one. Our experiments with removals of territorial males show that replacement males are capable of defeating the former residents if they have held the territory for more than four days, even though the former resident usually remains in the vicinity of his former territory, and can take it over and successfully defend it again if the new owner is removed. The greater value of a territory to a resident than to a challenger makes it highly unlikely that even vigorous challenges will be successful. This, combined with the risk of injury during fights, may be sufficient to explain the willingness of floating males to concede the dominance of territory holders without seriously challenging them.

Redwings apparently wandered extensively during their first year, but once they returned to our study area as yearlings, strong site fidelity developed, especially among males, even among individuals that failed to obtain territories. Most breeding females remained faithful to the general area, but not to specific sites. Our intensive study area was not large enough to reveal the full scale of female movements, but they commonly involved hundreds of metres.

Because there is no strong bond between the sexes, and because males in our study area, except for some feeding of fledglings, contribute only defence against predators to the breeding effort, it is unlikely that individuals of either sex would improve their nesting success by remaining with a former breeding partner. This is contrary to the situation in some strongly bonded monogamous species (Woolfenden & Fitzpatrick 1984). This may be why female Redwings were no more likely to return to a particular territory in subsequent years if the former owner was present than if he was not, and why females often changed territories within breeding seasons when a nest was destroyed. Improvements in breeding success with age are likely to be less in a polygynous species, such as the

Redwing, than among monogamous species, because the value of a coordinated pair is not added to the general value of prior experience in raising overall reproductive success. Male Redwings do occasionally improve their status by moving to nearby territories of higher quality (Beletsky & Orians 1987a), but only a small proportion do so. Although this behaviour is of adaptive significance and is presumably influenced by natural selection, it does not produce demographically detectable results.

In conclusion, the main factors affecting lifetime fledgling production in Redwings were longevity, harem size (males only), and the extent of nest predation. Because of the polygynous breeding system, there was greater variance in lifetime production among males than among females.

Acknowledgements

We wish to thank Lynn Erckmann for help during many parts of this study, and our field assistants, S. Birks, C. D'Antonio, E. Davies, M. Dunham, J. Erckmann, C. Halupka, T. Johnson, D. Mammen, C. Monnett, T. Olson, L. Patterson, L. Rotterman, S. Sharbaugh, R. Sulaiman, M. Titcomb, and S. Worlein. David Goeke kindly provided permission for us to work at the Columbia National Wildlife Refuge and supported us in many ways. This study was funded by several grants from the National Science Foundation to the first author, including BNS 8405486 and BSR 8614620. The second author was supported during the preparation of the manuscript by NSF post-doctoral fellowship BSR 8600123.

References

Beletsky, L.D. & Orians, G.H. 1987a. Territoriality among male Red-winged Blackbirds. I. Site fidelity and movement patterns. *Behav. Ecol. Sociobiol.* **20**: 21–34.

Beletsky, L.D. & Orians, G.H. 1987b. Territoriality among male Red-winged Blackbirds. II. Site dominance and removal experiments. *Behav. Ecol. Sociobiol.* **20**: 339–49

Bray, O., Kenelly, J.J. & Guarino, J.L. 1975. Fertility of eggs produced on territories of vasectomized Red-winged Blackbirds. *Wilson Bulletin* **87**: 187–95.

Dolbeer, R.A., Woroneck, R.B., Stickley, A.R. & White, S.B. 1978. Agricultural impact of a winter population of blackbirds and starlings. *Wilson Bulletin* **90**: 31–44.

Holcomb, L.C. 1974. The question of possible surplus females in breeding Red-winged Blackbirds. *Wilson Bulletin* **86**: 177–9.

Holm, C.H. 1973. Breeding sex ratios, territoriality and reproductive success in the Red-winged Blackbird (*Agelaius phoeniceus*). *Ecology* **54**: 356–65.

Hurley, T.A. & Robertson, R.J. 1984. Aggressive and territorial behavior in female Red-winged Blackbirds. *Can. J. Zool.* **62**: 148–53.

Hurley, T.A. & Robertson, R.J. 1985. Do female Red-winged Blackbirds limit harem size? I. A removal experiment. *Auk* **102**: 205–9.

Krebs, J.R. 1971. Territory and breeding density in the Great Tit (*Parus major*). *Ecology* **52**: 2–22.

Meanly, B. 1965. The roosting behavior of the Red-winged Blackbird in the southern United States. *Wilson Bulletin* **77**: 217–28.

Monnett, C., Rotterman, L.M., Worlein, C. and Halupka, K. 1984. Copulation patterns of Red-winged Blackbirds (*Agelaius phoeniceus*). *Amer. Nat.* **124**: 757–64.

Muldal, A.M., Moffat, J.D. & Robertson, R.J. 1986. Parental care of nestlings by male Red-winged Blackbirds.*Behav. Ecol. Sociobiol.* **19**: 105–14.

Orians, G.H. 1961. The ecology of blackbird (*Agelaius*) social systems. *Ecolog. Monog.* **31**: 285–312.

Orians, G.H. 1980. *Some Adaptations of Marsh-nesting Blackbirds.* Princeton: Princeton University Press.

Orians, G.H. & Willson, M.F. 1964. Interspecific territories of birds. *Ecology* **45**: 736–45.

Payne, R.B. 1969. Breeding seasons and reproductive physiology of Tricoloured Blackbirds and Red-winged Blackbirds. *University of California Publications in Zoology* **90**: 1–115.

Picman, J. 1980. Impact of Marsh Wrens on reproductive strategy of Red-winged Blackbirds. *Can. J. Zool.* **58**: 337–50.

Roberts. T.A. & Kenelly, J. J. 1980. Variation in promiscuity among Red-winged Blackbirds *Wilson Bulletin* **92**: 110–12.

Røskaft, E., Orians, G.H. & Beletsky, L.D. in press. Why do Red-winged Blackbirds accept eggs of Brown-headed Cowbirds? *Evol. Ecol.*

Searcy, W.A. 1986. Are female Red-winged Blackbirds territorial? *Anim. Behav.* **34**: 1381–91.

Searcy, W.A. & Yasukawa, K. 1983. Sexual selection and Red-winged Blackbirds. *American Scientist* **71**: 166–74.

Weatherhead, P.J. & Robertson, R.J. 1977. Harem size, territory quality and reproductive success in the Red-winged Blackbird (*Agelaius phoeniceus*). *Can. J. Zool.* **55**: 1261–9.

Woolfenden, G.E. & Fitzpatrick, J.W. 1984. *The Florida Scrub Jay: Demography of a Cooperative-breeding Bird.* Princeton: Princeton University Press.

Yasukawa, K. & Searcy, W.A. 1981. Nesting synchrony and dispersion in Red-winged Blackbirds: Is the harem competitive or cooperative? *Auk* **98**: 659–68.

Yasukawa, K. & Searcy, W.A. 1982. Aggression in female Red-winged Blackbirds: A strategy to ensure male parental investment. *Behav. Ecol. Sociobiol.* **11**: 13–17.

Part III. Co-operative Breeders

The term 'co-operative breeder' is used for bird species which live in social groups, the members of which co-operate to defend a communal territory. At least 3 % (300) of the world's bird species are co-operative breeders, and include members of several different families from a wide range of geographical locations and habitats. Four species are represented here, namely the Florida Scrub Jay *Aphelocoma caerulescens* of the south eastern United States, the Green Woodhoopoe *Phoeniculus purpureus* of Africa, the Splendid Fairy-wren *Malurus splendens* of Australia, and the Arabian Babbler *Turdoides squamiceps* of the Middle East. In all these species the group consists of a dominant breeding pair and a number of subordinates, usually offspring from previous years. The subordinates may stay in the group for several years, helping in territory defense, predator vigilance and care of young, and leaving only if driven out or if a better opportunity presents itself nearby. Some young eventually inherit their parents' territory, and those that disperse usually move very short distances, often to join a neighbouring group. In consequence some neighbourhoods are dominated for generations by a single family lineage. The combination of group living, territory inheritance and delayed dispersal has been seen as an adaptation to enchance offspring survival in a permanently crowded habitat. But it brings with it an unusually complex social system in which, beyond the 'co-operation', which often involves mutually beneficial alliances between particular individuals, there is intense conflict and competition for breeding opportunities.

All four studies illustrate the extent to which an individual's social environment can affect its reproductive success. The size of the group is important, as is the rank of the individual within the group. Some individuals live for years but never achieve breeding status. Such studies reveal how the optimal behaviour for the individual varies with circumstance, as some individuals pass from helping their parents, thereby gaining indirect fitness benefits, to becoming dominant breeders themselves.

Although all four species have much in common, they differ in interesting respects, for example in the extent to which subordinates participate directly in reproduction, and in whether incest is practised. The Splendid Fairy-wren shows considerable inbreeding, in that about a fifth of breeding partners are related at the level of sibling or closer. Arabian Babblers, in constrast, observe a strict incest taboo.

13. Florida Scrub Jay

JOHN W. FITZPATRICK & GLEN E. WOOLFENDEN

Since 1969 we have monitored the births, movements and deaths in a wild population of Florida Scrub Jays *Aphelocoma c. caerulescens*, a co-operative-breeding bird. These jays can live up to 14 years. Therefore, only recently have we accumulated a sample of entire lifetimes sufficient to characterize the patterns of lifetime reproductive success within this population. In this chapter we describe these patterns, and examine why some jays are successful while most are not. The habitat, social system, and demographic attributes of the Florida Scrub Jay have been described in prior publications (see especially Woolfenden & Fitzpatrick 1984, in press). Here we summarize a few pertinent life history traits.

Life history

Florida Scrub Jays are sexually monomorphic, crestless, blue and grey jays averaging 79 g. Compared with other corvids, Florida Scrub Jays are weak fliers. They forage in and under shrubs, and eat a variety of terrestrial arthropods, some small vertebrates, acorns and some other seeds. Predators of nestlings and young fledglings include a few corvids (*Cyanocitta, Corvus*), snakes (e.g. *Masticophis, Drymarchon*), mammals (e.g. *Lynx, Procyon*), and certain owls (*Bubo, Otus*) and hawks (*Buteo, Circus*). In addition, certain hawks (*Accipiter, Falco*) prey on adults.

The Scrub Jay in Florida is entirely restricted to the oak scrub of the peninsula. This is a xerophytic habitat found on ancient or recent sand dunes, hence of extremely patchy distribution. The low, shrubby habitat is dominated by several species of stunted oaks and sharply bounded by various other habitats (pine forests, grassy ponds, prairie, etc.) totally unused by the jays. The rare, patchy and sharply defined nature of the

LIFETIME REPRODUCTION IN BIRDS
ISBN 0-12-517370-9

habitat plays important roles in shaping the social organization of the Florida Scrub Jay.

As breeders, the jays live as monogamous pairs in large ($\bar{x}$ = 9.0 ha), all-purpose territories defended year round. Daily the resident jays defend their piece of oak scrub from neighbouring territory holders and from unfamiliar jays on dispersal forays. Breeder death rates are low (about 21 % per year) and divorce is rare, so the same pair often occupies the same area of scrub for several years. Territories grow or shrink at the expense of bordering territories, as all usable oak scrub is always occupied. Change in territory size usually corresponds to the presence or absence of prebreeding helpers.

Most jays first breed at age two years (males, 47 %; females, 58 %). Some jays, especially males, do not breed until 3–6 years old. Only a very few jays have bred at age one year. Prebreeders live in their natal territory, from which they foray into the neighbourhood searching for openings in the breeding population. Females search over longer distances than males. Every year more potential breeders exist than openings in the breeding ranks, so prebreeders compete intensely for the limited opportunities. Virtually all yearlings and older prebreeders that failed to become breeders remain at home. About 55 % of the breeding pairs share their territories with these prebreeders. The number of prebreeders for families that have them ranges for one to six, and averages about two.

Territorial groups almost never include more than one monogamous pair. Prebreeders normally assist breeders in territory defense, predator mobbing, sentinel behaviour, and care of dependent young. They do not build nests, incubate or brood young. Apparently as a result of their help, prebreeders increase the reproductive success of breeders and the survival of young juveniles as well as breeding adults.

The nesting season extends from March through June. By the first week in April most first clutches have been laid. Renesting after failed attempts occurs until about mid-May. With only a few exceptions (4 %) during unusually productive years, only a single brood is reared annually.

Most clutches are of three (50 %) or four (37 %) eggs, with a range of one to five. Predation accounts for virtually all nest failures (98 %). Prebreeding helpers increase reproductive success principally by decreasing predation rates at early nests. The probability that a nest will fledge young plummets from about 70 % if begun in March to 40 % in May (Woolfenden & Fitzpatrick 1984). Fledgling brood sizes most often are two or three.

Reproductive success fluctuates greatly from year to year. Over 18 years (1970–1987) average fledgling production varied three-fold, from

0.9 to 2.8 fledglings per pair ($\bar{x} = 1.97 \pm 0.55$ SD). Survival of fledglings to age one year (yearlings) fluctuated from 21 % to 44 % ($\bar{x} = 34$ %) during all but one year. In 1979 a presumed epidemic killed all but one of 93 juveniles, along with an unprecedented 45 % of the breeders.

Survival of yearling and older jays varies with social status, age, and sex. Yearlings die before breeding at a higher rate than older helpers, with 35 % of females and 20 % of males dying between their first and second birthdays. Annual mortality of older helpers is 26 % for females and 16 % for males. Earlier, longer and more frequent dispersal forays by females in search of breeding vacancies probably account for their higher prebreeding mortality. The converse, however, is that males remain as non-breeding helpers for longer periods.

Death rates of breeders are identical for males and females (21 %). Both sexes show higher mortality between mid-May and mid-July than in any other two-month period. Breeders with helpers show lower mortality. Mortality increases after nine years of breeding, and no breeder has survived more than 11 breeding seasons. Breeding success also declines at about age nine years (Fitzpatrick & Woolfenden 1988).

Helpers become breeders by: (a) replacement, following death of established breeders, (b) direct inheritance of the natal territory, or (c) through a process of territorial budding, in which male helpers inherit a portion of their natal territory, pair with a female from outside the family group, and thereby create a new territory. New territories established through budding usually are small in the first year, but approach average size thereafter. Incipient territories less than about 4 ha in size do not persist if they cannot expand within the first year.

The sample

The 140 focal individuals used in most of our analyses of lifetime reproduction consist of jays that first bred between 1969 and 1982, and for which all breeding was documented. This sample includes nearly identical numbers of males (69) and females (71). Six of these 140 individuals are alive at this writing (spring 1988), two males (each has bred for nine years) and four females (three have bred for seven and one for 10 years). Including these six jays allows us to incorporate into the sample all jays that first bred between 1979 and 1982 without introducing a bias toward shorter-lived breeders. Including a few living individuals in analyses of lifetime reproduction requires estimating future reproductive success for these individuals (as in Clutton-Brock *et al.* 1982). We do so by awarding a few additional years of breeding, with the average expected success rates, for the six living jays according to known survivorship rates for older breeders (Fitzpatrick & Woolfenden 1988).

For our overall estimate of lifetime reproduction (Fig. 13.1) we add 69 additional jays (35 males, 34 females) to the sample of 140 just described. These were jays banded as breeders early in our study, with unknown prior breeding histories. We know that most of these jays were early in their breeding careers (most breeders are at any given time), but the sample undoubtedly included a few older breeders as well. Mean breeding "lifespan" of these unknown-origin jays after banding (males: 4.5 ± 3.2; females: 4.2 ±3.1 years) was so similar to that of our known-age sample (males: 4.1 ± 2.9; females: 4.3 ± 3.1 years) that we feel justified in pooling the two samples to maximize the sample of lifetime reproductive rates thus far observed in our study.

Because Florida Scrub Jays are extremely sedentary (Woolfenden & Fitzpatrick 1986), we have a choice of currencies with which to measure and compare reproductive success. Production of fledglings, independent young, and yearlings are known precisely for all breeders in our sample. However, many offspring, including older ones, never become breeders because of the intense competition for breeding space (Fitzpatrick &

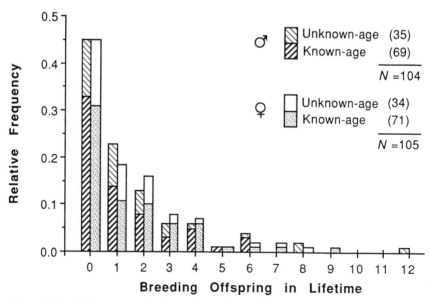

Figure 13.1 Lifetime reproductive success for 209 Florida Scrub Jay breeders, 104 males and 105 females. The unknown-age sample (n = 69) included some jays with prior breeding histories. The known-age sample (n = 140) includes only jays for which entire breeding histories are known.

Woolfenden 1986). Therefore we use production of breeding offspring as our most meaningful measure of reproductive success. We tested our decision to use post-dispersal jays as the currency by comparing the sex ratio of the breeding offspring we found. Because females disperse farther than males (Woolfenden & Fitzpatrick 1986,) the sex ratio of these replacement breeders should be biased toward males if we overlooked a significant number of dispersers. It is not (males, 73; females, 84; $\chi^2 = 0.77$, ns).

From 1969 through 1987, the 209 breeding jays (Fig. 13.1) produced 157 offspring known to have become breeders. To these we add 16, the expected number of currently living offspring that will soon become breeders, and five, the probable number of additional breeding offspring to be produced by the six living breeders in our sample. The resulting sum (178) still is 15 % smaller than our total sample of breeders (209). However, overall breeder density has declined about 15 % since our study began, indicating that recruitment has not quite kept pace with breeder mortality. Correcting for this fact, we are convinced that we locate virtually all breeding offspring produced within our study tract, given that a slight decline in overall population density has occurred during the study period (Woolfenden & Fitzpatrick, in press). Successful long-distance emigrations are rare, and we discover most of these.

Lifetime reproductive success

Nearly half of all jays that became breeders failed to produce breeding offspring of their own (Fig. 13.1). The largest number of breeders produced by an individual was 12 ($n = 1$), sired by an unknown-age male that bred for at least 11 years. As expected in a monogamous species with equal death rates of males and females, the sexes showed identical patterns of variance in reproductive output. Of the 178 breeding offspring represented in Fig. 13.1, 95 (53 %) were produced by only 13 (6 %) of the breeders.

Causes of variance in reproduction

We seek to explain why a few breeders are successful while most are not. Our most ambitious hope would be to isolate any heritable variation that might be involved, but we are not to that point. For the present we take two preliminary steps. First, after separating reproductive success into discrete life-history stages, we partition the overall variance to locate

Table 13.1 Correlations among components to lifetime production of breeders, and their relative contributions to the total variance.

| | Correlations | | | | | Multiple regression[a] | |
| | Breeding offspring $(L \cdot R \cdot S \cdot F)$ | Components | | | | Partial R^2 | Standardized β' |
		L	R	S	F		
Breeding lifespan (L)	0.64	—	−0.05	0.22**	0.23**	0.28***	0.65
Recruitment (R)	0.59		—	0.29**	−0.04	0.24***	0.39
Offspring survival (S)	0.46			—	0.12	0.13***	0.41
Fecundity (F)	0.37				—	0.11***	0.34

** $P < 0.01$; *** $P < 0.0001$
[a] Variables L, R, S, F are listed in descending order of their entry into the stepwise multiple regression model; standardized β' = standardized partial regression coefficient.

which (if any) stages provide the most important sources of variation. Second, we point out a few environmental and behavioural aspects that differ between successful and unsuccessful jays, and may help to explain some of the total variance.

PARTITIONING THE VARIANCE

Each individual's lifetime production of breeding offspring is the product of a series of components: annual fecundity, offspring survival through various stages, offspring recruitment, and all of these multiplied by the number of years of breeding. To examine whether individual variation in any of these components contributes more than the others to overall variance in lifetime success, we performed a stepwise multiple regression, using data on reproduction and offspring survival for the known-age sample of breeders (69 males, 71 females). For each individual we partitioned its lifetime reproduction into the following four stages: *fecundity* (average number of fledglings produced annually); *juvenile survival* (average survival of fledglings to age one year = number of yearlings/number of fledglings); *offspring recruitment* (average success by yearlings at becoming breeders = number of breeders/number of yearlings); *breeding lifespan* (number of breeding seasons survived, beginning with the first season in which a clutch of eggs was produced). For each individual, of course, the product of these four independent values was the integer number of breeding offspring produced in the lifetime, or lifetime reproductive success.

Correlations exist among these variables (Table 13.1). Not surprisingly, *lifetime reproductive success* is highly correlated wîth all four component variables, and most strongly with breeding lifespan. In addition, longer-lived breeders tend to show higher annual fledgling production, owing in part to the effects of helpers (which are offspring of previous breeding seasons), as well as increased fecundity and nest success with experience (Woolfenden & Fitzpatrick 1984). Longer-lived breeders also tend to show higher average survival of their offspring. This correlation may be a result of the contributions by helpers to the survival of breeders: breeders whose offspring survived to become helpers stand a significantly higher chance of surviving to the next breeding season (Woolfenden & Fitzpatrick 1984). Finally, breeders with high average offspring survival also tend to show higher success at recruitment by those offspring into the breeding ranks. We return to this most interesting correlation later.

We eliminated from the regression analysis all individuals with missing values for juvenile survival (because they produced no juveniles) or yearling recruitment (because they produced no yearlings). Potential biases result from including only those breeders that produced yearlings. For example, the sample (n = 89) is slightly biased toward longer-lived breeders (mean breeding lifespan = 5.2 years, compared with 4.4 years overall). However, similar results to those described below were obtained when all individuals were included and the missing values were assumed to be the average values for the population as a whole.

About 76 % of the overall variation in lifetime reproduction is explained by variation in the four parameters (Table 13.1). Breeding lifespan enters the multiple regression model first (R^2 = 0.28), followed by offspring recruitment (R^2 = 0.24). The effects of the remaining two variables are much lower after the first two are accounted for, but all four variables contribute significantly to the model.

Standardized partial regression scores show that one of the four parameters, *breeding lifespan*, is substantially more important than the other three in determining lifetime reproductive success. This result supports an earlier, similar analysis we made on a smaller data set (Fitzpatrick *et al.* 1988). The importance of breeding lifespan on lifetime reproduction comes as no surprise, as each year of breeding provides an independent opportunity to produce offspring that might succeed. The relative importance of offspring survival and offspring recruitment compared to fecundity has fundamental significance to the social regime of the Florida Scrub Jay. Mechanisms for increasing the survival and recruitment of older offspring include delayed dispersal, group living, a well-developed sentinel system, and territorial inheritance (Woolfenden & Fitzpatrick 1978).

Some correlates of reproductive success

The preceding analysis shows where in the life cycle the most important variation occurs for determining lifetime reproductive success. We are still far from answering why this variation occurs, or how much of it may be heritable as opposed to environmental or random. A few observations suggest that all three sources are involved.

RANDOMNESS

We conducted a computer simulation of lifetime reproductive outputs, based upon our overall observed probabilities of reproductive success and survival, and allowing a maximum of 12 breeding seasons. Two samples of 10,000 hypothetical breeding jays with slightly different annual survival rates both produced breeding offspring in frequencies nearly identical to those illustrated in Fig. 13.1 (Table 13.2). Therefore, our observed frequencies of reproductive output cannot be distinguished from those that would characterize a population of identical jays each of which shares identical probabilities of annual survival, reproductive success, and offspring survival, where those probabilities are set to the overall averages within our study population. It is important to stress that a frequency distribution not distinguishable from one produced by random processes does not confirm or support the importance of random processes in determining lifetime reproduction (e.g. Colwell & Winkler 1984).

HABITAT

Florida Scrub Jays are among the most habitat-specific land bird species in North America (Woolfenden & Fitzpatrick 1984, Cox 1987). Even within the relatively uniform oak scrub habitat historically supporting these jays, reproduction and survival can vary substantially. Habitat patches once supporting stable jay populations are known to become unsuitable after 30–40 years in the absence of fire, as the oaks grow tall and dense. One such population decline has been shown to be associated with decreases in reproductive success and survival of both adults and juveniles (Fitzpatrick & Woolfenden 1986). Table 13.3 summarizes the range of average values for reproduction and survival that we have observed in several grossly different kinds of habitat in or near our study area.

If gross differences in habitat can cause variation in reproduction and survival, then they certainly affect lifetime reproduction. Expected lifetime production of breeders varies between 1.1 and 0.4 among the habitats summarized in Table 13.3 (Fitzpatrick & Woolfenden 1986). We do not yet know whether such variation also characterizes local areas within our main study tract, much of which appears to be of high quality for jay reproduction. As a preliminary search for such spatial variation, we plotted the number of breeding offspring for all 104 male breeders at their activity centers on a map of the main study area (Fig. 13.2). Some clumping of successful jays is evident, suggesting (but not proving) that local habitat variation may play a significant role in determining lifetime reproductive output. Because local neighbourhoods may be dominated for generations by a single family lineage, locally higher success rates also may reflect genetic differences among lineages.

AGE AT FIRST BREEDING

Early theories of helping at the nest in birds postulated that helping provides practice, thereby improving later breeding efforts (Skutch 1961). However, in a preliminary analysis of lifetime fitness (Fitzpatrick &

Table 13.2 Observed and expected lifetime production of breeding offspring by Florida Scrub Jay breeders.

Breeding offspring in lifetime	Observed frequencies				Expected frequencies from null model[a]			
	Known-age only		Total		Breeder survival			
					0.82		0.79	
	N	(prop.)	N	(prop.)	N	(prop.)	N	(prop.)
0	67	(0.48)	94	(0.45)	4600	(0.46)	5036	(0.50)
1	27	(0.19)	43	(0.21)	2009	(0.20)	1962	(0.20)
2	18	(0.13)	30	(0.14)	1244	(0.12)	1267	(0.13)
3	9	(0.06)	14	(0.07)	898	(0.09)	761	(0.08)
4	11	(0.08)	13	(0.06)	574	(0.06)	435	(0.04)
5	2	(0.01)	2	(0.01)	333	(0.03)	258	(0.03)
6	4	(0.03)	6	(0.03)	185	(0.02)	157	(0.02)
7	1	(0.01)	2	(0.01)	90	(0.01)	79	(0.01)
8	0	—	3	(0.01)	39	(0.004)	29	(0.003)
9	1	(0.01)	1	(0.005)	14	(0.001)	9	(0.001)
10	0	—	0	—	10	(0.001)	6	(0.001)
11	0	—	0	—	4	(0.000)	0	—
12	0	—	1	(0.005)	0	—	1	(0.000)
	140	(1.00)	209	(1.00)	10,000	(1.00)	10,000	(1.01)

[a] 10,000 hypothetical breeding jays: annual survival = 0.82 (or 0.79) for maximum of 12 breeding seasons; annual probabilities of producing 0 to 3 breeding offspring: 0(0.74), 1(0.20), 2(0.05), 3(0.01) as observed in study population 1970–1987.

Table 13.3 Mean survivorships and reproduction of Florida Scrub Jays in several habitats, 1969–1986.

	Habitat			
	Periodically-burned, open-oak scrub	Unburned, overgrown scrubby flatwoods	Unburned Southern Ridge Sandhill (Slash Pine-Turkey Oak)	Mature citrus bordering unburned scrub
n (pair-years)	429	74	8	21
Seasonal nest attempts	1.38 (593/429)	1.49 (110/74)	1.50 (12/8)	1.11 (20/18)
Fledglings/pair	1.97 (843/429)	1.58 (117/74)	1.38 (11/8)	2.00 (38/18)
Independent young/pair	1.17 (500/429)	0.80 (59/74)	1.13 (9/8)	1.56 (28/18)
Yearlings/pair	0.60 (259/429)	0.36 (27/74)	0.50 (4/8)	0.61 (11/18)
First-year survival	0.307 (259/843)	0.231 (27/117)	0.364 (4/11)	0.289 (11/38)
Breeder survival	0.789 (697/883)[a]	0.723 (107/148)	0.688 (11/16)	0.619 (26/42)
Expected lifetime success/individual				
Breeding seasons	4.4	3.5	3.2	2.6
Fledglings	4.3	2.8	2.2	2.6
Independent young	2.6	1.4	1.8	2.0
Yearlings	1.3	0.6	0.8	0.8

[a] n = 883 breeder years for calculating breeder survival.

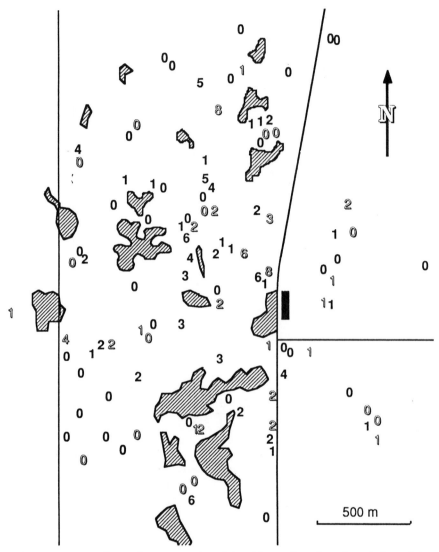

Figure 13.2 Breeding offspring produced by 104 males plotted at their activity centres within the study tract at Archbold Biological Station. Solid numbers (0–6) represent males for which lifetime breeding is known; open numbers (0–12) represent males for which some prior breeding may have occurred. Some clumping of successful jays is evident.

Woolfenden 1988), we found a *negative* correlation between age at first reproduction and lifetime reproductive success. Detailed study of this puzzling trend is underway, but its implications are so interesting that a summary is warranted here. It is now apparent that delaying breeding and helping for additional years beyond one does not improve an individual's subsequent performance as a breeder. Florida Scrub Jay breeders whose lifetime helping and breeding histories are known were split into two samples of roughly similar size, those that helped for one year (n = 25 males, 29 females) versus those that helped for two or more years (n = 33 males, 17 females). Survival and reproductive success of breeding females show no relationship with number of years they spent as helpers. Among males, however, striking effects exist. For males that helped only one year, survival as breeders was 0.86, compared to 0.72 for males that helped two or more years. Annual reproductive success was 2.03 vs. 1.74 fledglings per year, respectively. Eliminating the first breeding year (because novices always show lower reproductive success), the difference in reproductive performance is even more evident: 2.25 vs. 1.77 fledglings per year. Finally, fledglings produced by males that helped in only one year showed significantly higher probability of becoming breeders than those from males that helped in two or more years (0.19 vs. 0.12, χ^2 = 3.86, P < 0.5). Lower survivorship, lower annual fledgling production, and lower offspring survival combine to cause significantly lower lifetime reproductive success among male jays that help for more than one year (Fig. 13.3).

The biological implications of these results are not yet clear. We entertain the possibility that for males, number of years as a helper bears some relationship to the overall quality of the jay. Indeed, long-term helpers often are attracted to territories where a breeding vacancy exists, only to be found again in the home territory, having lost in the competition for breeding space. Wandering, inept, male prebreeders (WIMPs) are apparent to us because of the extremely sedentary behaviour of the species in general. We suggest that such poor-quality individuals exist in many avian populations as floaters, but remain unobserved because of their mobility, subordinance, and lacking a role in the breeding structure of the population.

Indirect component to fitness

Florida Scrub Jay helpers raise the reproductive success of the pairs they help, namely their parents or other close relatives. Production of additional, related individuals beyond what the parents might have

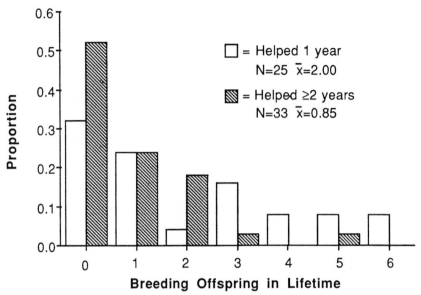

Figure 13.3 Lifetime production of breeding offspring by male breeders that helped for one year compared with male breeders that helped for two or more years.

produced adds an indirect component to lifetime reproductive success, a component characteristic of co-operative-breeding species (e.g. Koenig & Mumme 1987). Detailed analysis of this aspect of fitness is beyond the present scope. However, as a first-order attempt to estimate the relative magnitude of these indirect benefits, we tallied them for our known-history samples of jays (Fig. 13.4).

"Breeding offspring equivalents" were awarded to each jay (including those that helped but never became breeders) based on the number of breeders produced in its territory while that jay was a helper. Pairs with helpers produced an average of 0.40 breeders per year, while those without helpers produced 0.25 (unpubl. data). Therefore, about 38 % (0.15/0.40) of the offspring production by pairs with helpers accrued to those helpers. This percentage did not change with number of helpers, because additional helpers beyond one did not affect reproductive success (Woolfenden & Fitzpatrick 1984). Therefore, if more than one helper was present in a year when one or more breeders were produced, the 38 % was split evenly among the helpers. If the helper aided an unrelated breeder (or two), its accrual was reduced according to its average relatedness to the offspring produced. Thus, a lone helper that helped both its parents produce a breeder received 0.38 offspring equivalents, but only 0.19 offspring equivalents if one breeder was a step-parent.

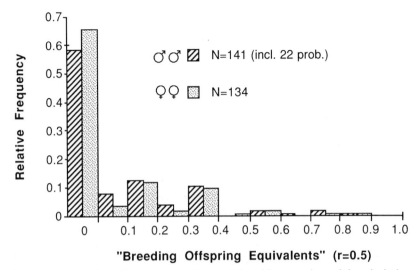

Figure 13.4 Breeding offspring equivalents attributable to male and female helpers. The samples include jays that became breeders and jays that died as helpers.

Individual jays, including those that never bred, gained from 0 to 0.89 breeding offspring equivalents via indirect accrual (Fig. 13.4). Because most annual breeding efforts fail to produce a breeder, and most helpers help for only one or two years, over 60 % of the individuals gained no indirect benefits whatsoever through helping to raise breeding kin. A significantly greater proportion of helpers that died before breeding gained no indirect benefits (86 of 121, or 0.71) than among those that later became breeders (84 of 154, or 0.55; $\chi^2 = 7.18$, $P < 0.01$). This difference results in part from a difference in duration as a helper. A significantly higher proportion of the helpers that failed to become breeders helped for only one year than among the sample of helpers that later bred (0.74 vs. 0.51, $\chi^2 = 14.7$, $P < .001$). Of those helpers that did receive a benefit, the average gain was 0.30 breeding offspring equivalents whether or not the helper eventually became a breeder. The difference in indirect benefits among these two samples of helpers reflects our previous finding (Fitzpatrick & Woolfenden 1986) that a helper's probability of becoming a breeder is enhanced by the presence of additional, subordinate helpers in the territory.

Discussion

Despite a social system characterized by monogamous pairs breeding within similarly-sized territories across a relatively simple and uniform habitat, lifetime reproductive success among Florida Scrub Jays is highly skewed. Considering only the individuals that achieve breeding status, nearly half fail to produce even a single offspring that becomes a breeder (Fig. 13.1). A much larger sample could be viewed as potential breeders, namely all those individuals that survived to age one year and experienced a breeding season as adult-plumaged jays ($n = 261$). Of this sample, only about 28 % (73/261) produced at least one breeding offspring and fewer than 18 % (46/261) replaced themselves genetically by producing two or more breeders.

Although we cannot distinguish the overall distribution of breeding offspring from an outcome of chance events among identical jays, several patterns suggest that variation does not play a role in skewing the reproductive success. Most important, production of breeding offspring is not uniformly, or randomly, distributed spatially across the study area. Several clusters of territories exist in which average success has been higher than elsewhere (Fig. 13.2; Fitzpatrick et al. 1988). This could reflect effects of fine scale habitat variation (currently being studied carefully in the field) or variation among family lineages that dominate these areas, or both.

Partial regression analysis shows that breeding lifespan plays an over-riding role in determining lifetime reproductive success, but that the fates of older offspring are not far behind in importance. Protection of the reproductive investment in older offspring may be central to the cooperative social system of the Florida Scrub Jay. Permanent defense of territories that are unusually large for birds of their body size allows Florida Scrub Jay breeders to provide safe haven for pre-reproductive offspring, allowing delayed dispersal to accompany delayed breeding. Perhaps more important, these large territories provide opportunity for certain offspring to gain direct access to limited breeding vacancies without leaving home. In a highly competitive population, where defense of relatively large territories is facilitated by open habitat structure, this form of parental investment can greatly increase the chances that a few of the offspring successfully compete to become breeders.

Long-term helpers can expect to gain indirect genetic benefits by helping to produce kin, and the accrual of these benefits shows a steady increase with number of years as a helper. Paradoxically, however, the long-term helper males show significantly inferior performances once they become breeders, with elevated mortality and reduced annual reproductive success. These are our first indications that helpers vary in quality. Predominant are those that help for one year, then begin making serious efforts to enter the breeding ranks, a process that can take another year or two even for high quality individuals. In addition, however, a fraction of the *male* helpers appears to lose at competitions for breeding, and even gives evidence of ineptitude at these competitions. These prebreeders may help for three to five years, gaining at least some form of lifetime reproductive success through their effects on collateral kin. After becoming breeders, their direct accrual of fitness—while considerably greater than that gained indirectly—is poor. These jays appear to show reduced ability to establish and maintain territories, and what offspring they do produce show reduced survival. Such a pattern among the less successful breeders provides further evidence of the great importance of territory as a parental, as well as an individual, investment to increase lifetime reproductive success.

Acknowledgements

Archbold Biological Station continues to provide extensive support for our study of Florida Scrub Jay ecology. We sincerely thank the Station personnel. We also thank Ronald L. Mumme, Robert L. Curry, and Ian Newton, who made useful comments on the manuscript and prepared the figures, and Jan Woolfenden, who assisted with the typing. Field work was supported by the Conover Fund of the Field Museum of Natural History, the University of South Florida, and by NSF Grant No. BSR-8705443.

References

Clutton-Brock, T.H., Guinness, F.E & Albon, S.D. 1982. *Red Deer, Behavior and Ecology of Two Sexes.* Chicago: University Press.
Colwell, R.K. & Winkler, D.W. 1984. A null model for null models in biogeography. In *Ecological Communities: Conceptual Issues and the Evidence*, eds. D.R. Strong Jr, D. Simberloff, L.G. Able & A.B. Thistle, pp. 344–59. Princeton: University Press.
Cox, J.A. 1987. Status and distribution of the Florida Scrub Jay. *Florida Ornithol. Soc. Spec. Pub.* **No. 3**, 110 pp.
Fitzpatrick, J.W. & Woolfenden, G.E. 1986. Demographic routes to cooperative breeding in some New World jays. In *Evolution of Animal Behavior*, ed. M.H. Nitecki & J.A. Kitchell, pp. 137–60. New York: Oxford University Press.
Fitzpatrick, J.W. & Woolfenden, G.E. 1988. Components of lifetime reproductive success in the Florida Scrub Jay. In *Reproductive Success*, ed. T.H. Clutton-Brock, pp. 137–60. Chicago: University Press.
Fitzpatrick, J.W., Woolfenden, G.E. & McGowan, K.J. 1988. Sources of variance in lifetime fitness of Florida Scrub Jays. In: *Acta Congressus Internationalis Ornithologici*, ed. H. Ouellet, pp. 871–891. Ottawa: University Press.

Koenig, W.D. & Mumme, R.L. 1987. *Population Ecology of the Cooperatively Breeding Acorn Woodpecker.* Princeton: University Press.

Skutch, A.F. 1961. Helpers among birds. *Condor* **63**: 198–226.

Woolfenden, G.E. & Fitzpatrick, J.W. 1978. The inheritance of territory in group-breeding birds. *BioScience* **28**: 104–8.

Woolfenden, G.E. & Fitzpatrick, J.W. 1984. *The Florida Scrub Jay, Demography of a Cooperative-Breeding Bird.* Princeton: University Press.

Woolfenden, G.E. & Fitzpatrick, J.W. 1986. Sexual asymmetries in the life history of the Florida Scrub Jay. In *Ecological Aspects of Social Evolution,* ed. D.I. Rubenstein & R.W. Wrangham, pp 87–107. Princeton: University Press.

Woolfenden, G.E. & Fitzpatrick, J.W. In press. Florida Scrub Jays: a synopsis after 18 years of study. In *Cooperative Breeding in Birds: Long-term Studies of Ecology and Behavior,* ed. P.B. Stacey & W.D. Koenig. Cambridge: University Press.

14. Green Woodhoopoe

J. DAVID LIGON & SANDRA H. LIGON

The Green Woodhoopoe *Phoeniculus purpureus* is a co-operatively breeding member of the coraciiform family Phoeniculidae, a group restricted to sub-Saharan Africa. Within this extensive region, the species is widespread, occurring from East to West Africa and south to southernmost South Africa. These woodhoopoes occupy a wide range of habitats—savanna, open woodland, palm groves, riverine forest in arid thornbush—and are found from near sea level to over 2000 m. They are, however, absent from dense forest, such as the rainforests of West Africa. In general, their habitat requirements seem to be open woodland with at least some trees large enough to provide nest and roost cavities (Ligon & Davidson 1988).

Green Woodhoopoes, like other members of their family, are primarily black, with an iridescent metallic green and blue-purple sheen. Bills, tarsi and toes of adults of both sexes are red-orange while those of juveniles are black. Unlike several other co-operative breeders, Green Woodhoopoes are strongly sexually dimorphic in weight and in bill length (males at around 78 g are about 20 % larger than females at around 64 g), and, in addition, all vocalizations of adults are sexually diagnostic. We suspect that the larger size of males is related to intense male–male competition for territories and females. Whatever its adaptive basis, large body size appears to carry a cost to males of all ages, in the form of greater annual mortality relative to females (see below).

Foraging woodhoopoes move as a group as they search tree trunks, branches and twigs for arthropods. Although the tail is long and flexible, it is used for support as the birds hang beneath or on the sides of branches. Males and females tend to forage in different parts of the trees (our unpublished data), according to differences in their body size and bill length. Thus females forage more on small terminal branches and

LIFETIME REPRODUCTION IN BIRDS
ISBN 0-12-517370-9

twigs, while males more often feed on the main trunk and large branches, as well as on the ground at the bases of trees.

Territorial social units contain from two to 16 individuals (60 % number 3–5, $n = 189$), but only one breeding pair, regardless of group size. Other flock members, usually relatives of one or both breeders, serve as nest helpers, providing food to the incubating female and later to nestlings.

Breeding in each year normally begins in late May or early June, following the "long rains" of March–May, and generally ends in December or earlier. The number of successful nesting efforts per year appears to be initially controlled by food availability, in turn determined largely by patterns of rainfall (Ligon & Ligon 1982, 1988). Two broods during the period June–December are common, and if conditions are extremely favourable up to three broods can be produced by one breeding pair during this interval. If conditions are poor, however, breeding is limited, and only one, or even no nesting attempt may occur. In poor years, there is considerable territory-to-territory variation in the time of breeding as well as in the number of nests attempted.

Another important factor influencing year-to-year variation in breeding success is predation. Known predators of nestlings include driver ants, Tribe Dorylini, Harrier Hawks *Polyboroides radiatus*, and Pearl-spotted Owlets *Glaucidium perlatum*. In addition, we suspect that an arboreal agamid lizard ate eggs at a few nests.

Clutches normally contain two to four eggs. Four eggs is a common clutch size in all groups, including those with only one female. More than one female probably laid eggs in no more than three of 73 nests for which we have appropriate information. Another line of evidence supporting the suggestion that only one female normally contributes to the clutch is the pattern of incubation. The breeding female is the sole incubator of eggs. Of 51 nests watched for extended periods during incubation, two females appeared to incubate in only one, with four eggs. These observations led us to conclude that the alpha female, like the alpha male, normally is the sole reproductive of its sex.

Unlike most co-operative breeders for which long-term data are available, Green Woodhoopoes exhibit high annual rates of mortality, about 40 % per year for males and 30 % per year for females (Ligon 1981, Ligon & Ligon 1988). Woodhoopoes without exception use cavities or other cover for roosting, and a large cost apparently is associated with this behaviour. All our evidence suggests that most predation on these birds occurs at the roost holes. Major nocturnal predators are driver ants, Tribe Dorylini, and Large-spotted Genets *Genetta tigrina*. Birds occupying cavities in weak wood, or cavities that are shallow and/or with large entrance holes, are vulnerable to genets and possibly to wild and feral cats (*Felis libyca* and *F. domesticus*). All cavities are vulnerable to invasion by raiding driver ants. It seems paradoxical that Green Woodhoopoes should inevitably roost in cavities where they are so susceptible to capture, but perhaps they are unable to remain endothermic in the absence of insulation afforded by the roost cavities (Ligon *et al.* 1988).

Green Woodhoopoes of both sexes tend to be extremely philopatric. Eighteen of 38 females banded as nestlings or fledglings, and that we later recorded as breeding, did so in their natal territories and 14 did so in an adjacent territory. Three of the remaining six birds bred only two or three territories away from their natal territory (mean number of territories females moved = 1.24). We have similar data for 33 males: nine attained breeding status in their natal territories, 20 bred in an adjacent territory, and four bred two territories from their birth sites (mean number of territories males moved = 0.85). The limited dispersal pattern of male woodhoopoes is similar to that of males of several other co-operative breeders; however, female woodhoopoes differ from the usual pattern in that about half bred in their natal territories and 85 % bred 0–1 territories from their hatching site. Female Green Woodhoopoes thus employ the full range of dispersal possibilities, from no movement to long-distance (ca. 13 territories) emigration.

At our study area near Lake Naivasha, in the central Rift Valley of Kenya, woodhoopoes occupy open woodland consisting almost entirely of one tree species, the Yellow-barked Acacia *Acacia xanthophloea*. These trees, growing on former lake bed, form a belt of varying width that surrounds the lake. Our primary study site was Morendat Farm, with 18 marked flocks (not all present at one time) in contiguous territories. We also studied woodhoopoes along the south side of Lake Naivasha, near Crescent Island, where seven social units were banded, and between the Morendat River and North Lake Road along the Naivasha–Nakuru highway where we banded eight flocks.

Our study began in mid-1975 and the main body of the project ended in January 1982. JDL returned to the study site in mid-1984. We ultimately marked for individual recognition a total of 386 woodhoopoes in 33 flocks; 269 birds were banded on or just adjacent to Morendat Farm. By January 1982, 93 % of the birds on Morendat Farm were of known parentage. This was possible both because mortality was high and because most woodhoopoes of each sex are extremely philopatric.

Results

INDIVIDUAL VARIATION IN LIFETIME REPRODUCTIVE SUCCESS

In our Green Woodhoopoes, as in many other species of birds, most individuals that lived long enough to attain independence never bred successfully (79 % of 143 females and 78 % of 131 males). Offspring that themselves survived to breed are a far better estimate of relative fitness (Ligon 1981, Fitzpatrick & Woolfenden 1988), and by this more restrictive measure, 91 % of all females and 89 % of all males in our study failed to produce any direct descendants.

Of those few birds that did attain breeding status, many failed to produce even one independent offspring and many more failed to produce any breeders. Fig. 14.1 illustrates that about two-thirds of all woodhoopoes that became breeders left no breeding offspring. About 20 % of male

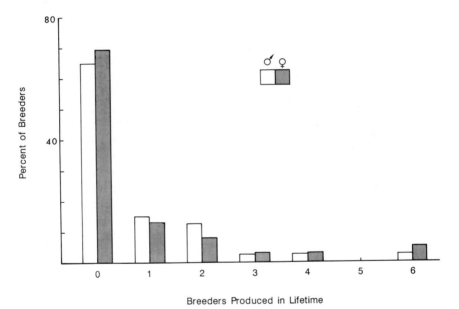

Breeders Produced in Lifetime

Figure 14.1 Frequency distribution of the number of breeding offspring produced by 41 male and 40 female Green Woodhoopoes that either were breeders at the onset of the study in 1975 (assigned 0 previous offspring) or that obtained breeding status in 1975 or later, and that died prior to 1984. In a few cases, breeders present in 1975 almost surely had previously produced offspring, some of which became breeders; in addition, long-distance dispersal of some birds that subsequently became breeders may have occurred. Both of these factors may have affected the precision of these values to some degree.

and female breeders produced about 80 % of the young birds that eventually attained breeding status. Three birds, two female and one male, produced six breeding offspring. The male and one of these females were mates.

Both sets of values, all birds and breeding birds, are to some extent biased toward conservatism (range of variation decreased), since breeding birds present at the beginning of the study in 1975 were assigned zero prior offspring, regardless of the number and ages of helpers in their groups. For example, one female, known to have produced 18 independent young in 1975–1981, was also the probable mother of four yearling helpers present in her flock in 1975. Breeders still alive at the end of the study in 1984 were excluded from consideration.

SURVIVORSHIP AND REPRODUCTIVE SUCCESS

We have produced survivorship curves for birds of known age (Fig. 14.2), although samples such as this are nearly always biased to some degree toward individuals that died early (Clutton-Brock *et al.* 1983, Fitzpatrick and Woolfenden 1988), because individuals living to the end of the study are excluded from analysis. Thus we also generated survivorship curves for all birds that died or disappeared prior to 1984. For this analysis breeders present at the beginning of the study were assigned an age of three years (mean age of first breeding for 19 males and 15 females of known age was 3.6 years). Non-breeding helpers were assumed to be

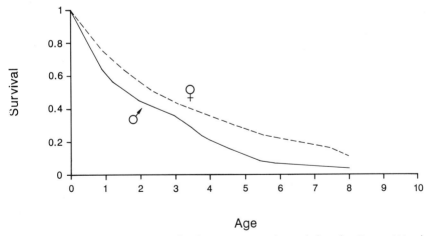

Age

Figure 14.2 Survivorship curve for known-age male and female Green Woodhoopoes. These values may underestimate mean survival for each sex by as much as a year (see text).

one year of age, the minimum value possible. This added about one year to the expected lifespan of each sex (Table 14.1).

Most of the 386 woodhoopoes that we banded failed to live for many years. At the time of our last period of field work in mid-1984, eight known-age female breeders averaged 6.1 years of age and six known-age male breeders averaged 4.8 years. Two other male breeders still alive in 1984 were banded in January 1977 and June 1978 and were probably nine and eight years old, respectively. Our longevity record is of a female hatched in 1975 and seen by us as a breeder in her natal territory in 1984 and by H. Dinkeloo in 1985.

Annual survival of known-age Green Woodhoopoes was 72 % ($n = 63$) for females and 61 % ($n = 67$) for males, a difference that was significant (logrank Chi-square = 4.52, $P = 0.03$). The difference probably was a reflection of the larger body size of males, which often required them to use less secure cavities for roosting, with resulting higher predation (Ligon & Ligon 1988, in press).

The breeding lives of male and female Green Woodhoopoes showed no differences. Both sexes showed the general relationship between number of breeding years and lifetime number of young birds produced (40 females, $P < 0.0005$; 41 males, $P < 0.03$). In itself, however, this relationship does not provide much useful insight into the social and ecological factors that influence individual variation in lifetime reproduction. For species such as the Green Woodhoopoe three obvious variables that might influence individual lifetime reproductive success are the environmental conditions, the effects of helpers and the territory quality.

Table 14.1 Lifespan and lifetime reproductive success by sex in Green Woodhoopoes.

Variable	Males		Females	
Mean lifespan, all known-age birds	2.30[a]	(93)[b]	2.22	(97)
Mean lifespan all birds,[c] including pre-1975 hatched birds	3.27	(160)	3.49	(173)
Mean age of first breeding, known-age birds	3.3	(19)	3.9	(15)
Mean breeding lifetime	2.62	(48)	2.72	(50)
Mean no. young produced[d]/lifetime by birds attaining breeding status	2.90	(41)	3.28	(40)

[a] Years.
[b] Number individuals.
[c] Minimum estimate. See text.
[d] Number young birds that survived to end of calendar year.

BREEDING SUCCESS, RAINFALL AND HELPERS

Since the great majority of woodhoopoes that live long enough to attain independence never breed, it might be assumed that most individuals contribute to their lifetime inclusive fitness by providing aid to close relatives. And, indeed, most woodhoopoes do act as nest helpers, if they have lived long enough to be present when breeding takes place in their social group.

Although most social units of Green Woodhoopoes include nonbreeding helpers the mean number of young produced per year does not differ significantly between pairs and groups with helpers ($\bar{x}$ pairs = 0.94, $n = 17$; $\bar{x}$ groups = 1.36, $n = 1.55$; Kruskal–Wallis Test, $\chi^2 = 1.53$, $df = 1, P < 0.20$); nor does the probability that no young will be produced in a given year differ between pairs and groups ($n = 155$ groups, 17 pairs, Kruskal–Wallis Test, $\chi^2 = 0.846$, $P = 0.36$).

The year-by-year data (Tables 14.2 and 14.3) suggest that the pattern of rainfall (timing and amount) might account for much of the variation recorded for each group size (Fig. 14.3). Our analyses revealed that, perhaps surprisingly, the quantity of precipitation during the "long rains", which occurs March or April–May or June, just prior to the onset of breeding, does not significantly influence the production of young woodhoopoes on an annual basis ($r = 0.063$, $P = 0.42$, $n = 171$). Rather, *dry season* (January–March) rainfall is inversely associated with reproductive success ($r = 0.335$, $P < 0.0001$, $n = 171$). This finding is explained by the biology of the many species of moth larvae, which comprise the vast majority of food items brought to nestlings and eaten by both adult and young birds during the potential 6–7 month nesting

Table 14.2. Precipitation patterns and annual reproductive success in the Green Woodhoopoe.

Year	Dry season precip. (cm)	Long rains precip. (cm)	Mean no. young produced/flock[a]	Mean flock size
1975	0.41	18.06	1.5	5.8 (22)[b]
1976	2.29	14.00	1.9	5.6 (22)
1977	10.06	43.15	2.4	4.8 (22)
1978	17.22	45.72	0.7	6.3 (25)
1979	27.21	17.70	0.5	5.3 (24)
1980	5.67	36.70	1.2	4.0 (24)
1981	2.11	36.05	2.5	3.9 (21)

[a] Number of offspring surviving to the end of the calendar year.
[b] Number of flocks.

Table 14.3 The relationship between patterns of rainfall and predation to annual reproductive success in the Green Woodhoopoe.

Year	Timing and quantity of rainfall	Mean no. nesting attempts/ flock	Per cent nesting attempts producing at least one fledgling	Mean no. fledglings produced/ successful nest
1979	Unfavourable[a]	1.00 (21,0–2)[b]	48[c] (21)[d]	1.67 (9,1–2)[e]
1980	Favourable	2.40 (20,1–5)	44 (48)	1.65 (20,1–4)
1981	Favourable	2.00 (19,1–3)	82 (38)	1.81 (31,1–4)

[a] See Table 14.2.
[b] Number of flocks, range in number of nest attempts.
[c] Low percentage of successful nests reflects high rate of predation.
[d] Number of nests.
[e] Number of successful nests; range in number of fledglings produced/successful nest.

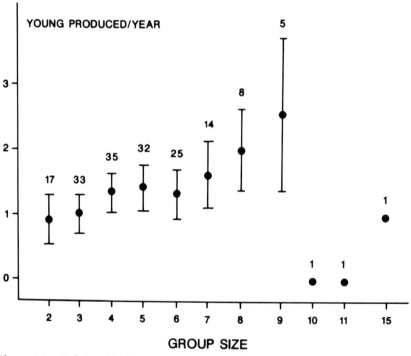

Figure 14.3 Relationship between group size at the onset of breeding in each year and number of young surviving to the end of the calendar year. Circles indicate means, vertical lines show one standard error. Numbers above points indicate sample sizes.

season (Ligon & Ligon 1982). Numbers of these caterpillars through the year are related most directly to the amount of precipitation that falls during the putative dry season. During this period, the moth caterpillars pupate in the ground and survive well only if the soil remains dry. If much rain falls during the "dry" season, mortality of the pupae is very high (M. Clifton, National Museum of Kenya, pers. comm.), with few adults emerging to give rise to the generation of caterpillars that appears following the long rains.

A multiple regression analysis considering the long rains, dry season precipitation, and number of helpers, provides an indication of the relative importance of each of these factors in the annual production of young woodhoopoes. Dry season rainfall had by far the greatest effect ($r^2 = 0.112$, $n = 171$, $P < 0.0001$), while precipitation during the long rains was of negligible importance (multiple $r^2 = 0.01$, $P = 0.16$). Once the effect of precipitation during the dry season is controlled, the relationship between number of helpers and number of young wood-hoopoes produced per year becomes significant (multiple $r^2 = 0.024$, $P = 0.03$). Thus, although helpers have the potential to increase the production of young birds, their effects are usually over-ridden by environmental factors.

LIFETIME REPRODUCTIVE SUCCESS AND TERRITORY QUALITY

Where a woodhoopoe breeds is by far the most important factor governing its lifetime reproductive success. The critical ultimate measures of territory quality, especially in species occupying permanent territories, are survival and reproductive success (Ligon & Ligon 1988). On this basis we used three measures of territory quality: (a) A Natality/Mortality ratio for each territory to provide two categories—territories with a N/M ratio of greater than one, and those with a ratio of less than one. (b) Number of descendants in the two classes of territories over a ten year period. (c) Territories in which the same genetic lineage persisted over the 10 years of study, and those where either the lineage changed or ceased to exist. In addition, we found that trees in the territories with an N/M ratio > 1 were larger than in the other territories to a degree almost significant statistically. Safe cavities for roosting seemed to be the single most critical resource determining territory quality (Ligon & Ligon 1988, in press) and, other things being equal, cavities are much more likely to be present in larger trees.

All of these measures led to the same conclusion: Green Woodhoopoe territories on Morendat Farm fell into one of two categories, either very productive ("High Quality") or virtually non-productive ("Low Quality").

About one-third of the territories were responsible for almost all the young birds produced on the study site. Breeders on the other territories did not produce sufficient young to replace themselves; such territories continued to exist only by virtue of continual immigration of new birds from elsewhere. When this did not occur, the territory disappeared and the space was incorporated into neighbouring territories.

Territory quality affected lifetime reproductive success of breeders in several ways: (a) Overall, annual survival was significantly greater for all birds in the high quality territories (Wilcoxon Chi-Square = 15.65). (b) Breeding life was also significantly greater in the high quality territories (Fig. 14.4). This alone could produce the relationship discussed earlier between breeding lifespan and lifetime number of offspring. However, significantly more young birds were produced per year, independent of individual breeders, in the high quality territories (Ligon & Ligon 1988). (c) Production of more young, plus their greater survival, together led to the great disparity in the N/M ratios of the territories. (d) Individual breeders in the five high quality territories produced significantly more breeding offspring than did breeders in the other territories.

A relationship also emerged between territory quality and age of first breeding. On Morendat Farm, those females hatched on high quality territories, and that remained there and eventually inherited breeding status, were somewhat older when they first bred than were females attaining breeding status on the other territories (mean age high quality = 4.3 years; low quality = 3.8 years). This delay was promptly made

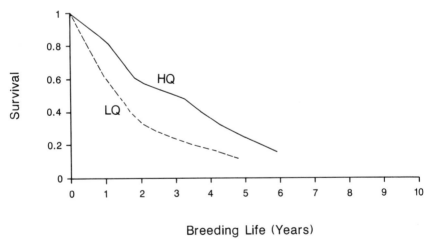

Breeding Life (Years)

Figure 14.4 Survivorship curve for breeding lifespans of Green Woodhoopoes occupying high quality (HQ) and low quality (LQ) territories on Morendat Farm.

good, in that *first-time* female breeders produced an average of 3.7 young ($n = 10$) in their first year on high-quality natal territories, compared with 0.8 young ($n = 16$) on other territories. These values illustrate that *where* a woodhoopoe first breeds is far more important to its lifetime reproductive success than when it first breeds.

All breeders alive in years when environmental conditions were optimal (e.g. 1977 and 1981, Table 14.2) did better overall than those birds that bred in average or below-average years, regardless of territory quality. Breeders in both high and low quality territories produced about twice as many young in exceptionally good years as in average or poor years (Table 14.4). However, territory quality still exerted a powerful effect: twice as many young were produced in high quality territories in both good and bad years. Thus, the important effect of rainfall patterns did not alter the *relative* productivity of high and low quality territories.

Discussion

Individual differences in the lifetime reproductive success of Green Woodhoopoes were great. Most birds never attained breeding status, and of those that did, about two thirds left no breeding offspring. This picture has important implications concerning effective population size and inbreeding in the woodhoopoe population of Lake Naivasha (Ligon & Ligon 1988).

The most critical factor affecting relative reproductive success was the territory in which the individual bird bred. On our main study site, Morendat Farm, only five of 16 territories were producers or "sources," of new woodhoopoes, the remaining 11 were net importers of birds from elsewhere or "sinks." This places a great premium on breeding in certain territories. In four of the five productive, or "high quality" territories, over a 10-year period, breeding status was passed from mother to daughter and sister to sister, whereas in the other 11 territories complete turnover of genetic lineage occurred one or more times during the study, or in four cases the territory went out of existence.

Although there was a relationship between territory quality, as we have identified it, and group size, the number of helpers available to a breeder did not significantly affect annual reproduction; the potential effect of helpers was strongly over-ridden by the timing and amount of rainfall. The increased survival of non-breeders in the high quality territories may, in large part, have accounted for the relationship, weak though it was, between number of helpers and number of young produced per year. Moreover, since individual breeders in high quality territories live

Table 14.4 The effects of annual rainfall patterns on annual reproductive success in high and low quality territories of Green Woodhoopoes on Morendat Farm.

	Favourable years[a]			Average or below average years[a]					
	1977	1981	Mean	1975	1976	1978	1979	1980	Mean
High quality territories	3.4 (5)[b]	3.6 (5)	3.5	1.0 (4)	2.8 (4)	1.0 (4)	1.4 (5)	3.2 (5)	1.9
Low quality territories	1.7 (11)	1.3 (7)	1.5	1.6 (10)	1.2 (10)	0.5 (11)	0.4 (11)	0.6 (9)	0.9

[a] See Tables 14.2 and 14.3.
[b] Mean number of young surviving to end of calendar year and number of territories included.

significantly longer than those in low quality ones, they are more likely to breed in one or more of the years when conditions are especially favourable. For a variety of reasons, the chances of success strongly favour breeders occupying certain territories.

Acknowledgements

We want especially to acknowledge the contributions to this chapter made by P.B. Stacey and D.A. McCallum. Our field studies of Green Woodhoopoes in Kenya were made possible via the support of R.E. Leakey and G.R. Cunningham van Someren of the National Museums of Kenya. E.K. Ruchiami, of the Office of the President of Kenya, kindly granted research permits. S. Zack, D.C. Schmitt and J.C. Bednarz assisted with the field work and W. and R. Hillyar and R. and B. Terry provided logistical support. This project was generously supported by the National Science Foundation, The National Geographic Society, The American Museum of Natural History, The National Fish and Wildlife Laboratory, and The University of New Mexico. We thank all of these individuals and institutions.

References

Clutton-Brock, T.H., Guiness, F.E. & Albon, S.D. 1982. *Red deer, behavior and ecology of two sexes.* Chicago: University Press.

Fitzpatrick, J.W. & Woolfenden, G.E. 1988. Components of lifetime reproductive success in the Florida scrub jay. In *Lifetime Reproductive Success*, ed. T.H. Clutton-Brock, 305–320. Chicago: University Press.

Ligon, J.D. 1981. Demographic patterns and communal breeding in the Green Woodhoopoe (*Phoeniculus purpureus*). In *Natural Selection and Social Behavior*, ed. R.D. Alexander & D.W. Tinkle, pp. 231–43. Ann Arbor, MI: Chiron Press.

Ligon, J.D., Carey, C. & Ligon, S.H. 1988. Cavity roosting, philopatry and cooperative breeding in the Green Woodhoopoe may reflect a physiological trait. *Auk* **105**: 123–7.

Ligon, J.D. & Davidson, N.K. 1988. Order Coraciiformes, Phoeniculidae, woodhoopoes. In *The Birds of Africa*, Vol. III, eds. E.K. Urban, C.H. Fry & S. Keith, pp. 356–70. London: Academic Press.

Ligon, J.D. & Ligon, S.H. 1982. The cooperative breeding behavior of the Green Woodhoopoe. *Sci. Am.* **247**: 126–34.

Ligon, J.D. & Ligon, S.H. 1988. Territory quality: key determinant of fitness in the group-living Green Woodhoopoe. In *The Ecology of Social Behavior*, ed. C. Slobodchikoff, pp. 229–53. Flagstaff, AZ. London: Academic Press.

Ligon, J.D. & Ligon, S.H. In press. Green Woodhoopoe: life history traits and sociality. In *Cooperative Breeding in Birds: Long-term Studies of Behavior and Ecology*, ed. P.B. Stacey & W.D. Koenig. Cambridge: University Press.

15. Splendid Fairy-wren

IAN ROWLEY & ELEANOR RUSSELL

The Splendid Fairy-wren *Malurus splendens* is one of 13 species in the genus *Malurus*, which with four other genera makes up the endemic Australasian Family Maluridae (Schodde 1982). Despite their common names, the Maluridae are not closely related to the wrens of Eurasia and the Americas (Troglodytidae), but are a highly specialized family which diverged early from the main Australian passerine radiation (Sibley & Ahlquist 1985).

Adults of the genus are strongly sexually dimorphic; in *M.splendens*, the males are brilliant blue and black, while the females are grey-brown; juveniles and immatures of both sexes are brown like the females, and the males become blue in the first spring after they are hatched; there is a slight size dimorphism (male 10.2 ± 0.5 g; female 9.6 ± 0.6 g; $n = 74$). These small birds are weak fliers; they are insectivorous and forage on the ground or in low shrubs, moving about by hopping on their strong legs. They generally pounce on their prey from a standing position, but may fly the occasional sortie into a swarm of termites or other slow-moving prey.

M. splendens occurs as three recognized sub-species across Australia south of the Tropic of Capricorn, except in the south-east corner where it is replaced by *M.cyaneus*. Near Perth in Western Australia, where we have studied them for more than 14 years, *M. splendens* lives in a scattered woodland of eucalypts, *Eucalyptus calophylla* and *E.wandoo*, with a dense and varied heathey understorey of 1–2 m tall xerophytic shrubs of the families Proteaceae and Myrtaceae. The study began when eight groups were colour-banded in 1973 and now has more than 30 groups and over 100 colour-banded adults in 120 ha. Each year all young produced are colour-banded as nestlings or fledglings.

M.splendens is a year-round resident, living in groups of 2–8 members (mean 3.3 ± 1.2) in contiguous territories of $1.6 - 9$ ha (mean 3.8 ha),

LIFETIME REPRODUCTION IN BIRDS
ISBN 0-12-517370-9

which persist in much the same location year after year. Like many other small Australian passerines, *M. splendens* live a fairly long time. Annual survival of breeding males is 72 % and of females 65 %; the longest survival from the start of breeding recorded for a male is 12 years, and for a female eight years. Mean lifespan for birds which become breeders is 5.0 years for males and 4.2 years for females, compared with that calculated from survivorship of 707 birds banded as nestlings: 1.9 years for males and 1.7 years for females. Age at first breeding is similar in both sexes—about half breed first at one year and half at two years, with three males breeding for the first time at three, four at four and one at eight years (55 females; 49 males). Clutch size is almost always three, producing three fledglings (Table 15.1). When nests are subject to predation or parasitism, all nestlings are usually lost. Variation in annual reproductive effort is not through clutch size, but in the number of nests per year. Individual performance reaches its peak in the third year of breeding and does not appear to decline subsequently. For 13 females that bred for at least five years, including seven that bred for six years, there was no significant difference in production of fledglings per year from year three onwards (Kruskal–Wallis One-way ANOVA, $P > 0.05$).

The basic breeding unit is the pair, which remains together as long as both survive. Where there are more than two birds in the group, the

Table 15.1 Summary of reproductive success of *Malurus splendens*, Gooseberry Hill 1973–1986.

A. Overall success in terms of individual nesting attempts.

Number of nests found	561
Number (%) in which eggs were laid[a]	520 (93)
Number (%) in which young were hatched[b]	316 (61)
Number (%) from which young fledged	264 (84)
Mean clutch size (±SD)	2.90 ± 0.34
Mean brood size (±SD) at hatching[b]	1.52 ± 1.41
Mean brood size (±SD) at fledging	1.28 ± 1.36
Mean young to independence (±SD) per clutch laid	0.85 ± 1.13

B. Overall success in terms of yearly production per group.

Mean (±SD) nests per female per year	2.04 ± 0.92
Mean (±SD) fledglings per female per year	2.86 ± 2.02
Mean (±SD) young to independence per female per year	1.88 ± 1.58

[a] Percentage of previous figure.
[b] Assumes a nest which hatched a cuckoo hatched no wrens.

oldest male and the oldest female are presumed to be paired on the basis of mate-guarding during nest-building and egg-laying. However, electrophoretic analysis (Brooker *et al.* 1989) has shown that genetic parentage (as compared to visible, social parentage) is not always what is expected, and many offspring are sired by a male other than the putative father. At high population density, some plural breeding (*sensu* Brown 1987) occurs, with more than one female in a group nesting concurrently (12 % (30/257) of group years). Progeny tend to remain in the family group after they reach independence, and many stay as "helpers" for one or more years after they reach sexual maturity, taking part in territorial defence, feeding nestlings and shepherding and feeding fledglings; *M. splendens* is thus a co-operative breeder. Some 98 % of helpers in this study were progeny of the group in which they helped. The size and composition of groups depends on breeding success in previous seasons; most possible combinations of males and females have occurred at some time (Russell & Rowley 1988), with simple pairs being the commonest (34 % of group years), and one male the most frequent extra (20 % of group years). In most years the sex ratio was male-biased, ranging from 0.97 to 1.81 males per female, with a mean of 1.3:1. Only in one year (1981) were there more females than males. Males tend to be the more philopatric sex; 57 % of males whose full history we knew, and which became breeders, did so in their natal territory, while 57 % of females became established in other than their natal territory. Few birds dispersed across many territories; 89 % of known males and 80 % of known females which became breeders did so in their natal group or in a group next door.

In our study area, with its Mediterranean climate, the breeding season lasts from September to January, giving time to raise two or even three broods. Over 14 years, females have had a mean of 2.04 nests per year, but the number of nests in a season was very much affected by a female's breeding experience and number of helpers. If she had helpers, she might lay the second clutch earlier, leaving the final stages of raising the first brood to the rest of the group. Brood parasitism reduced productivity, and affected 0–44 % of nests each year (overall 21 % of 562 nests found were laid in by Horsfield's Bronze-cuckoo, *Chrysococcyx basalis*). Although fledgling cuckoos were dependent for less time than wrens, they still occupied a large portion of the breeding season. It was rare for a group to raise two cuckoos in a year, but having raised one cuckoo, they often had time for only one brood of their own, unless there were helpers.

Since *M.splendens* has a very short nestling period of 10 days, followed by 3–4 weeks when the fledglings are dependent on their associated adults, production of both fledglings and independent young is given in Table 15.1. About 33.5 % of fledglings (51.2 % of independent young) reached one year old, and approximately 22 % became breeders (Fig. 15.1). Thus about 11 % of eggs gave rise to breeders.

Those birds which bred at one year old had no period as adult helpers (17 % males and 29 % females which reached one year old). Forty per cent of males and 49 % of females helped for only one year, and 34 % of males and 18 % of females helped for two years or longer, before achieving a breeding vacancy or disappearing. One male helped for seven years before moving next-door as a breeder. Thus significant numbers of birds which reached sexual maturity died before they had a chance of achieving breeding status (more males than females).

LIFETIME REPRODUCTIVE SUCCESS

In estimates of LRS, we included only birds which began breeding in 1982 or earlier to avoid bias towards short-lived individuals. For five

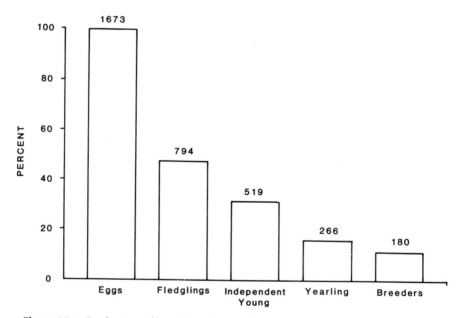

Figure 15.1 Production of breeding offspring from eggs in *Malurus splendens*. Data from 562 nests, 1973–1986. Per cent of breeding offspring is an estimate (10.7 %) and could range from 9 to 13 % depending on whether none or all of helpers currently alive become breeders. Allowance is made for offspring which dispersed and became breeders by assuming that this number is matched by the number which moved in.

males and four females that are still alive, we calculated expectation of further life at their present age and added the average values for offspring that would be produced by experienced breeders (after Clutton-Brock *et al.* 1982). For two males and one female of unknown age that bred in 1973 and for at least four more years, we estimated production in one previous year. This gave 36 males and 46 females for analysis, with average lives as a breeder of 3.9 ± 2.7 years and 3.2 ± 2.2 years respectively (Fig. 15.2). Many females bred for only one or two years, with one year the modal value. In males, the modal value was three years. For all these birds we calculated the lifetime production of eggs, fledglings, independent young, yearlings and breeding offspring (Table

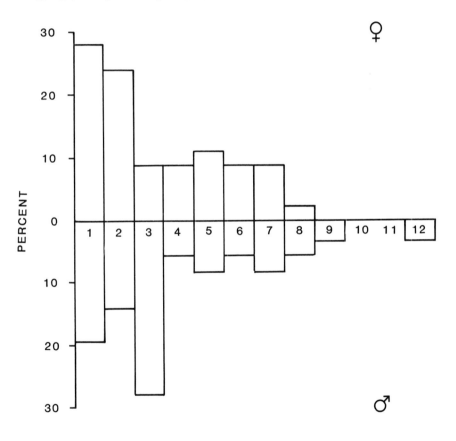

Years as Breeder

Figure 15.2 Breeding lifespan of male and female *Malurus splendens*.

15.2a, Fig. 15.3). For this analysis, we assumed that the oldest male in the group was the father of the offspring, although electrophoresis has shown that this is not always so.

Variation among individuals in mean LRS becomes greater as success is measured in terms of older offspring, particularly for females, where 28 % bred for only one year. The mean LRS of breeding males was

Table 15.2 Lifetime reproductive success of male and female *Malurus splendens*. (a) Actual data for established breeding birds: 36 Males and 46 Females which were established breeders and which began breeding in 1982 or earlier. (b) Estimated, for birds reaching sexual maturity: 61 Males and 60 Females which represents the above established breeders plus an estimate of the corresponding number of birds which reached sexual maturity at one year old but died before becoming established breeders.

Figures for LRS are given as the mean ($\bar{x}$) with standard deviation (SD) and standardized variance (I = variance/square of mean, Wade & Arnold 1980). P is probability (one-tailed) from Mann–Whitney U-test that male and female LRS are equal.

(a) Known breeding birds

Lifetime production of	Males (n=36)			Females (n=46)			
	$\bar{x}$	SD	I	$\bar{x}$	SD	I	P
Eggs	23.9	19.8	0.68	18.8	16.5	0.77	0.074
Fledglings	12.6	9.5	0.57	10.0	9.4	0.88	0.056
Independent y	8.3	6.1	0.54	6.6	6.5	0.96	0.056
Yearlings	4.6	3.7	0.65	3.5	3.9	1.3	0.042
Breeders	2.3	2.0	0.76	1.7	2.0	1.4	0.044

(b) All birds reaching sexual maturity[a]

Lifetime production of	Males (n=61)			Females (n=60)			
	$\bar{x}$	SD	I	$\bar{x}$	SD	I	P
Fledglings	7.4	9.6	1.7	7.7	9.3	1.5	> 0.05
Breeders	1.3	1.9	2.0	1.3	1.9	2.2	> 0.05

[a] From the estimate of how many birds which disappeared and did not die but became breeders elsewhere, we calculated the ratio of non-breeders to breeders for males and females separately and used this to calculate how many non-breeders corresponded to our sample of 36 males and 46 females.

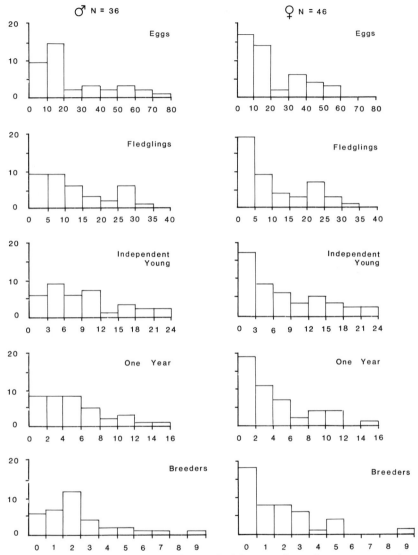

Figure 15.3 Lifetime production of eggs, fledglings, independent young, yearlings and established breeders by male and female *Malurus splendens*.

higher than that of breeding females for all classes of offspring; the differences in mean production of yearlings and breeders are significant, and those in fledglings and independent young nearly so. The variance for females in the production of yearlings and breeders is almost twice that for males. Although there were significant correlations between production of breeding offspring and of eggs ($r = 0.83$), fledglings ($r = 0.81$), independent young ($r = 0.84$) and yearlings ($r = 0.92$) (for females, $n = 46$; $P < 0.001$ for all values of r), the considerable amount of unexplained variation ($1 - r^2 = 0.34$, for fledglings) means that numbers of eggs, fledglings or independent young provide only an approximate measure of lifetime production of breeding offspring.

Some 30 % of breeding females produced 75 % of future breeders, and 16 % produced 50 % of future breeders (Fig. 15.3). About 39 % of females produced no breeding offspring at all. Males showed less variation: 44 % produced 75 %, 25 % produced 50 % and only 17 % failed completely. These figures are minimal, as they are based only on offspring which were located breeding. Some independent young which disappeared before they were one year old, and some helpers which disappeared after one or more years, would have dispersed to breed elsewhere. About half the territories in our study area were peripheral, from which young were likely to disperse beyond our searching. The number of unknown dispersed breeders was estimated in two ways: (a) by assuming that the number of unbanded birds which moved into the study area as breeders (25 males and 34 females) equalled the number which dispersed out; (b) from the proportion of dispersing birds expected to have died or to have become breeders (according to the method of Woolfenden & Fitzpatrick 1984, Appendix M). These estimates suggested that the numbers of breeding offspring should be increased by 48 % (estimate 1) or by 35 % (estimate 2). To give some idea of the scale of this effect, we calculated mean LRS for the 36 males after assigning an extra 34 breeders (from a mean of estimates 1 and 2) in proportion to the number of independent young produced. Mean LRS (breeding offspring) increased from 2.3 ± 2.0 ($I = 0.76$) to 3.2 ± 2.6 ($I = 0.66$). For females ($n = 46$) mean LRS calculated in the same way, with an extra 31 breeders assigned, increased from 1.7 ± 2.0 ($I = 1.4$) to 2.3 ± 2.5 ($I = 1.2$). From the proportion of fledglings that became breeders (Fig. 15.1), and the relative proportion of males and females whose age at first breeding was known, we could calculate that approximately 24 % of females and 21 % of males that left the nest became breeders, and that 11 % of birds that left the nest reached one year old but did not breed.

Birds which reached sexual maturity but did not achieve breeding status were excluded from the above calculations of LRS for breeding birds. We estimated the overall ratios of non-breeders to breeders for males and females which reached one year old, from the data (1973–1986) on how many birds became breeders in the study area and from the estimate of how many birds which disappeared became breeders elsewhere or died without breeding. We used these ratios to calculate the number of non-breeders appropriate to our sample of the 36 males and 46 females, and recalculated the standardized variance (*I*) in Lifetime Production of Breeders. This changed from 0.76 to 1.98 for males, and from 1.43 to 2.17 for females (Table 15.2b).

ENVIRONMENTAL FACTORS AFFECTING REPRODUCTIVE SUCCESS

Mean annual productivity varied considerably between individuals and between years (Table 15.3). In good years, when most individuals did well, mean reproductive success was high and variance was low. Any year with a mean production per territory of three or more fledglings may be called a "good" year, when many females raised more than one brood. In bad years not only was mean reproductive success low, but variance tended to be high, with a few birds doing well but many failing completely. The main environmental factors involved were the unpredictable ones of fires and cuckoos. With many birds breeding in only one or two years, the actual year in which an individual bred had a considerable influence on its LRS.

Although the study area, on the edge of a city, is more fire prone now than before the arrival of European man, the adaptations of the vegetation suggest that fire has long been frequent. Incredibly, most adults and independent juveniles survived a very intense fire at the end of the

Table 15.3 Variance among years in annual mean productivity by male and female *Malurus splendens*, Gooseberry Hill 1973–1986. Annual variance is measured by variance/square of mean. For each year, productivity was measured in terms of eggs (e), fledglings (fl), yearling (1yo) and number of young which subsequently became breeders (br).

		Annual mean productivity			Annual variance	
	No. of years	Class of offspring	$\bar{x}$ of $\bar{x}'s$	Range	$\bar{x}$ of annual variances	Range
Female	14	e	5.86	4.00–7.92	0.19	0.05–0.34
	14	fl	3.00	1.78–5.20	0.55	0.18–1.03
	13	1yo	0.96	0.22–1.82	1.58	0.47–3.94
	10	Br	0.50	0–1.18	2.22	0–8.00
Male	14	e	6.05	4.00–9.77	0.19	0.05–0.34
	14	fl	3.06	1.78–5.20	0.53	0.23–1.09
	13	1yo	0.96	0.22–1.67	1.57	0.55–3.94
	10	Br	0.46	0–1.00	2.45	0–9.00

breeding season in January 1985 (Rowley & Brooker 1987). Although the shrubby vegetation regenerated rapidly, it took several years to recover its former density and during this time, nests were more vulnerable to cuckoos and predation, fledgling production was relatively low, and survival to one year old of the few fledglings raised was poor. Productivity was depressed for at least two years after this major fire. Between 1973 and 1984, the long-term mean for nests/female/year was 1.8. After the fire, in 1985, this figure was 2.5 and in 1986, it was 2.8 nests/female/year. At the start of the 1986 and 1987 breeding seasons there were more vacancies for female than for male breeders (1986, 15:7, 1987, 14:9) as against the long-term (over 12 years) figure of 1.1 : 1, which suggests that the increased energy demand on the females after the fire, when they had to re-nest frequently after failure, helped to shorten their lifespan.

Territory quality also had a significant influence on LRS. The boundaries of territories in "good" habitat changed little from year to year. In other places, vegetation regrowth during the fire-free period from 1978 to 1984 changed some areas from being unsuitable to marginal or suitable habitat. The few marginal areas were frequently occupied by dispersing birds breeding for the first time. Such attempts were rarely successful, and turnover of breeding females was high, accounting for some of the short-lived unsuccessful females in Figs 15.2 and 15.3.

In years of heavy cuckoo parasitism (1975, 1979, 1986), some females were parasitized twice and produced no young of their own, others lost one nest to predation and another to a cuckoo. This meant not only nil production in that year, but in the next year, no (or few) helpers, which lowered productivity and survival chances.

Year to year variations in weather appeared to be a minor factor. Cold wet weather in August meant a late start and less time available in which to fit in two broods if delays due to cuckoos or predation occurred. The survival of juveniles may have been depressed by very hot weather in late summer, often associated with the late appearance of autumn rains, and survival may also have been reduced by winter conditions colder and wetter than average. However, there was no significant correlation between rainfall and production of fledglings or their survival to one year old.

SOCIAL FACTORS AFFECTING LRS

Several aspects of the co-operative breeding social organization of *M. splendens* affect LRS.

Achieving a breeding opportunity

Many birds which survived to breeding age disappeared before they found a breeding vacancy, and many presumably died. Assuming that in each year all suitable territories were occupied, then there was a surplus of non-breeding adults available for any vacancy that occurred. Competitors included helpers of one or more years experience, as well as young birds of the previous breeding season which had never helped. Over 13 years, the mean number of candidates per vacancy was 3.5 ± 1.2 for males and 2.1 ± 1.9 for females. This difference in opportunity was reflected in the fact that all females were breeding by the age of three and none helped for more than two years. In contrast, some males helped for many years and finally disappeared without ever having bred. In a group with more than one male helper, it was the oldest male which took over if a vacancy occurred. For *M. splendens*, a breeding vacancy was partly a matter of luck and partly a matter of success in competition. It depended on the availability of competitors surviving from earlier years in an individual's natal and neighbouring territories, and on the occurrence of vacancies in the right place—at home or next-door. Of the 14 longest-lived females in our sample of breeders (five years or more), two bred in their natal territory and nine dispersed into a well-established territory nearby.

For some females, an alternative route to breeding status was to breed in their natal group as a second (y) female. In about 80 % of cases the y female had helped in the group for one previous season and in 20 % she was a yearling who helped at the first nest and then built her own. The second female was in most cases the daughter of the older (x) female. After fledging, x and y broods coalesced and were cared for by the whole group. Some y females later became sole breeder in the same territory, some dispersed to become sole breeder elsewhere, a few bred as y females a second time, and some disappeared. Seven out of the 36 males had a second female in 11/146 group years, laying in 14 nests. Six of these nests produced fledglings (three of which became breeders); for convenience, these are attributed to the LRS of the oldest male in all cases, even if there was another male in the group. In no group was the matter of paternity assured.

y females produced the same as other novices, 1.7 fledglings per year, without the cost of dispersing. A second female breeding in the group could impose a cost on the primary female, who lost her help. It rarely subtracted other helpers from the primary female's nest, but if that nest failed, even the primary female sometimes helped at the y female's nest. It could enhance the LRS of a male to have a second female breeding in his group. From Table 15.3, mean production by males was 3.1

fledglings per year; for all males in plural breeding groups in the years 1973–1986 (n = 22; 28 male-years), production of fledglings was 4.0 ± 1.7 per year, giving an increased chance of producing breeding offspring. For six of these 22 males, plural breeding occurred in two years. Since plural breeding occurred in 12 % of all group years, it was a significant factor affecting LRS. It added to the success of an already successful male, since it was generally in large, long-established successful groups that it happened.

Extra-Pair Fertilizations

It is becoming increasingly common to find fertilization being accomplished by individuals other than the assumed mated pair (Gowaty & Karlin 1984, Quinn *et al.* 1987). In *M.splendens*, stolen copulations resulting in extra-pair fertilizations (EPF) could arise either from male helpers within the social group or extraneous "philandering" males from other groups (Rowley 1981).

Recent electrophoretic studies (Brooker *et al.* 1989) have indicated that more than half the progeny tested (n = 91) could not have been fathered by a male within their natal social group. Males trespassing in neighbouring territories were frequently seen early in the breeding season, often performing a display in which the bird carried a bright purple or pink petal. Occasionally, these "philanderers" were seen to attempt (and achieve) copulations. Without more precise data on parentage, it is not possible to estimate the exact significance of EPF to the LRS of an individual. For the present, the best we can do is to acknowledge that estimates of LRS for males must be treated with caution.

The importance of helpers and experience

The presence of helpers may lead to an increase in production by the breeding pair, and may also improve a breeding female's chances of surviving to the next season. Of females with helpers (n = 128), 76 % survived to the next breeding season, compared with 55 % of those without (n = 75; χ^2 = 10.7, $p < 0.001$). Of 26 females breeding for the first time, with no helpers, 35 % did not survive to breed in a second year, and 31 % produced no young that survived to one year old, and thus had no helpers for a second year, with mortality again higher than for females with helpers. This is one source of the skew towards unsuccessful females in figures for LRS. Helpers had no effect on the annual survival of breeding males, which was 72%.

Offspring production was also affected by the age and experience of the breeding female (Table 15.4; Russell & Rowley 1988). A female breeding for the first time as a member of a simple pair fledged only half as many young per year as a female with two years experience as a breeder, for whom the presence of helpers further increased production. It made little difference to the number of fledglings produced from a single nest, but reduced the time to renesting, and allowed more broods to be raised. Females with two years' experience and no helpers had a mean of 2.2 nests per year, and 47 % renested after successfully fledging a brood. With the same experience and at least two helpers they had 2.8 nests per year, and 67 % renested after a successful nest. The effect of experience as a helper may also have influenced productivity as a breeder, but this effect cannot be separated from that of age. Females with one year's experience as a helper breeding for the first time as two year olds with helpers produced 2.8 fledglings ($n = 11$), compared with 2.0 fledglings for females that first bred at one year old with helpers ($n = 7$) (Russell & Rowley 1988). Juveniles from early broods sometimes attended the later nestings of their parents (Rowley 1981). One-third of all progeny had the opportunity to help, but our observations suggest that their "help" was insignificant. Any experience that juveniles might have gained did not affect their productivity as one-year old breeding females.

Inbreeding

The level of inbreeding in *M.splendens* appears to be high, and in about 20% of pairs male and female are related at $r = 0.5$ (as sibs) or greater.

Table 15.4 Effects of helpers and experience on annual production of fledglings by female *Malurus splendens*.

	Fledglings produced per year by			
Female	Pair alone	Pair plus ≫ 1 helper	Pair plus 1 helper	Pair plus > 1 helper
Novice	1.5 ± 17 (n = 46)	2.2 ± 1.5 (36)	—	—
One year experience as breeder	3.2 ± 2.4 (18)	—	3.1 ± 1.4 (18)	2.4 ± 1.8 (16)
Two years experience as breeder	3.2 ± 1.5 (19)	—	3.1 ± 2.0 (19)	4.0 ± 2.1 (46)

This arises partly due to the filling of breeding vacancies from within a group, and partly due to plural breeding groups, where the second female frequently mates with her father or brother. No reduction in the production of fledglings by inbred pairs has been found (Rowley *et al.* 1986). Any cost (so far undetected) may be outweighed by the importance of taking a breeding opportunity when it arrives. We cannot be confident that the senior male always fathered all the brood. Even allowing the possibility of stolen matings by subordinate males or near neighbours, the level of inbreeding is unlikely to be much decreased. The average helper male is closely related to the breeding female (64 % were helping mother or sister, and only 25 % helping an unrelated female). Of males which dispersed to breed elsewhere, more than 75 % went no further than a territory next to their natal territory.

Indirect Fitness Benefits

In a co-operative breeding species, there is the potential for non-breeding helpers to gain an indirect fitness benefit from the increased production of young by the breeders which they help (Brown 1987). In *M.splendens*, this indirect benefit has two parts, a benefit from increased production in the current year by the breeding female who was helped, and a future benefit arising from the increased survival of that breeding female. Such kin benefits are enhanced by inbreeding. Russell & Rowley (1988) have calculated that in one year, the total indirect benefit from helping may be close to the direct genetic benefit from breeding as an unaided novice female.

INDIVIDUAL LIFE-HISTORY FACTORS AFFECTING LRS

One factor in particular, lifespan as breeder (L), had a marked effect on LRS, and was in turn affected by several other factors. Lifetime production of fledglings showed a strong correlation with L (for males $r = 0.86$, $P < 0.001$; for females, $r = 0.90$, $P < 0.001$). The correlation between L and lifetime production of breeders was significant, but weaker (males, $r = 0.66$, $P < 0.001$; females $r = 0.81$, $P < 0.001$), reflecting the competition for breeding vacancies from a reservoir of non-breeders, some of whom never achieve breeding status. If a long L was important for LRS, then to breed as soon as possible should be important, but this was constrained by lack of opportunities. Moreover, not all breeding opportunities were equivalent. To move into a good territory with helpers as a two-year old female was likely to be at least as productive as to disperse as a novice at one year.

By expressing LRS as the product of several components ("episodes of selection", Arnold & Wade 1984a,b), the contribution to total variance in LRS (opportunity for selection, $= I$) of each component can be calculated. The expression used was: LRS $= L \times F \times S_1 \times S_2 \times S_3 \times S_4$. The components considered were (1) Survivorship, measured by years as a breeder (L); (2) Fecundity (F) expressed as eggs laid (or sired) per year; (3) a series of survival measures: proportion of eggs that gave rise to fledglings (S_1); fledglings to independent young (S_2); independent young to yearlings (S_3); yearlings to breeders (S_4). This expression did not include indirect components of LRS which arose from helping, nor did it separate competition for breeding vacancies, which was included in component S_4. At each stage, the standardized variance in reproductive success and its components was expressed as the variance in individual fitness divided by mean fitness squared (I). The use of I has been criticized by Downhower *et al.* (1987) because I is not independent of the mean, and because mean fitness values of less than one will dramatically increase I. However, for our data, where mean fitness values of less than one have variance also less than one, there was no such effect, and we considered it worth using I in attempting to estimate the relative importance of factors affecting LRS.

Table 15.5 lists for females the opportunity for selection associated with component (w_i), the running subtotals for the total opportunity for selection $(w_i, w_j, \ldots w_n)$ calculated as each component is introduced, the opportunity for selection associated with correlations between fitness components (the "cointensity" terms) and the total opportunity for selection (W). The table also lists the percentage of the total opportunity for selection represented by each component, and subtotals as each is added. The uncertainty of male parentage makes a similar calculation for males of doubtful value.

As successive components were added, the value of I increased. The components which had most effect on I as they were added to the expression were years as breeder, survival of offspring to one year and proportion of yearlings which became breeders. Factors affecting fecundity and survival of eggs to fledglings and fledglings to independence were least important. The large negative covariance components result largely from correlations between various classes of offspring.

Discussion

Our study of *M.splendens* suggested that potential factors contributing to LRS were: (a) Competition for breeding vacancies; (b) Fecundity; that is, the number of eggs laid per season; (c) Survival of eggs and fledglings; (d) Offspring success at getting breeding vacancies; (e) Individual survival

Table 15.5 Partitioning of the opportunity for selection (= variance) on lifetime reproductive success of female *Malurus splendens* (n = 46 individuals). Lines separate running subtotals.

Source of variation in relative fitness	Symbol	Contribution to total opportunity for selection	
		Value	Percentage
Years as breeder (w_1)	I_1	0.45	30
Eggs/year (w_2)	I_2	0.24	16
Cointensities (12)	*	0.08	5
Lifetime production of eggs ($w_1 w_2$)	I_{12}	0.77	51
Survival, eggs to fledgling (w_3)	I_3	0.37	24
Cointensities (123)	*	−0.24	−16
Lifetime production of fledglings ($w_1 w_2 w_3 w_4$)	I_{123}	0.88	59
Survival, fledgling to independence (w_5)	I_4	0.32	22
Cointensities (1234)	*	−0.24	−16
Lifetime production of independent young ($w_1 w_2 w_3 w_4$)	I_{1234}	0.97	64
Survival, independence to yearling (w_5)	I_5	0.70	47
Cointensities (12345)	*	−0.41	27
Lifetime production of yearlings ($w_1 w_2 w_3 w_4 w_5$)	I_{12345}	1.26	84
Survival, yearling to breeding (w_6)	I_6	1.01	67
Cointensities (123456)	*	−0.76	−51
Lifetime production of breeders ($w_1 w_2 w_3 w_4 w_5 w_6$)	I_{123456}	1.50	100

* Cointensities include covariance terms between the current episode and all prior episodes, e.g., "Cointensities (12)" = COI (1,2) + COI (12/2) + COI (12,2/1) − COI (12,2). For formulae, see Arnold & Wade (1984a).

(years as breeder); and (f) Indirect fitness benefits. These factors applied to males and females, although (a) may take a different form for females.

The primary factor which determined all others was getting a breeding vacancy. Without that, LRS was zero or nearly so in the case of a helper or a secondary female. The data reported here begin after birds had achieved breeding status. Allowing for birds that bred outside the study area, we calculated that approximately 22 % of fledglings eventually bred, but only 15 % of females and 17 % of males fledged produced breeding offspring.

Variation in individual quality, as discussed by Coulson & Thomas (1985), was an important factor in the production of eggs per season by both males and females. Variation in clutch size was small, and for all females except novices, production per nest was similar (Russell & Rowley 1988). Differences in annual reproductive success arose mainly by ability to renest quickly. Multiple broods were particularly advantageous when the incidence of cuckoo parasitism was high, as they provided a greater chance of fledging at least one young than a single larger clutch would (May & Robinson 1985). Selection could decrease the interval between nests either by improving the efficiency of individual females or by evolving a social system that enhances rapid recycling. The data were too few to show whether individual females differed in how quickly they could renest, but the presence of helpers did speed the process. Production of eggs per season for a male was influenced by female quality, by the number of breeding females in his group, and by his opportunity to achieve EPF.

The actual number of eggs laid was affected by cuckoos and predation, and had little effect on LRS. It was the survival of fledglings to become breeders which was important, and this was influenced by unpredictable factors such as cuckoos, fire and predators. The delayed dispersal that operates in most co-operative breeders should improve the survival of offspring and their chances of breeding.

Another important factor in success was survival to breed over a number of years. For most of their breeding lifespan females had a constant survival rate. However, the presence of helpers increased the survival of breeding females (novice and experienced), so for a female to disperse as a breeder into a group with helpers should be a better strategy than to set up on her own with no helpers. To breed at one-year old would give a longer L only if it entailed no extra risks. Since females which bred at one-year old with no helpers and no experience as helpers were even less successful than other females, they were less likely to have helpers in year two, and also suffered higher mortality. Since about half of female breeders started at one-year old and half at two, the balance was not wholly in favour of early breeding. For a female which did not breed at one-year old but helped for a year, the indirect fitness gained by helping kin was close to that she would have gained by breeding as an inexperienced novice with no helpers (Russell & Rowley 1988).

Thus a successful female *M.splendens* was one who weathered the first two less productive years and by the time she was three or four years old, was established in a group with at least two helpers. To live so long it was clearly better to become a breeder in an established group. A successful male was one who lived a long time in a group with helpers;

it was to his advantage that he had helpers, since they improved female survival (and productivity when she was older), and decreased the chance that he would be mated with a succession of novices. Most males were breeding by two years old—to remain in the natal territory meant a good chance of becoming a breeder by the start of the third year, with only one year of possible breeding lost. Annual survival for established breeding males was high (72 %).

Most of the 13 longest-lived females successfully produced fledglings year after year. In their first year two were parasitized by cuckoos and did not renest. In most cases where older females did not produce any wren fledglings in a year, this was because two successive nests were parasitized. Whether the reason for high lifetime success was quality of the individual, the presence of helpers or the quality of the territory, these long-lived females produced some fledglings each year, despite fires and cuckoos, most of them having two successful broods in at least two of their years as breeders. All 21 females established as breeders for three years or more produced 3.5 ± 1.0 fledglings per year. This suggests that it was not just a few successful birds which managed to raise three or four fledglings per year, but that, because of the multi-brooded strategy, cuckoos and invariant clutch size, any female who became *established* and *experienced* had the potential to perform at this level. The most important factor determining lifetime production was lifespan as a breeder (L). Furthermore, the longer the L, the greater the chance of compensating for the poor production of bad years.

Since the mating system of *M. splendens* is hard to define until the parentage of nestlings can be exactly identified, comparisons between males and females are difficult. However, mean values for apparent LRS in males are significantly higher than in females for individuals which became established in breeding roles. There were no differences in actual variance, but standardized variance (I) for production of yearlings and breeders was twice as high for females as for males (Table 15.2). The incidence of EPF in males could either increase or decrease mean and variance in male LRS. If apparently unsuccessful males sired some offspring, mean and variance would decrease. If a few males (perhaps already established as senior in a group), achieved most EPF, the variance would increase.

Female production of offspring was skewed towards the unsuccessful (Fig. 15.3), due to two factors. Firstly, mortality in females was slightly higher and mean L shorter than in males; more females than males bred for only one or two years, and thus never got past the first two less productive years. Secondly, all females that breed must go through these first two less productive years, irrespective of whether they were mated with novice or experienced males. In contrast, not all males suffered this depression of reproductive success—males breeding for the first time had a probability of at least 0.4 of being mated with an experienced female, who produced equally well when mated to a novice or an experienced male. When all individuals that reached one-year old were considered, more males than females survived for some time but did not become established. The LRS of these was nearly zero unless they achieved a significant number of EPF. Females that attempted to breed (24 % of those fledged) produced a mean of 1.7 known breeding offspring in a lifetime. Including females which survived to one-year old but did not get a chance to breed gave a figure of 1.3 breeding offspring in a lifetime. This was not very close to the two breeding offspring needed to maintain the population, but presumably some offspring dispersed and bred elsewhere, their loss being balanced by immigrants to the study area.

References

Arnold, S.J. & Wade, M.J. 1984a. On the measurement of natural and sexual selection: Theory. *Evolution* **38**: 709–19.

Arnold, S.J. & Wade, M.J. 1984b. On the measurement of natural and sexual selection: Applications. *Evolution* **38**: 720–34.

Brooker, M.G., Rowley, I., Adams, M. & Baverstock, P.R. 1989. Promiscuity: an inbreeding avoidance mechanism in a socially monogamous species? *Behav. Ecol. Sociobiol.* **25**.

Brown, J.L. 1987. *Helping and Communal Breeding in Birds: Ecology and Evolution.* Princeton: University Press.

Clutton-Brock, T.H., Guinness, F.E. & Albon, S.D. 1982. *Red Deer: Behavior and Ecology of Two Sexes.* Edinburgh: University Press.

Coulson, J.C. & Thomas, C. 1985. Differences in the breeding performance of individual kittiwake gulls, *Rissa tridactyla* (L.). In *Behavioural Ecology: Ecological Consequences of Adaptive Behaviour,* ed. R.M. Sibly & R.H. Smith, pp. 489–503. Oxford: Blackwell.

Downhower, J.F., Blumer, L.S. & Brown, L. 1987. Opportunity for selection: an appropriate measure for evaluating variation in the potential for selection? *Evolution* **41**: 1,395–1,400.

Gowaty, P.A. & Karlin, A.A. 1984. Multiple maternity and paternity in single broods of apparently monogamous Eastern Bluebirds (*Sialia sialis*). *Behav. Ecol. Sociobiol.* **15**: 91–5.

May, R.M. & Robinson, S.K. 1985. Population dynamics of avian brood parasitism. *Amer. Nat.* **126**: 475–94.

Quinn, T.W., Quinn, J.S., Cooke, F. & White, B.N. 1987. DNA marker analysis detects multiple maternity and paternity in single broods of the Lesser Snow Goose. *Nature (Lond).* **326**: 392–4.

Rowley, I. 1981. The communal way of life in the Splendid Wren, *Malurus splendens. Z. Tiepsychol.* **55**: 228–267.

Rowley, I. & Brooker, M.G. 1987. The response of a small insectivorous bird to fire in heathlands. In *Nature Conservation: the Role of Remnants of Native Vegetation*, ed. D.A. Saunders, G.W. Arnold, A.A. Burbidge & A.J.M. Hopkins, pp. 211–218. Sydney: Surrey Beatty.

Rowley, I., Russell, E.M. & Brooker, M.G. 1986. Inbreeding—benefits may outweigh costs. *Anim. Behav.* **34**: 939–41.

Russell, E.M. & Rowley, I. 1988. Helper contributions to reproductive success in the Splendid Fairy-wren *Malurus splendens*. *Behav. Ecol. Sociobiol.* **22**: 131–40.

Schodde, R. 1982. *The Fairy-Wrens: a Monograph of the Maluridae*. Melbourne: Lansdowne.

Sibley, C.G. & Ahlquist, J.E. 1985. The phylogeny and classification of the Australo-Papuan passerines. *Emu* **85**: 1–14.

Wade, M.J. & Arnold, S.J. 1980. The intensity of sexual selection in relation to male sexual behaviour, female choice, and sperm precedence. *Anim. Behav.* **28**: 446–61.

Woolfenden, G.E. & Fitzpatrick, J.W. 1984. *The Florida Scrub Jay: Demography of a Cooperatively Breeding Bird*. Princeton: University Press.

16. Arabian Babbler

AMOTZ ZAHAVI

The Arabian Babbler *Turdoides squamiceps* is a member of the Paleotropical family Timallidae. Several closely related species from the same genus replace each other geographically, along the desert belt from India to Morocco (Meinerzhagen 1954). The Arabian Babbler occurs in the Arabian peninsula, the hot desert parts of Israel and the Sinai. In Israel it is common along the Rift Valley, north to Jericho and in several of the large wadies in the West and Central Negev.

The babbler is 65–85 g in weight and 280–290 mm in length, of which 145–150 mm is tail. Its cryptic coloration matches the desert background. The iris is light grey in young birds, but changes during the first year of life to light yellow in males and dark brown in females. This is the only obvious difference between the sexes, although the base of the bill is usually more yellow in females and the white of the eye-ring is usually more pronounced.

Babblers fly slowly and usually near to the ground. They always keep close to cover, into which they fly at the slightest alarm. They hop and walk more than they fly, searching for food on the ground and among bushes and trees. They dig in the ground and peel legumes and bark from branches in search of food. They eat all kind of animals, including small reptiles, and also vegetable matter, such as berries, nectar (e.g. *Loranthus*), fleshy leaves (e.g. *Rumex*) and flowers.

My study site extends over 25 km² in the Rift Valley, around the Field Study Center of Hatzeva, 30 km south of the Dead Sea. In 1971 we colour-ringed 20 groups of babblers, containing 125 individuals. Since then, we have followed the population of that area and to some extent of nearby areas, monitoring all the birds hatched in the area and their subsequent life histories. Since 1976 all the birds have been tamed by occasional feeding with bread titbits. The tame babblers allow us to observe them while we stand in the middle of the group.

LIFETIME REPRODUCTION IN BIRDS
ISBN 0-12-517370-9

The climate is very arid, and the annual rainfall averages 35 mm. Perennial vegetation is mostly confined to dry river beds. The special geological features of the area, which consist of sandstone and intercalations of clay beds from a former freshwater lake, preserves much water under ground and consequently the area has more bushes and trees than do other parts of the desert with similar rainfall. The nearby villages, vegetable gardens and garbage dumps provide additional cover and food for the babblers. These man-made additions have major effects on the population, but probably no greater than those created by an oasis. On the other hand, babblers die more frequently around villages through road accidents and other misfortunes. But part of the study area is far from gardens and roads, so the groups living there provide a control.

The population

Babblers live in groups of 2–20 individuals. The groups are resident and territorial. About 5–10 % of the population are non-territorial birds, living alone or in small groups. They are chased by the residents whenever they are encountered. Some of them stay in small home ranges which they usually cannot defend against their neighbours. But sometimes, through a series of fortunate events (Fig. 16.1, MTMC), such marginal groups develop into new territorial groups. Others of these refugees, especially females, succeed in entering a resident group as breeders. I shall not deal further with the refugee population in this chapter.

The number of birds in the study area fluctuated between a maximum of 220 (in 23 groups) in 1983 after a succession of good breeding years and a minimum of 65 (in 15 groups) in the beginning of the 1979 breeding season after two dry years.

The history of one group over 17 years is presented in Fig. 16.2. In early 1979, the group consisted of only three individuals. At this stage the death of the dominant breeding male could have caused the extinction of the group. In 1983, however, the same group was up to 15 individuals. This group, like other groups, varied through time in the ages, sexes and relatedness of its members (Fig. 16.2).

Within each group a strict dominance hierarchy prevails, in which older birds dominate younger ones of the same sex. Young from the same brood fight with one another during the first few months of their life and the winner dominates the others whilever they stay together in the same group. Adult males dominate all adult breeding females which always come from another group. Male offspring often grow to dominate older female sibs.

Breeding

There is only one nest at a time in the territory of a group and all group members attend it. Clutches are laid during February to July, rarely in other months. In good breeding years, with over 30 mm of rain spread over the winter months to allow the growth of annuals, a group may fledge up to three broods. Dry years may end without any breeding, as in 1973, 1977, 1978, 1984 and 1987. Clutches contain three to five eggs. Clutches of five eggs occur only in the middle of the breeding season, while clutches of three are more common in the early and late stages of the season (Zahavi in press (b)).

Babblers do not mate with individuals born within their own group, so there is no conflict over breeding between parents and offspring so long as the group is composed of parents and their offspring. Babblers can breed at two years but some reach six or even eight years without breeding if they stay in their natal group and their parents are still alive. These non-breeders, which are found in many species of co-operative breeding birds, have been termed helpers at the nest (Skutch 1961). They participate in incubation, feeding and guarding the brood. The presence of several helpers does not increase the breeding success of the group, beyond the level achieved with only one to two helpers (Zahavi in press). The major advantage to all group members, in having helpers, is in their ability to defend the territory against other groups, thus helping the offspring to survive and have a base from which to disperse in order to breed. The advantage to the individual helper is in raising its social status (Zahavi in press).

The death of the breeding female is followed by the arrival of one or more females, usually sibs from a nearby group (Zahavi in press). These new females usually drive all the female line from the group and mate with all the adult males in the group (Zahavi 1988). Dominant males copulate more and dominant females lay more eggs. We have observed up to four females mating and laying at the only nest in the territory. Such multiple clutches occur only once after the arrival of the new females. These multiple clutches, which may contain up to 13 eggs, never fledge more, and often fledge fewer, offspring than single clutches (Fig. 16.2). Once a coalition of sib females has been successful in fledging young, their collaboration stops and the dominant female is the only one which continues to lay. The rest of the new females either disperse or stay and help without breeding. Males continue to share in reproduction. The subordinate males continue to mate with the laying female, although many of their copulations are in the days before the laying of the clutches, sometimes weeks before, and if they copulate during the laying period

```
Name        Sex    1974   75   76   77   78   79   80   81   82   83   84   85   86   87   88

            Ma            ********************************X
            Mb                      ---- ===== r

AMMT                                                          5    6    0    3    0    7    8    0
            M                       ----- ===== =============*******************X
            Mc                      ----- ===== ====X
            Md                      ----- ===== =============*********
            Fa            >++++++++X++
            Fb                                   >+++++++++++++++X
            Fc                                   >++++++++++++++X


Name        Sex    1974   75   76   77   78   79   80   81   82   83   84   85   86   87   88

                                               8    4    7    3    6    7    0    6
            M                      ---->*****************************************
            Ma                          -->====X
            Mb                          ->=X
            Fa            >+++++++++++++++++++++++++++++++++++++++++++++++
SMTA        Fb                          ------ =====r


Name        Sex    1974   75   76   77   78   79   80   81   82   83   84   85   86   87   88

                                     4    8    1    3    0    4    2    2    0
            M            --------->***********>r  >****>
            Ma                     ------>rX
            Mb                     ------  >r>=>=>===>===X
            Fa            +++++++++++++++
MTMC        Fb            -----==========::
            Fc                               >++++>r
            Fd                                    >++++++>r
            Fe                                         +++++++X
            Fe                                         ============+++++++++++++
```

Figure 16.1 Life history diagram of three males. Numbers show numbers of nestlings ringed. Ma = Male a; Fa = Female a; r = Refugee; X = missing. The various males and females shown in the diagram are breeding individuals associated with the life history of the key individual. – – – – – Dominant breeding female; = = = = = = Breeding helper; ****** Dominant breeding male; +++++++ Dominant breeding male; lived with father (Ma) and a sib (Mb) of the same brood. His mother (Fa) was replaced in 1977 by an alien female (Fb) and he became a breeding helper. His dominant brother (Mb) was chased from the territory, and he became the dominant male after his father's death. In 1982 there was a change in females (Fc) and consequently no breeding. His sons (Mc and Md) from the former female became breeding helpers. He chased one out and stayed with the other until his death. As far as I know, the helper son that remained had no access to the breeding female. AMMT spent all his life in the same territory, waited for his turn to be dominant, lived for 12 years and had 29 offspring as a dominant breeder.

SMTA hatched in 1979 in TMR territory. When two years old he moved together with two sibs (Ma and Mb), which were one year old, to take over the adjacent POL territory in which he mated with a mother and her daughter. In the first year one of his collaborators disappeared and two years later the other disappeared. During the time he had 12 offspring, most of them probably his own. The daughter (Fb) of his dominant mate (Fa) did not lay. She became a refugee in 1982. The two dominants continue to breed to the present. They share 40 offspring.

MTMC hatched in 1974 in MZG territory. Together with a sib (not shown) he took over the TMR territory, where he bred with Fa, having 12 offspring before he was chased from it by his subordinate sib. He lived as a refugee for a year, but was joined by one (Ma) and later by another of his male offspring (Mb). With the second son he formed a small group with a female (Fc) from a nearby territory on a marginal area. Their four offspring did not survive. In 1983 the breeding male in a nearby territory died, and father and son joined that group to mate with a mother (Fd) and her daughter (Fe). His son disappeared in 1984 and the (Fd) mother in 1985. MTMC and Fe still survive and breed. He is now approaching 15 years old. His 12 offspring from the first territory survive and some breed, as do some from his third territory.

```
Name    Sex   1972  73   74   75   76   77   78   79   80   81   82   83   84   85   86   87   88

aatz    M     *************X
mtmt    M     --=========********************************************X
actz    M     ---X
cctz    M     ---==X
aztz    F     +++X
tvcc    F     ---------X
catz    F     ->d
avtz    F     ->h
zhvt    M     N----====X
zvvt    F     N----X
cact    F            >+X
htmc    M     N-----========X
mtmz    M     N-----========>d
tamh    F     N------------X
ahzt    F            >+X
tltz    F            >+X
atva    M            >==X
zmht    F            >++++++++++++++++++++++++++++++++++++++>d
mmtm    F            >==>d
halt    F            >=X
cltz    L            NX
hlct    L            NX
zzat    M            N---------->d
tsch    M            N----------->hd
hlst    M            N------------>d

Name    Sex   1972  73   74   75   76   77   78   79   80   81   82   83   84   85   86   87   88
```

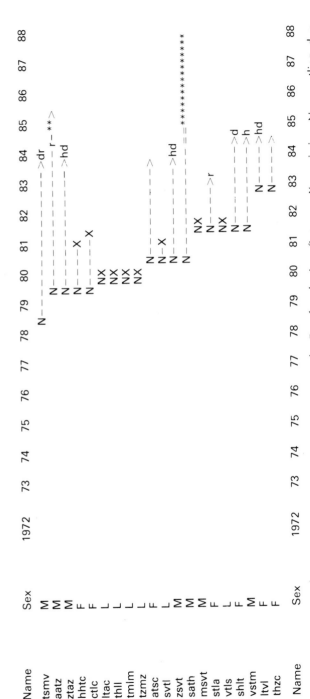

Figure 16.2 History of group ZEH during 1972–1989. M = male; F = female; L = first year; X = missing; N = nestling; d = dominant; h = helper; r = refugee; >= joined or moved out of group; - - - - = non-breeding offspring; = = = = = helper which might breed; +++++++ dominant breeding female; ********** dominant breeding male. (Continued over page)

Figure 16.2 Continued.

they do so less often than the alpha male breeder and usually later in the day (Zahavi in press (a)).

Breeding coalitions among females last only during the first successful brood. Male breeding coalitions may survive over several years. Eventually all breeding coalitions end up with a pair of dominant breeders, together with the offspring of the former coalition. The offspring consider the dominants as parents, even though this may not be genetically so. The death of a male in a group with several sons ends with the eviction of the female line from the group and the arrival of alien breeding females (one exception). The social changes which accompany the arrival of new females may take time, and groups have lost a good breeding year because of intragroup conflicts; e.g. AMMT in 1982 (Fig. 16.1). The conflicts may be between the daughters of the group (which often tend to stay) and the new females, or among the new breeding females to determine which of them lay, or among the few breeding males to determine which of them, how often and when they copulate with the laying females. These conflicts often result in egg destruction (Zahavi 1988 and in press).

Babblers do not always wait for breeding vacancies to occur. Groups of males or females evict or kill a breeder, or breeders, usually in adjacent groups, and mate with the members of the other sex. It is always members of larger groups, usually the offspring, which supplant the breeding members of small groups.

Dispersal

Babblers use their parental group as a base for finding a breeding opportunity (Zahavi in press). Still, a very few have moved when only a few months old (Fig. 16.2, year 83), either alone or with an older member of the group.

About 30 % of adult males inherit breeding status by staying in their natal group (Zahavi in press); e.g. AMMT (Fig. 16.1). About 50 % find a breeding opportunity in a next door territory (e.g. SMTA, Fig. 16.1) or one territory further away (Zahavi in press (b)). Most females disperse at two to four years old, but some females remain in their parental group without breeding for five to six years. Females may inherit dominant status in their natal territory only in small groups which either lost their breeding male or in which the male line was displaced by another (e.g. MTMCs mate in the third territory. Fig. 16.1). In large stable groups, in which there are always male offspring, females nearly always move out. Females disperse up to four to five territories from their natal place.

The multiple pathways which can lead to a breeding option result in life histories of breeders which are very different from each other (see Fig. 16.1 for three successfully breeding males). For each successful case, many others follow a similar way of life which ends by the disappearance of the bird somewhere along the way.

Lifetime reproductive success

I have selected for this analysis 261 babblers, all the offspring of 11 groups and ringed as nestlings in 1979–1983. I did not include the whole population because of disturbances caused by man to some of the groups. Some groups living near roads were also excluded because they lost many of their offspring by road accidents. However, decimation of a population also occurs naturally during a succession of dry years, or under high predation. Five of the chosen groups lived away from the vegetable gardens and roads. The other six groups, although visiting man-made habitats, were not very different in their history from the five and so were included in the sample.

Survival

One hundred and twenty-two out of the 261 babblers died before they reached 1 January of their first year of life (Table 16.1). We do not know if they were males or females because we cannot sex babblers in confidence before that time. But mortality in the first year of life is clearly higher than in later years.

Mortality in 1984 was lower than in other years (Table 16.2). This was a dry year without breeding, and I attribute the reduced mortality to the fewer attempts of dispersal and the fewer conflicts over breeding. 1987

Table 16.1 Survival of first year Babblers. From the time they were ringed at the nest to 1 January the following year.

Year	Birds ringed	Number survived	% Survived
1979	37	25	68
1980	50	24	48
1981	88	57	66
1982	40	15	38
1983	46	18	40

Table 16.2 Survival of male and female Babblers hatched 1979–1983.

a. Males	Numbers alive in:								
Year hatched	1980	1981	1982	1983	1984	1985	1986	1987	1988	1989
1979	18	16	15	15	12	9	5	5	2	2
1980		10	8	7	6	5	3	0	0	0
1981			32	30	26	22	17	13	10	9
1982				4	2	2	2	1	1	1
1983					5	5	4	3	2	2

b. Females	Numbers alive in:								
Year hatched	1980	1981	1982	1983	1984	1985	1986	1987	1988	1989
1979	7	6	4	3	1	1	1	1	1	1
1980		14	10	7	5	4	4	4	4	3
1981			26	21	12	10	6	2	1	0
1982				11	8	8	4	1	1	1
1983					13	11	8	6	4	1

was also a dry year without breeding but, by 1987, our sample population was already greatly reduced and no longer dispersing. Therefore the effect of the drought on survival is not clear in that sample. But the set of birds born in 1985 (not analysed in this paper) also survived somewhat better in 1987 than in 1988.

Most females disappear and probably die earlier than males. I have included in Fig. 16.3 only birds hatched in 1979–1983 in order to show the survival of the same cohorts over six years. The shorter average lifespan of females may be caused partly by our failure to find a few females after their dispersal. But in most groups the dominant breeding males are older than the breeding females: in 1988 the average age of breeding males was 7 years (3–14 years, $n = 21$) and of females 6.3 years (3–13 years, $n = 19$). This supports the view that females die earlier than males.

Some babblers survive more than 14 years. Obviously it is impossible to see in 1989 such old birds in a sample of babblers born in 1979–1983. But 13 males and six females of the sample are still alive and some of them may reach that age. At present only one female is over 12 years and two males are over 14 years in the entire population. The relative scarcity of old birds in the present breeding population may be due to the lack of breeding in the dry years of 1977 and 1978. First year survival of the cohorts of 1979 and 1981 was better than that of the other cohorts, 68 % and 66 % as compared to 38–48 % (Table 16.1). In the males, the long term survival of the 1979 and 1981 cohorts was also better than for

A. Zahavi

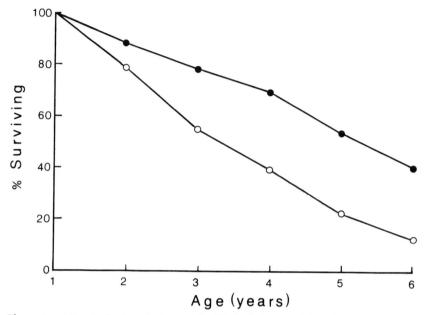

Figure 16.3 Survival of male (*n* = 69, filled circles) and female (*n* = 72, open circles) Babblers hatched in 1979–1983.

the 1980 and 1982 cohorts, 40 % and 50 % compared with 25 % and 30 % at the end of the sixth year (Table 16.2). Male survival seems more variable among cohorts (Table 16.2) than female survival but samples are small.

Attaining a breeding option

Babblers breed mostly when they attain the rank of alpha dominant, but subordinates may also achieve some breeding. Three females in our sample succeeded in depositing eggs in a common clutch, while 46 subordinate males were often observed to copulate with breeding females. Some 25 of these males later reached the status of alpha breeders. As it is difficult to guess their contribution to the breeding of the groups (Zahavi 1988), I here consider all the offspring of the group in a particular year to derive from the dominant male. It is reasonable to assume that some, probably not many, are offspring of the subordinate males.

Babblers attain dominant breeding status between their second and eighth year (Fig. 16.4), females generally earlier than males (females: *m*

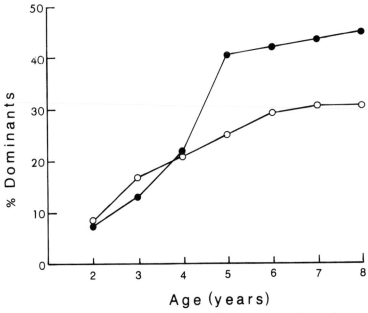

Figure 16.4 Ages at which males (*n* = 69, filled circles) and females (*n* = 72, open circles) became dominant breeders.

= 4.0 years, range 2–7 years, *n* = 20; males *m* = 4.4 years, range 2–8 years, *n* = 31). Since females disperse further than males, we may have missed 1–2 breeding females, but we certainly did not miss any of the males.

The age of first breeding is affected to a large extent by the size and structure of the population at the time the babbler matures, as these factors influence the availability of vacant breeding slots. Earliest breeding was encountered among males hatched in 1979 (Fig. 16.5), when the total population was small and composed of many small groups. Five of the 18 babblers, identified as males at the age of 1 year, started to breed at 2 years old. The average age of first breeding in this cohort was 3 years for the males and 2.5 years for females. This is much lower than that of the 1981 cohort which averaged 5.3 for males and 4.6 for females. None of the 1981 males bred before its fourth year. The oldest female to attain dominant breeding status was 7 years old and the oldest male 8 (Fig. 16.4).

By 1989 all surviving babblers in our sample have become dominant breeders. Altogether 31 out of 69 one-year old males and 20 out of 72 females attained this status. We may have missed a few breeding females,

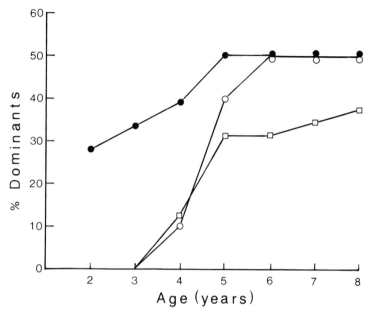

Figure 16.5 Ages at which three cohorts of males became dominant breeders (1979: $n = 18$, filled circles; 1980: $n = 10$, open circles; 1981: $n = 32$, open squares).

but none of the males for reasons given earlier. We cannot yet summarize the entire lifetime reproductive success for all these birds because 13 males and seven females are still alive.

Variation in the lifetime reproductive success of males is clearly seen in the cohort born in 1979 (Table 16.3). Of 18 babblers identified as males when one year old, nine (50 %) became dominant breeders, and two are still alive. These are the two top breeders of our sample, having produced 53 and 40 nestlings respectively. A third male had 21 offspring and the rest 3–9. The pattern is even more skewed, when considering the survival and recruitment of offspring into the breeding population. The top breeder has 15 living offspring, three of which are already breeding as dominants and a few others have a good chance of becoming dominants. The second male, also still alive, has 12 live offspring, four of which are already dominants. The third had nine offspring, one of which is already dominant. Only two of the offspring of the remaining six male breeders still survive, both as breeding females. Hence, most of the breeding success, as measured by the number of recruits to the breeding population, in the 18 males of the 1979 cohort, was achieved by 2–3 individuals (Table 16.3).

Twenty-one of the 72 one-year old females have bred as dominants, but have produced fewer offspring than the males. The top breeder had 24 offspring and three others had more than 10. However this picture for females may be somewhat distorted partly because of the 11 groups selected for analysis. Although most of the breeding males are offspring of large groups, several breeding females came from small groups which did not last for many years and were excluded from the analysis.

A more realistic picture of the breeding success of females comes from analysis of the present population in the whole study area. Among the females not included, one, the mate of the second top male, produced 43 offspring (three with another male), and another female, the present mate of the top dominant male, produced 41 offspring with him alone. A third female produced 18 offspring. All these females came from outside the study area. Another set of good breeders were daughters of small short-lived groups, the territories of which were occupied by males from nearby groups. One of them produced 14 offspring, another produced 17. Both are still alive.

The breeding of small groups

Every group may pass through a stage when it is small, either because the group has not bred (owing to dry years or social disturbances) or because of misfortunes and accidents which killed their offspring. As small groups do not have several adult males to defend the boundaries, they may disintegrate following the death of the dominant male. They are also an easy prey to strong neighbours which look for a breeding opportunity. The offspring of small groups which have disintegrated live as refugees until they find a breeding option. They have to search for it by themselves or with any collaborators they might find among the non-territorial babblers. A daughter of a small group which has lost its dominant male, or was supplanted by a nearby male, has a chance to remain at home and breed, but the sons of small groups are chased from the territory. Solitary males have no way of penetrating into large breeding groups, but solitary females may establish themselves within a large group which has lost its breeding female.

The difference in the strategies by which males and females attain breeding status is clearly demonstrated in the present breeders. Out of 22 males, seven have inherited their status, 11 have occupied neighbouring territories with the help of their parental group, and the other four have gained such status by solitary dispersal into an existing slot. Out of 22 breeding females, eight became breeders when neighbouring males took

Table 16.3 Lifetime reproductive success of males hatched in 1979 and 1980.
? = bred outside study area, S = surviving at time of writing.

Name	Age when went missing	Age became dominant	Breeding duration (Years)	Offspring		
				No. fledged	No. surviving	No. becoming dominant
a. Males hatched 1979						
TCVS	1		0			
ACTS	1		0			
MCTS	2		0			
HSLT	4		0			
HSVT	4		0			
ASMT	4		0			
HSTA	5		0			
SVTV	5		0			
ALST	5	2	3	8	?	?
CSTA	6		0			

HLST	6	2	4	6	1	1
TSMZ	6	3	3	7	1	1
TASA	6	5	1	3	?	?
ATSA	8	2	6	9	?	?
TSCH	8	4	4	21	9	1
TSMV	8	5	3	5	0	0
CTSC	5	2	6	53	15	3
SMTA	5	2	6	40	12	4

b. Males hatched 1980

TLMH	1		0			
CTLZ	1		0			
TZAL	2		0			
ZTLZ	3		0			
HTAZ	4		0			
LTHC	5	5	1	3	1	1
AATZ	5	5	1	5	3	0
CTLL	6	4	2	3	?	?
ZTAZ	6	5	1	6	0	0
TCHM	8	6	2	7	3	0

over their parental territory, nine supplanted the females in a nearby
territory with the help of female sibs, while four moved by themselves
into open slots. This sex difference in strategy provides the advantage to
females over males in small groups. Since daughters of large groups are
much more numerous than daughters of small groups, a one-year old
daughter of a small group could have a better chance to breed than the
daughter of a large permanent group, but we have not checked this.

Breeding of subordinate birds

Three subordinate females in our sample succeeded in laying and raising
young along with the dominant female of their group. They were not
included among the breeding birds in our sample and their offspring were
attributed to those of the dominant females. It is not difficult to know
whether one or more females have laid, but it is hard to ascribe the
surviving nestlings to a particular female. As females collaborate only in
one brood, the contribution of subordinate females to the reproduction
of the population is small.

Some 38 males were adults and subordinate members of the group
when an unrelated female came to their territory, enabling them for the
first time to participate in breeding. Twenty-two of these males never
bred as dominants, while 16 other males remained for several years as
subordinate breeders before they became dominants. Many of the
subordinates mate with the laying female but, without genetic fingerprinting, it is not possible to assess their breeding success. Even if it is low
they still curtail the reproduction of the dominant males below what it
would be if they helped without copulating with the dominant female.
Hence the number of offspring ascribed to a particular dominant male in
our analysis is inflated above what it really is. Nevertheless, the top
breeders who bred for many years shared less with subordinate males
than did males which bred for fewer years. The male which had 53
offspring never had to share with a subordinate breeder. The same is
true for the one with 21 offspring. The second-best male had a male
helper breeder only during the time the first 11 offspring were produced,
but as far as I know his subordinates were not very successful in copulating
with the breeding female.

Discussion

Lifetime reproductive success varies greatly among babblers. About 50 %
of the adults (i.e. two-year old males and females) never become dominant

breeders. Although some females (about 5 %) may lay as subordinate helpers and about 25 % of males may copulate as subordinate males, their contribution to reproduction is small. If reproductive success is measured by the number of offspring which attain breeding status, then the number of babblers which succeed in breeding is reduced to about 30 % of the adult population and most of the reproduction is achieved by about half of the 30 % which succeed in producing large family groups. In the 1979 cohort of males (Table 16.3), nine out of 16 adult males became breeders. Three of these produced 114 offspring, as compared with 38 produced by the other six breeding males, and 36 of the surviving offspring as compared with two. By 1989, eight of the offspring of these three males had already attained breeding status, compared with two for the rest of the cohort. Moreover, two of these three males are still alive and may produce more young, while several more of their offspring will almost certainly become dominant breeders. The skew in the reproduction of females is not much smaller, as most of the successful male breeders have produced most of their young with one breeding female, and females also live up to 14 years.

In babblers much of the success of the individual is determined by the social circumstances within its own and nearby groups. Whatever other circumstances are, the fact that a babbler may have, or may not have, a big brother, or a set of young helpers may decide its fate as a breeder. Babblers seem to have a flexible phenotypic response to the complex social circumstances into which they hatch.

The lifetime reproductive success of co-operative breeding birds, especially when measured in terms of recruits to the breeding population, is dependant largely on the help provided to them by their parental group (Woolfenden & Fitzpatrick 1984). In babblers male offspring from large groups succeed more often than offspring from small groups in entering the breeding population. A breeder which succeeds in accumulating a large number of offspring has in proportion more of its offspring in the following generation (Table 16.3). A large group helps the offspring to survive and also helps the young males to inherit the territory or to conquer an adjacent territory (Zahavi in press), or to bud off on part of their parental territory. They use the natal territory as a safe base from which to disperse and into which they may return if they fail, or they use it as a place to rest while fighting for a neighbouring territory.

Although females may not inherit the territory of a large and successful group, they still enjoy the safety of the parental territory as a base from which to monitor the breeding options in the neighbourhood. Zack & Rabenold (in press) found that female Stripe-backed Wrens *Campylorhynchus nuchalis* succeed in fighting for a breeding slot when their parental territory is near the territory for which they fight. Female babblers may set out from their parental territory, with several of their sibs, to win a breeding slot in other groups.

Daughters of small groups have an advantage over sons of such groups, when male neighbours conquer their parental territory; they may stay and breed in natal territories. Females also succeed in dispersing by themselves more than males do. Nevertheless daughters in large groups prefer to disperse in sib groups and solitary females often make a coalition with a non-kin, non-territorial female, to take over a breeding option.

WHY HELP?

The benefit of helping is not simply that an individual thereby gains membership of a large group and all that that offers. There is a subtle argument of group selection here (Zahavi in press), and such a system would be vulnerable to social parasites which might benefit from being members of a large group, but would not invest in helping. The same is true for the argument of kin selection. When more than one member of a group may provide the help, and when there is no need for more than one or two helpers, the individual which helps gains less than the individual which does not help. But all babblers, even in groups with many helpers, help and even compete over the option to help. My suggestion that the helper gains by increasing its social status is based on a simple model of individual selection, which is immune to social parasitism and fits the competition we observe among helpers.

The rank system within a group of adult babblers is very stable. A dominant may enjoy all the associated advantages without having to enforce its position by overt aggression, including precedence in mating. It is also similar to that in other co-operative breeders in which older birds dominate younger group members. Revolts do occur. A babbler may dispute its inferior rank, but only once, as there is a cost. The loser of such a fight dies or is chased away from the group. Thus a decision to rebel should be preceded by an accurate assessment of the possible outcomes, especially because a defeated dominant is also lost. I think that this is the reason why the rank system is so stable (Zahavi in press (b)). Since a higher rank provides better breeding opportunities, it is reasonable to assume that babblers will invest much to advertise their claim for social status (Carlisile & Zahavi 1986) by investing in the welfare of their group, including the "help at the nest". A higher social status may function to intimidate a subordinate from rebellion and to allow a dominant or a female to obtain a larger share in reproduction.

Not all offspring have an equal opportunity to breed. The dominants among the offspring seem to have a better chance. Dominance is a matter of age and a measure of the babbler's strength against its sibs of the same brood during its first year of life. It is always the dominant son of a group which disperses, with the help of its younger sibs, into a conquered territory, and the second son always stays at home. Hence the accident of being a first-, or second- or third-ranking son (the third usually disperses with the first) largely determines the breeding fortunes of the individual. In females it is the dominant daughter which gains the help of her sisters to disperse and establish herself as the dominant breeder. The youngest daughter often stays at home for another 1–3 years until another set of females is ready to disperse. If the parental territory is occupied in the meantime by alien males, she may join her mother in breeding with them. These differences between the breeding strategies of the sexes are due to the group size and the complexity of the babblers' breeding system, not seen in simpler co-operative breeding systems, like that of the Scrub Jay (Chapter 13). The extreme philopatry of males seems to be a consequence of the greater help they receive from their parental group and their option to inherit their territory. Unlike females, it is the males which dominate and defend the territory. The need to defend the territory was probably the selection pressure, on males, to form long-term coalitions. Males often share in breeding over several years, while females share over one brood only until they establish themselves as breeders.

WHY LIVE IN SMALL GROUPS?

Babblers often form groups of non-kin individuals. There are always non-territorial babblers (refugees) which are willing to join resident groups. If it is so good to be a member of a large group, why do not solitary babblers and small unsuccessful groups join to form larger groups? I suggest that the risk of having one additional competitor in the group often outweighs the gain from having an additional collaborator. The risk is smaller when continuing a coalition with a sib known over several years, since the dominant has already assessed the value of its collaborator and the risk that it may face from its previous experience with that individual. In the absence of kin, babblers collaborate with the smallest number of alien birds that may provide them with a chance to succeed in establishing a new group.

FACTORS AFFECTING LRS

The best social circumstances for breeding success is in having a group composed of a single breeding pair and a set of young babblers which consider both members of the pair as parents. They do not have to be the real genetic parents, because young babblers consider any older babbler which was in their group at the time they were nestlings as a parent with whom they should not mate. Such groups do not have conflicts over breeding, and the breeding pair enjoys the help of the young in defending the territory.

An established breeding female seems to be important to the breeding success of the dominant male, since no male offspring has ever rebelled against his father while his mother was alive. The same is true for the female which may continue to breed in the territory only as long as her mate survived. I suggest that it is the risk of fighting against the dominant pair which deters babblers from such a rebellion. The presence of a mother should not otherwise deter the rebel since mothers leave the territory when their mate dies. It is thus not surprising to find that the best success is achieved by breeding with the same mate over several years (Fig. 16.2 years 1979–1984).

Even if his father has got a new mate, with whom the son may also mate, the son had better wait before he rebels until the group has a few male offspring to help him to defend the territory. In the meantime, the son can only share copulations as a subordinate. Chasing his father or elder brother at an earlier stage will leave him as the sole defender of the territory, while the alternative leaves the territory with a good breeding group in exchange for the loss of one breeding year (Fig. 16.2).

The desert habitat with its frequent droughts, introduces one more element of chance into the breeding success of babblers. It is clear that a babbler with a breeding lifespan of 5 years, 2–3 years of which were not suitable, is at a big disadvantage, compared to an otherwise similar babbler which was a dominant breeder in a group during five successive good years. It is also reasonable to assume that prospects for fledglings born in a dry year are not as good as for those born in a good breeding year. This was manifest in our sample in the higher survival of the 1981 cohort in relation to the 1980 cohort (Table 16.2).

To summarize, the lifetime breeding success of a babbler is dependant on many social variables within its natal group which are consequences of a series of social accidents. Such variables include its group size, its position among its sibs, the age of its parents and their survival, the number of sibs of its own sex, etc. It is further dependant on the size of neighbouring groups and on the chance of getting a good mate which will survive to breed with it over several years.

The extreme philopatry of babblers makes them dependant, more than most other bird species, on local variation in the abundance of food and predators. I suggest that their philopatry is most probably a consequence of their special social system, in which their survival and breeding fortunes depend to a large extent on their parental group.

Some other co-operative breeding birds have smaller groups with a simpler social structure. Individuals do not share in reproduction. Breeding females may stay and accept a new mate into the territory following the death of their previous mate. Offspring do not disperse in sib groups, but individually. In spite of all these differences, however, they are not dissimilar from babblers in the variance of their lifetime reproductive success.

Acknowledgments

My wife, A. Kadman-Zahavi, helped much in discussing the life strategies of the babblers, in interpreting the data and in preparing the figures. Many thanks are due to Ian Newton, who encouraged me to invest in the preparation of the manuscript, and also edited it.

References

Carlisile, R.T. & Zahavi, A. 1986. Helping at the nest, allofeeding and social status in immature Arabian Babblers. *Behav. Ecol. Sociobiol.* **18**: 339–51.
Fitzpatrick, J.W. & Woolfenden, G.E. 1988. Components of lifetime reproductive success in the Florida Scrub Jay. In *Reproductive Success*, ed. T.H. Clutton-Brock, pp. 305–20. Chicago: University Press.
Meinertzhagen, R. 1954. *Birds of Arabia*. London: Oliver & Boyd.
Skutch, A.F. 1961. Helpers among birds. *Condor* **63**: 198–226.
Woolfenden, G.E. & Fitzpatrick, J.W. 1984. *The Florida Scrub Jay, Demography of a Cooperative-Breeding Bird*. Princeton: University Press.
Zahavi, A. 1988. Mate guarding in Arabian Babbler. *Proc. Int. Orn. Congr.* **20**: 420–427. Ottawa, Canada.
Zahavi, A. In press. The Arabian Babbler. In *Cooperative Breeding in Birds*, ed. P.B. Stacey & W.D. Koenig. Cambridge: University Press.
Zack, S. & Rabenold, K.N. In press. Assessment, age and proximity in dispersal contests among cooperative wrens: field experiments. *Anim. Behav.*

IV. Birds of Prey

The four species that comprise this section include two diurnal raptors, the Sparrowhawk *Accipiter nisus* and Osprey *Pandion haliaetus*, and two nocturnal ones, the Screech Owl *Otus asio* and Ural Owl *Strix uralensis*. They thus include both short-lived and long-lived species, all of which are monogamous. As in most other raptors, the males are smaller than the females: for most of the breeding cycle the males provide the food, while the females stay at their nests and tend the eggs and young. For the biologist, this behavioural difference makes it difficult to trap and identify the males, and only for the Osprey, studied by Postupalsky in Michigan, is information provided on the LRS of both sexes.

In his study of Sparrowhawks in southern Scotland, Newton found changes in reproduction and survival through the lifespan which were more marked than in most other birds, including a decline in breeding performance and survival in old age. Otherwise, both aspects of performance (and hence LRS) were greatly affected by territory quality.

In his study of Screech Owls in Texas, Gehlbach found marked differences in mean LRS between two study areas; in one area the population was not self-sustaining and evidently persisted only as a result of continual immigration. In contrast, the most striking feature of the Ural Owl, studied by Saurola in the conifer forests of southern Finland, was the extreme annual variation in breeding rate, corresponding with cyclic fluctuations of rodent prey. In years of prey scarcity, most owls made no attempt to breed, and LRS values depended largely on the number of rodent peak years in the lifespan.

17. Sparrowhawk

IAN NEWTON

The Sparrowhawk *Accipiter nisus* is a small woodland raptor which preys primarily upon other birds. It breeds throughout the Palearctic region, from Britain to Japan, in boreal, temperate and Mediterranean zones. Wherever it has been studied, it breeds monogamously, and the pairs are well spaced out, mostly using the same nesting territories year after year.

Sexual dimorphism in plumage is slight, but, as in other birds of prey, the female is the larger sex. The Sparrowhawk is extreme in this respect, and females (at 290 g) weigh nearly twice as much as males (at 150 g). In consequence, the sexes differ in ecology, with the female hunting more in open country than the male, and taking some larger prey. As in most other raptors, the sexes divide duties in the breeding season. The male does practically all the hunting from before egg laying until the young are about half grown, while the female stays at the nest, incubates the eggs and broods and feeds the young. From the mid-nestling period both sexes share the hunting.

In Scotland, where this study was made, Sparrowhawks are resident year round, centred on woodland and forest. They start nest building anytime from February to April, constructing a large nest of sticks in the lower canopy. They lay their eggs (1–7) in May. The eggs hatch in June after 32–34 days of incubation and the young leave the nest about 30 days later in July. The young then remain in the nest vicinity, fed by their parents, for another 20–30 days, before leaving. Their dispersal normally ceases in late September, and thereafter most individuals stay in the same general area for the rest of their lives, though not necessarily in the same nesting territory. In general, females move further between their natal and subsequent breeding sites than do males (median distances 27 and 14 km respectively).

LIFETIME REPRODUCTION IN BIRDS
ISBN 0-12-517370-9

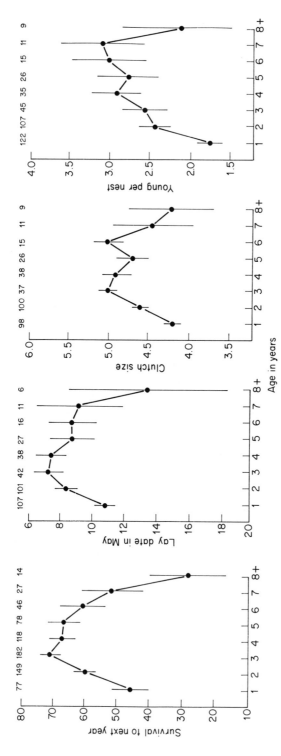

Figure 17.1 Annual survival and breeding success in females of different ages. The small figures along the top show sample sizes. In survival, laying dates and clutch sizes, females performed best in middle age, but in overall production they did not decline until after the seventh year. Survival data were based on the annual trapping of breeding females at their nests. They give minimum estimates of annual survival, because a few individuals may have left the area or been missed within it, and would thus have been counted as dead. In each set of data a bird may appear more than once, the unit of observation being one bird-year. The figures for overall production (young per nest) include nil values.

Study areas and procedure

This chapter is based on data from two areas of south Scotland, Annandale (1971–1980) and Eskdale (1972–1988), some 17 km apart at their nearest points. The landscape is similar in both areas, with small mixed farms and woods on the low ground, and open grassland and large conifer plantations in the hills. In both areas I attempted each year to find all the nests, record the breeding performance, ring the young, and trap as many breeders as possible for individual identification. Females were easier to catch than males, and data were obtained for 194 individual females whose entire breeding lives fell within the study period. Birds not seen for two or more years were counted as dead. In this chapter I shall be concerned with females, enlarging the sample from previous analyses (Newton 1985, 1988a, 1988b), and exploring some additional points.

No females were included if they started breeding after 1984. Only one of the 34 females present in that year was still alive in 1988 (the cut-off point), but in earlier years occasional individuals bred for up to eight years. In their breeding lifespans, therefore, the females of the sample were reasonably representative of the female population as a whole (see also Newton 1985). Although the Annandale breeding population declined by 45 % during the period of study, the Eskdale population remained stable. Mean lifetime fledgling production was slightly lower in Annandale (mode = 4.0, mean 4.82 ± SE 0.42) than in Eskdale (mode = 4.0, mean = 5.26 ± SE 0.52), but, as this difference was not significant, records from both areas were pooled for the analyses below.

Survival and longevity

The survival of female Sparrowhawks was previously estimated at 49 % in the first year, and at 64–71 % per year thereafter, depending on the method of analysis (Newton 1986). The equivalent figures for males were 31 % in the first year and 67–69 % thereafter. As the sex ratio at fledging was equal, the greater first-year mortality of males led to a surplus of females among birds of breeding age (Newton 1988a). However, this sex difference in mortality was not apparent among the national ring recoveries of Sparrowhawks in Britain, and may have been peculiar to south Scotland.

To judge from recaptures of breeders, the annual survival of females was not constant after the first year, but increased to a peak at 3–6 years, and declined thereafter, markedly after seven years (Fig. 17.1). In practice, the improvement in early life may have been less than the data in Fig. 17.1 suggest, because breeders more often changed territories between years 1 and 2 than later in life (Newton & Marquiss 1982), so were more likely to have left the study area then (and thus counted as dead). The decline in recorded survival later in life may have been exaggerated in a different way, if some individuals ceased to breed at that age, thus precluding their capture at nests, and lived on as non-breeders. There was, however, no evidence for this, and no ring recoveries were obtained from the general public of birds older than the oldest breeder (in tenth year). The late-life decline in annual survival was a new finding over a previous analysis (Newton 1988a), evident only because of an increase in the number of older females available. The entire survival data were best fitted by a quadratic regression model in which annual survival(s) = 0.253 + 0.255 age − 0.028 age^2 (r = 0.990, P < 0.0001). This model explained 98 % of the variance in age-related survival, whereas a linear model explained only 15 %.

Annual breeding success

Overall, only about 57 % of nests produced young, varying annually between 47 and 75 %. The percentage was correlated with the weather in the pre-lay period, being high when March–April was warm and dry, and low when it was cold and wet (Newton 1986). Weather in these months influenced the breeding of songbirds, and hence the subsequent food-supply for Sparrowhawks. In fact, the main cause of nest failure in all years appeared to be food shortage, which was manifest chiefly in failure to lay and in egg desertion, but occasionally in starvation of young (Newton 1986). Predation on nest contents was infrequent and may also have been facilitated by food shortage, which caused the female to hunt and leave the nest contents unguarded. In those circumstances, Jays *Garrulus glandarius* and Red Squirrels *Sciurus vulgaris* ate eggs and Tawny Owls *Strix aluco* ate young. Major predators of young Sparrowhawks in other regions, namely Goshawks *Accipiter gentilis* and Pine Martens *Martes martes*, were absent from the study areas, as a result of human action long ago. The Sparrowhawks were also contaminated with organochlorine pesticide residues, but at too low a level to have a major impact on their breeding in the years concerned. Most successful clutches consisted of 4–6 eggs, the majority of smaller clutches being deserted soon after laying; in many successful nests, one or more eggs failed to hatch, so that most broods consisted of 2–5 young. On limited data, less than 10 % of young died between fledging and independence.

No more than one brood was raised by each pair in a year. The few cases (less than 1 % of all failures) of a repeat clutch, laid after the first had failed at an early stage, followed an accidental loss, such as nest collapse. Repeats never followed desertions, presumably because, if a pair abandoned one nest through food shortage, they were unable to start another that year.

Age and breeding success

Breeding performance did not remain constant through life. Between the first and third years, the mean trends were for egg laying to become earlier and clutches larger (Fig. 17.1). As early nests were more successful than late ones, and large clutches produced more young than small ones (Newton 1986), these changes could be regarded as improvements in breeding. These features then changed little until about the sixth year, after which laying became later again and clutches smaller. A third aspect of performance, the number of young produced per attempt, continued to rise, at least until the seventh year of life; thereafter it seemed to decline, but too few birds were alive by then to be sure.

Such trends in mean performance emerged from the pooling of records from all individuals of each age class. Further analysis revealed that the mean trends could be attributed entirely to changes in the performance of individual birds as they aged (unpublished data). They were in no way attributable to birds with different performance levels entering or leaving the breeding population at different ages. There was, for example, no greater tendency for birds which bred badly to disappear from the breeding population at an early age.

Lifetime breeding success

Lifetime breeding success was calculated for the 194 females, on the assumption that all their breeding attempts occurred within the study area. For the majority, this was a reasonable assumption, but a minority may have had one or more attempts outside the study area and hence unknown to me. This meant that a small proportion of birds may have raised more young than recorded. A further assumption was that the female trapped at a nest (usually during incubation) was the mother of all the young therein. For the purpose of analysis, egg laying was counted as a breeding attempt, and nest building without laying was not.

Table 17.1 Lifespan, age of first breeding and years of breeding for reproductive females of known life history. Figures show number of females in each category.

	Number of years										*Mean*
	1	*2*	*3*	*4*	*5*	*6*	*7*	*8*	*9*	*10*	
Lifespan[a]	43	41	33	30	17	8	10	9	2	1	3.34
Age of first breeding[b]	65	65	61	3	0	0	0	0	0	0	2.01
Years of breeding[c]	97	41	17	14	8	10	5	2	0	0	2.25

[a] Taken from year of birth to year of last recorded breeding inclusive.
[b] Taken as age of first egg-laying (some birds paired and built a nest at an earlier age).
[c] Taken as years from first to last recorded breeding inclusive, and including any non-breeding years in between.

Females studied throughout their lives spent an average of 2.3 years (range 1–8) as breeders and raised from 0 to 24 young (Table 17.1, Fig. 17.2). The distribution of individual productions was skewed, with a long tail to the right. The peak at 3–5 young was because most birds raised only one brood during their lives, and single broods usually contained 3–5 young. Some 17 % of all females which attempted to breed produced no young. These included 27 birds which laid in only one year, four that laid in two years, one that laid in three years and one that laid in four.

The numbers of fledglings produced (y) was in turn closely dependent on the numbers of eggs laid (x) during a lifetime ($y = -0.152 + 0.567x$, $r = 0.825$, $P < 0.0001$). Overall, some 68 % of the variance in numbers of young raised could be explained in terms of the numbers of eggs laid. The correlation is not perfect because of variation between nests in survival from the egg to fledging stages.

The data in Fig. 17.2 refer to fledglings. A better measure of lifetime production would have been the number of young which survived and bred. Between them, the females in the sample raised 976 young, but only 40 were later found nesting in the study areas, other survivors presumably having dispersed elsewhere to breed. Nonetheless the number of local recruits from individual females was correlated with the numbers of fledglings they had raised (excluding females which produced no fledglings, $n = 161$, $r = 0.44$, $P < 0.001$); so the fledgling productions of individuals could be taken as a reasonable index of their contributions to future breeding populations. The regression line depicting this relationship passed close to zero. This implied that, on the average,

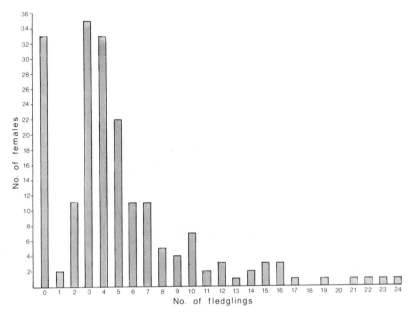

No. of fledglings

Figure 17.2 Lifetime fledgling productions of 194 female Sparrowhawks. Mean per female = 5.03 young.

individual fledglings had the same chance of breeding, whether they came from more or less productive mothers (Newton 1988a).

Much of the variation in lifetime fledgling production could be attributed to variation in longevity between individuals, and to variation in age of first breeding, which together determined the length of breeding life (Table 17.2). On linear regression analyses, some 43 % of the variance in lifetime productions could be explained in terms of variation in lifespan (Fig. 17.3), 10 % in terms of variation in age of first breeding, and 48 % in terms of length of breeding life. On a multiple regression analysis, age of first breeding and lifespan did not explain appreciably more of the variance in lifetime production than did lifespan alone (Fig. 17.3 caption). Productions of up to five young were recorded from some birds which lived for only one year, but all production greater than 10 young were from birds which lived for more than four years. There was considerable variation within age groups: some birds lived for five years and still raised no young; others lived eight years and raised only five young; and the longest lived bird, which died at 10 years, produced only 12 young.

Some of the birds in the sample started breeding in their first year of life (10 months old), others in their second, third or even fourth year

Table 17.2 Lifetime fledgling production according to lifespan and age of first breeding. n = number, M = mean, SE = standard error.

Age of first breeding	Lifespans of following numbers of years										
	1			2			3			4	
	n	M	SE	n	M	SE	n	M	SE	n	M	SE
1	43	2.79	0.24	9	4.67	0.80	2	7.00	3.00	11	8.18	1.83
2				32	2.38	0.32	11	5.73	1.01	22	10.86	1.39
3+							20	2.00	0.49	44	6.64	0.65

Significance of differences. Horizontal comparisons: age of first breeding 1: lifespans 1 and 2, $t = -2.97$, $P < 0.01$; lifespans 1 and 3, $t = -3.47$, $P < 0.01$; lifespans 1 and 4, $t = -5.29$, $P < 0.001$; age of first breeding 2: lifespans 2 and 3, $t = -4.18$, $P < 0.001$; lifespans 2 and 4, $t = -7.02$, $P < 0.001$, lifespans 3 and 4, $t = -2.45$, $P < 0.01$; age of first breeding 3: lifespans 3 and 4, $t = -4.51$, $P < 0.001$.

Vertical comparisons; lifespan 2 years: first breeding 1 and 2, $t = 3.11$, $P < 0.01$; lifespan 3 years: first breeding 1 and 3, $t = 2.90$, $P < 0.01$; first breeding 2 and 3, $t = 3.76$, $P < 0.001$; lifespan 4+ years: first breeding 2 and 3, $t = 2.75$, $P < 0.01$.

Only significant differences are mentioned above; all other differences are non-significant.

(Table 17.1). In general, among birds at any given lifespan, those that started breeding early in life raised more young than those that started later (Table 17.2). By implication, each year's delay in first breeding meant an average loss of about 2.7 young from the lifetime total.

The advantage in breeding at an early age was in fact greater than the figures in Table 17.2 suggest. Those figures referred only to birds which survived long enough to breed, and took no account of other individuals which died before they could reach two, three or four years. On average, a bird which failed to breed in its first year had only a 46 % chance of surviving from then to its second year, 27 % of surviving to the third year, and 19 % to the fourth (calculated from Fig. 17.1). Thus, taking survival chances into account, there was an even bigger advantage in starting to breed as early in life as possible.

Factors preventing some individuals from breeding early in life included shortage of food (implied by weights), shortage of good nesting territories, and (in the case of females) shortage of mates (Newton 1986). In other words, some individuals were prevented from breeding at an early age by competition with other individuals for resources in short supply. There was no evidence from their lifetime productions that the loss in production

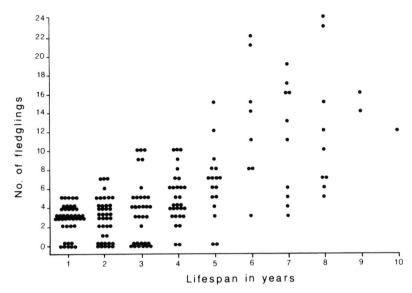

Figure 17.3 Lifetime fledgling production of females in relation to their lifespan. Each spot represents the number of fledglings raised by an individual female. Relationship between lifetime production (y) and lifespan (x₁): $y = 0.127 + 1.426x_1$, $r = 0.652$, $P = 0.0001$. Relationship between lifetime production (y) and age of first breeding (x₂): $y = 3.954 + 0.536x_2$, $r = 0.098$, $P = 0.175$. Multiple regression relationship between lifetime production (y) lifespan (x₁) and age of first breeding (x₂): $y = 2.569 + 1.774x_1 - 1.718x_2$, $r = 0.706$, $P = 0.0001$. Relationship between lifetime production (y) and duration of breeding life (= lifespan minus age at first breeding, x₃): $y = 0.854 + 1.850x_3$, $r = 0.706$, $P = 0.0001$.

from delayed breeding was offset by greater production in later life (Table 17.2).

What distinguished those individuals which started to breed when young from others of their cohort which waited? I could find nothing in early experience which predisposed a bird to start breeding at one age rather than another. Thus no relationship was apparent between age of first breeding and year of birth, fledging date within the season, natal brood size, age of mother or grade of natal territory.

However, as shown previously, females which bred at an early age tended to be larger than those which waited (Newton 1988b). Wing-chord was used as an index of body size, and was correlated with lifetime production, mainly because of an inverse relationship with age of first breeding. Entry to the breeding population was competitive, at least for females, and larger individuals evidently had an advantage early in life.

Relationship between habitat (territory quality) and lifetime production

In a previous analysis, no difference emerged in lifetime success of birds nesting on hills or in valleys, the two main environments of the study areas (Newton 1985). However, in both situations territories varied in quality, that is, in the chance they offered for successful breeding. Territories in the study area were initially graded in five classes (1 = poor, 5 = good), according to frequency of use in a five-year period (Newton & Marquiss 1976). This grading preceded the period in which most of the present data were collected, and was thus largely independent of them. In general, territories which were used most often in this period showed the best nest success, and those used least often showed the poorest success (Newton & Marquiss 1976). The same conclusion held, and individual territories were generally consistent in their grades, when the study was extended to 10 years (Newton 1988b).

In order to find whether territory quality affected lifetime production, birds were scored according to the grades of territories they used. Individuals which changed territories during their lives were given a weighted score, depending on the number of years they spent on territories of different grade. Females which bred in only one year could occupy only one territory, but those which bred in two years occupied an average of 1.3 territories each, and those which bred in three or more years occupied an average of 1.7 territories each.

A clear relationship emerged between lifetime success and mean territory grade (Fig. 17.4): on average, lifetime production increased by 1.4 young for each step in territory grade, and by 5.5 young over the whole five classes. Further analysis revealed that this was due mainly to greater longevity on the better territories (Fig. 17.4): on average, lifespan increased by 0.55 years for each step in territory grade, and by 2.2 years over the whole five classes. Annual production was also slightly better on the good territories, increasing by about 0.15 young for each step in territory grade, and by 0.60 young over the whole range, but this trend was not significant statistically.

Age of first breeding was also somewhat higher on good territories than on poor ones. This was largely because most young birds were relegated to poor territories, the survivors moving to better territories later in life. Greater longevity on better territories was consistent with an earlier analysis, showing greater year-to-year persistence of females on good territories than on poor ones (Newton 1988b). Hence, territory quality emerged as a major factor associated with longevity and lifetime reproduction.

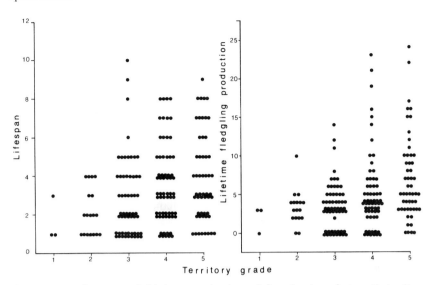

Figure 17.4 Lifespan and lifetime production of females in relation to territory grade. Territories were graded in five categories, 1–5 (poor–good), based on occupancy and nest success, and individual females were allotted a weighted score depending on all the territories they occupied during their breeding lives. Regression relationship between lifespan (*y*) and mean territory grade (*x*): *y* = 1.263 + 0.551*x*, *r* = 0.261, *P* = 0.0002. Regression relationship between lifespan (*y*) and grade of territory at first breeding attempt (*x*): *y* = 2.263 + 0.083, *r* = 0.088, *P* = 0.224. Regression relationship between lifetime production (*y*) and mean territory grade (*x*): *y* = −0.141 + 1.376*x*, *r* = 0.297, *n* = 194, *P* = 0.0001. Regression relationship between lifetime production (*y*) and grade of territory at first breeding attempt (*x*): *y* = 3.453 + 0.229*x*, *r* = 0.111, *P* = 0.125. Relationship between age of first breeding (*y*) and mean territory grade (*x*): *y* = 1.652 + 0.095*x*, *r* = 0.913, *P* = 0.116. Relationship between age of first breeding (*y*) and grade of territory at first breeding attempt (*x*): *y* = 1.753 + 0.066*x*, *r* = 0.090, *P* = 0.215. Relationship between duration of breeding life (*y*) and mean territory grade (*x*): *y* = 0.684 + 0.420*x*, *r* = 0.241, *P* = 0.0007. Relationship between duration of breeding life (*y*) and grade of territory at first breeding attempt (*x*): *y* = 1.975 + 0.042*x*, *r* = 0.053, *P* = 0.461. Relationship between mean brood size (*y*) and mean territory grade (*x*): *y* = 1.785 + 0.153*x*, *r* = 0.100, *P* = 0.166. Relationship between brood size (*y*) and grade of territory at first breeding attempt (*x*): *y* = 1.999 + 0.055*x*, *r* = 0.069, *P* = 0.341.

Some females had good territories throughout their breeding lives, while others had a poor territory to begin with and later moved to a better one. Birds which had good territories from the start tended to survive longest and raise most young overall, but neither tendency was significant statistically (Fig. 17.4, caption). It seemed that the fate of a bird was not entirely sealed by the quality of its first territory, and that later ones were important too.

Relationship between year of first breeding and lifetime production

A further factor which may have influenced the productions of at least the once-only breeders was the calendar year in which nesting occurred. Breeding success varied significantly between years, with 47–75 % of nests producing young in any one year. Thus a once-only breeder which happened to nest in a poor year was less likely to produce young than one which bred in a good year. However, the effect of calendar year was not apparent in the sample of once-only breeders available for analysis (ANOVA: $F_{13,83} = 0.87$, ns). Nor was it apparent in the lifetime productions of birds which bred in more than one year ($F_{13,83} = 0.91$, ns).

Factors affecting recruitment

Analysis has so far been restricted to those individuals which attempted to breed. In an earlier analysis, I calculated that such individuals formed only 28 % of birds which fledged, and that the remaining 72 % of fledgings died before they could attempt to breed in their 1st–4th year of life (Newton 1988a). These figures were calculated from the combined data for both study areas, in one of which numbers declined. In a stable population pre-breeding mortality is likely to be lower, perhaps around 60 %. But whichever figure is taken, survival between fledging and recruitment to the breeding population emerges as a crucial component of individual fitness.

To find whether early experience had any influence on recruitment chances, I compared the years of birth, fledging dates, brood sizes, age of mother and grades of natal territory of birds that were recruited to the local breeding population with similar data for birds that were not. The latter would have included many birds which died before they could breed, and some others which bred elsewhere. Analysis was again restricted to females because too few breeding males were caught to properly assess recruitment.

Only one significant relationship emerged: young fledged earliest in the season were more likely to breed in the local population than were young fledged later (Table 17.3). Ring recoveries from members of the public showed no difference in the dispersal distances of early and late fledged young (Newton & Marquiss 1982), so the greater recruitment from early young could only have resulted from their better survival, compared to late young.

Table 17.3 Recruitment in relation to fledging date, based on the pooled records from all years.

Fledging dates	Number of young females ringed	Number later found breeding in study areas (%)
−10 July	77	10 (13.0)
11–20 July	461	49 (10.6)
21–30 July	300	14 (4.7)
31 July–	34	1 (2.9)

Note: Significance of variation between categories: $\chi_3^2 = 10.5$, $P < 0.02$.

A tendency for females raised by yearling mothers to show lower recruitment than those raised by adult mothers was almost significant statistically, but this may have been partly because, on average, yearling mothers laid later in the season than adult mothers. The effect of fledging date on recruitment held among young from adult mothers alone, however.

Mother–daughter correlations

For a small number of locally-bred females, which settled to nest in the study areas, it was possible to compare their ages at first breeding, lifespans and lifetime productions with those of their mothers (Table 17.4). In each case, the correlations were poor and not statistically significant, as were the corresponding heritability estimates (Falconer 1981). These findings fit the expectation that characters which are strongly related to fitness should show little or no heritability, for otherwise they would have been incorporated by selection long ago (Fisher 1930).

Table 17.4 Correlations between mother–daughter values for various components of fitness, together with estimates of maximum heritability (h^2). No relationship was statistically significant.

	n	Correlation coefficient	h^2
Age of first breeding	15	0.40	0.11 ± 0.07
Lifespan	13	−0.19	0.04 ± 0.06
Lifetime fledgling production	12	−0.08	0.06 ± 0.28

Non-breeding

Females which had not previously bred formed about two-fifths of the total female population during the breeding season (Newton 1985). Radio-tracking revealed that most such birds lived singly in large home ranges, away from prime breeding sites (Marquiss & Newton 1982). However, some became associated with a male on a nesting territory in spring, and a proportion got as far as nest building. On average, about 15 % of the nests found each year were not subsequently laid in (Newton 1986). Females trapped at such sites were largely first-year birds, but included some second year and older birds which had not previously been recorded breeding (Table 17.5). Other females which had bred before occasionally had a non-breeding year. Again some were caught at nesting territories and others were not. Such skipping of breeding years was also more frequent among younger age groups, and no case was recorded among females older than five years. Overall, it seemed that about one half of the females present in the study areas bred (i.e. laid) in any one year.

Experimental provision of supplementary food showed that food-supply influenced the stage of breeding reached; in general, only well fed females produced eggs (Newton & Marquiss 1981).

Table 17.5 Age composition of different components of the female Sparrowhawk population in spring. Based on birds trapped on nesting territories 1971–1984.

	Number (%) of females aged		
Nesting places where:	*1 year*	*2 years*	*3+ years*
No nest built	21 (58)	7 (19)	8 (22)
Nest built, no eggs laid	16 (44)	8 (22)	12 (33)
Nest built, eggs laid	16 (14)	112 (21)	351 (65)

Significance of variation: $\chi_4^2 = 166.7$, $P < 0.001$.

Discussion

ASSESSMENT OF DATA

On average, the females studied produced only 5.0 fledglings during their lives, or 1.4, if birds which died before they could breed were included. This figure fell short of the 2.0 young (one male and one female) which each female must produce, on the average, to maintain population stability. For reasons given earlier, mainly involving movements, the lifetime productions of some females may have been underestimated, but in addition the population of Annandale declined during the study period, so that a mean production of less than two young per female was not unexpected. The slightly higher lifetime success of the Eskdale over the Annandale females was not significant, and stability of the Eskdale population was achieved largely by greater immigration (Newton & Marquiss 1986). An increase in mean lifetime production, needed to stabilize the Annandale breeding population, could have been achieved by an improvement in annual survival, a lowering of age of first breeding, or an increase in the number of young produced at individual attempts.

AGE AND PERFORMANCE

Sparrowhawks did not perform equally well throughout their lives. Annual survival and some aspects of breeding performance reached their peak at ages 3–6, deteriorating thereafter. The improvement in performance in early life could be attributed to a rise in experience, social status and competitive ability, enabling some individuals to move to better territories. If experience was important, it was probably that resulting from hunting and other everyday activities, rather than from breeding itself. Thus among second-year females, individuals which had bred previously showed no better success than those that had not, and the same was true among third-year birds (Newton 1989). Likewise any benefit gained from familiarity with a particular territory or mate was clearly offset by the advantage gained from a move to a better territory (which also entailed a change of mate) (Newton & Marquiss 1982, Newton 1988b). The decline in survival and breeding performance in later life was not associated with a move to a poorer territory and could only be attributed to senescence.

Interestingly, although some aspects of breeding performance peaked at 3–6 years, actual productivity, as measured by the mean number of young produced per clutch laid, continued to rise progressively, at least to the seventh year. Perhaps this aspect of performance depended more on experience, and was less subject to effects of physiological ageing than were laying late and clutch size.

TRADE-OFFS BETWEEN CURRENT AND FUTURE REPRODUCTION

Studies on some bird species have implied that reproduction has costs, which are manifest in reduced survival and future reproduction (e.g. Bryant 1979, MacGillivray 1983, Ekman & Askenmo 1986). Two aspects were investigated in female Sparrowhawks, namely age of first breeding (this chapter) and production at individual attempts (Newton 1988b). In age of first breeding, no advantage was apparent in waiting, because birds which started in their first or second year raised significantly more young in total than those which waited till their third. Shortage of resources meant that age of first breeding was competitive, and delay occurred in birds unable to obtain good territories, food supply or mates at an early age. Delayed breeders gained in no obvious way, but the analysis compared different groups of females which first bred at one, two or three years. If each individual bred at an age which was best for it, the advantage of delay could not have been detected.

From individual attempts, no evidence was found that small broods were advantageous to future reproduction. No inverse correlation was apparent between brood size in one year and brood size the next, nor between brood size and subsequent maternal survival (Newton 1988b). Evidently these aspects of performance varied independently of reproductive effort in the previous year. Hence, no evidence was found from the natural situation that the rearing of young reduced future reproductive potential in the Sparrowhawk. But this conclusion suffers from the same weakness as that on age of first breeding, that if an individual breeds at its chosen rate, and raises the brood size which is best for it in prevailing conditions, then no cost of reproduction need emerge. A more rigorous test would entail an experimental manipulation of brood sizes on a random basis. Such manipulations on other species have often demonstrated reproductive costs (Askenmo 1979, Nur 1984, and other studies reviewed by Partridge, Chapter 25) and are currently underway in Sparrowhawks.

LIFETIME PRODUCTIONS

With a lifespan up to 10 years, and up to six young in a brood, a female Sparrowhawk could raise 60 young in its life. As none of the 194 birds studied raised more than 24 young, none raised anywhere near this maximum. The most striking feature of lifetime production was the great spread of variation between individuals, with most breeders raising 0–5 young during their lives and few raising more than 20. From knowledge of mortality at different ages, it was possible to work out the lifetime production of a whole generation of fledglings, including those which died before breeding (Newton 1988b, Chapter 26). In the study areas 72 % of female young died before they could breed in their first to fourth year of life, 5 % attempted to breed but failed, and only 23 % produced young, but in greatly varying numbers. On the pattern observed, less than 5 % of fledglings (or 20 % of breeders) produced 50 % of the next generation, and 7 % of fledglings (or 30 % of breeders) produced 90 % of subsequent young. The relative contributions of individuals to future populations was therefore enormously skewed, a pattern repeated in each successive generation.

The chance of recruitment to the breeding population was influenced by the laying date of the mother, a feature over which the individual itself had no control. The age of recruitment was related to wing-chord (body size), with a tendency for larger females to start breeding earliest in life. Entry to the breeding population was a competitive process, in which large body size may have helped, at least in young birds. Wing-chords were correlated between mothers and daughters ($n = 34$, $r = 0.58$, $P < 0.001$), so this may have been one way in which inheritance might have influenced the lifetime productions of those individuals which survived long enough to breed. In practice, however, on a small sample there was no correlation between mothers and daughters in age of first breeding, lifespan or in lifetime production (Table 17.4).

Individual lifespan, and hence lifetime breeding success, was greatly influenced by territory quality. Again good territories were contested (Newton 1986), so this was another aspect in which competitive ability was important.

The performance of individuals was also influenced by chance events. Thus many birds suffered from accidents of various kinds. Some died from collisions with obstacles, such as windows and wires, while others failed in their breeding through similar unpredictable events, such as nest collapse in strong wind.

Highly productive females showed no significant tendency to perform better early in life than other females (Newton 1988b), and merely by living longer they usually acquired good territories and mates, and accrued more breeding attempts than others. I found no way in which high-performance birds could be identified as such early in their lives. None of the correlates of high performance, such as date of birth, wing-chord or territory grade, were sufficiently precise to be applied to particular individuals, as the variance was large. Still, the really successful female was one which was raised early in the season of its birth, which had a relatively large body size, which started breeding in its first year of life on a good territory, and which maintained a good territory throughout. It was also a lucky individual, avoiding year after year all the chance events which might shorten its life or destroy its nesting attempts.

Acknowledgements

I am grateful to the many landowners in the study areas for granting access to their woodland, to Dr M. Marquiss for his contribution to the study over the years, and to Drs J.P. Dempster and M. Marquiss for helpful criticism of the manuscript.

References

Askenmo, C. 1979. Reproductive effort and the return rate of male Pied Flycatchers. *Amer. Nat.* **114**: 748–53.

Bryant, D.M. 1979. Reproductive costs in the House Martin (*Delichon urbica*). *J. Anim. Ecol.* **48**: 655–75.

Ekman, J. & Askenmo, C. 1986. Reproductive cost, age-specific survival and a comparison of the reproductive strategy in two European tits (Genus *Parus*). *Evolution* **40**: 159–68.

Falconer, D.S. 1981. *Introduction to Quantitative Genetics*, 2nd edn. London: Longman.

Fisher, R.A. 1930. *The Genetical Theory of Natural Selection*. Oxford: University Press.

McGillivray, W.B. 1983. Intra-seasonal reproductive costs for the House Sparrow (*Passer domesticus*). *Auk* **100**: 25–32.

Marquiss, M. & Newton, I. 1982. A radio-tracking study of the ranging behaviour and dispersion of European Sparrowhawks *Accipiter nisus. J. Anim. Ecol.* **51**: 111–33.

Newton, I. 1985. Lifetime reproductive output of female Sparrowhawks *J. Anim. Ecol.* **54**: 241–253.

Newton, I. 1986. *The Sparrwhawk*. Calton: T & A.D. Poyser.

Newton, I. 1988a. Age and reproduction in the Sparrowhawk. In reproductive success, ed. T.H. Clutton-Brock, pp. 201–219. Chicago: University Press.

Newton, I. 1988b. Individual performance in Sparrowhawks: the ecology of two sexes. *Proc. Int. Orn. Congr.* **19**: 125–154.

Newton, I. & Marquiss, M. 1976. Occupancy and success of Sparrowhawk nesting territories. *Raptor Research* **10**: 65–71.

Newton, I. & Marquiss, M. 1981. Effect of additional food on laying dates and clutch-sizes of Sparrowhawks. *Ornis Scand.* **12**: 224–9.

Newton, I. & Marquiss, M. 1982. Fidelity to breeding area and mate in Sparrowhawks *Accipiter nisus. J. Anim. Ecol.* **51**: 327–41.

Newton, I. & Marquiss, M. 1986. Population regulation in Sparrowhawks. *J. Anim. Ecol.* **55**: 463–80.

Nur, N. 1984. The consequences of brood size for breeding in Blue Tits. 1. Adult survival, weight change and the cost of reproduction. *J. Anim. Ecol.* **53**: 479–96.

18. Osprey

SERGEJ POSTUPALSKY

The Osprey *Pandion haliaetus* is a large, distinctive, fish-eating raptor, which breeds over much of the northern hemisphere and the Australasian region. In boreal and temperate regions, the species is migratory, moving to tropical and sub-tropical areas for the winter (Henny & Van Velzen 1972, Österlöf 1977). It occupies both coastal and freshwater habitats, and in breeding areas builds a huge nest of sticks, usually on a tree, snag, or rock pinnacle near water. The nests are used year after year over long periods, as individual pairs and their successors return repeatedly to the same sites. They raise up to three young each year.

The sexes are similar in appearance but, as in most other raptors, females are larger than males (mean weight of 50 Michigan females: 1798 g, and of 29 males: 1427 g). Once a pair is established, the male provides nearly all the food from before egg-laying until the young leave the nest (Levenson 1979). Up to that stage the female stays at the nest, and only when the young start flying does she help with the hunting. Thus young Ospreys—even well-feathered ones—are seldom left unguarded.

This chapter is based mainly on my work in Michigan, but I also draw comparisons with other studies elsewhere. In Michigan Ospreys return to their breeding areas about the second week in April, departing again from late August to October. To judge from band returns, the birds winter in Central and northwestern South America.

Study area and methods

My study is centred on two principal Osprey breeding areas in the northern Lower Peninsula of Michigan: Fletcher Pond, a water storage reservoir (3600 ha, max. depth 3.5 m) in Alpena and Montmorency

LIFETIME REPRODUCTION IN BIRDS
ISBN 0-12-517370-9

counties, which now supports the largest Osprey "colony" (>20 pairs) in the state (Postupalsky 1977, 1978), and a group of smaller (50–500 ha) wildlife floodings established by the Michigan Department of Natural Resources near Houghton Lake in Roscommon County. Between one and five pairs are currently nesting at each flooding. Ospreys fish on the larger of these impoundments, as well as on Houghton Lake (8100 ha, max. depth 7 m). These two breeding areas are 75–110 km apart. Juveniles raised in one breeding area regularly nest in the other.

Most Osprey nests in the study area occur on man-made platforms (Postupalsky & Stackpole 1974, Postupalsky 1978). On Fletcher Pond these platforms reversed a declining population trend, caused in part by loss of natural nest sites (dead trees and stubs). The number of nesting pairs grew from 11 in 1966 to 21–23 in 1985–1987. In the Houghton Lake area, despite the presence of unused platforms, no increase in occupied nests was noted until after Osprey numbers started increasing statewide in 1977, following cessation of DDT use. The number of pairs increased from seven to nine prior to 1978, to 20 in 1987. While these platforms did attract new pairs into the study area, other platforms were installed for new pairs which had moved in and tried to nest on decaying snags and low stumps.

The accessible and durable man-made nest sites enabled me to undertake a long-term study. Banding of nestlings started in 1963 (few young were produced then at the height of the DDT era). Trapping of adults started in a small way in 1969–1970, with an artificial owl decoy and noose-bonnet (Scharf 1968), and accelerated in 1971 with a dome-shaped noose-carpet tied over the nest. Colour-banding commenced in 1972, since when 97 % of all young produced in the study area have been banded.

Each adult was marked uniquely with a numbered US Fish and Wildlife Service band and two or three colour bands. Nestlings received the USFWS band and a single colour band to identify hatching year. When these birds returned to nest three or more years later, their age could be seen at a glance. Trapping them as breeders enabled me to determine their identity (band number, natal site), and to apply additional colour bands to provide a unique combination. In this manner over the years I built up a population of about 60 individually marked adults, which included about 86 % of the females and 50 % of the males in the study area. Several additional adults were marked to hatching year only. Each spring, while checking nest contents, I viewed the adults (through binoculars), as they circled overhead, to determine their identity. I then tried to trap any unbanded breeders, once incubation was underway.

Osprey reproduction in Michigan, which was much reduced during the

DDT era, began recovering in the late 1960s and reached the estimated replacement level (see next section) by 1970 (Postupalsky 1977), as DDT use in the United States declined and ceased. In 1974 productivity rose well above replacement level and the population started growing in 1977. My records relating to reproduction of individually marked and known-age adults cover the period 1970–1987, with most during the last 14 years; in other words, following recovery from DDT effects.

The man-made nests have been more productive than natural nests. This was expected, as the platforms had been placed selectively to maximize reproduction, and excess platforms were occupied by young recruits (Postupalsky 1978). In recent years this difference has narrowed, as success in natural nests has increased also.

Inasmuch as a study of this type should ideally extend at least over one maximum lifespan for the species, the results presented here should be viewed as preliminary.

General lifestyle

Ospreys are fairly long-lived. The oldest North American Ospreys reported to date are one 22-year old bird of unknown sex (Clapp *et al.* 1982) and a 25-year old male (Spitzer 1980). The oldest Osprey in my study, a female, first banded in 1971 as a young breeding adult on a new territory, was still present there in 1987, when she was at least 19 years old.

Annual survival rates were estimated by Henny & Wight (1969) from the recoveries of birds banded in New York and New Jersey. Recoveries of birds "found dead" and combined recoveries of birds "found dead" and "shot" yielded maximum and minimum estimates of survival rates: 48.5–42.7 % for the first year of life and 83.8–81.5 % for each year thereafter. Values based on birds "found dead" are believed to be the most valid. My data from annual returns of individually marked Michigan Ospreys yielded an annual survival rate of 85.0 % for adults of three years and older.

Reproductive data given below are from the whole Lower Peninsula of Michigan (brood size, productivity) or from my study area alone (clutch size, proportion of eggs hatching and resulting in fledglings, proportion of non-breeders in the population). Ospreys in the study area comprised 75–80 % of the total Lower Peninsula population and could thus be regarded as representative. Clutch size varied from one to four, with the mean of 2.92 eggs ($n = 537$, 1967–1987); clutches of three eggs were most common (76.4 %). Renesting occurred only if the first clutch was lost very early in the cycle. Mean brood size at fledging was 2.14 young per successful nest (n = 370), and overall productivity averaged 1.27 young per occupied nest (n = 623), i.e. including failed and non-breeding pairs. About 9.6 % of pairs (n = 469) were non-breeders, but they were all associated with nests. Their proportion increased somewhat in recent years, associated with increased density and the possibility that Osprey numbers may have reached their limit in some areas. Non-breeders were mostly young adults undertaking their first attempts at nesting, but some older adults with previous breeding experience failed to lay eggs in certain years. First-time breeders typically laid 2–4 weeks later than established breeders, and their clutches and broods tended to be smaller. Overall in my study area, at least 66.6 % of eggs laid (n = 1,161) hatched and some 54.5 % resulted in fledglings. Except for clutch size, which was unchanged during the DDT era, all reproductive data given are for 1974–1987.

Since 1974 most nest failures were attributable to inexperience of the numerous first-time breeders in the population and to environmental factors, such as food supply and weather, affecting the birds' foraging success.

Using their range of survivorship estimates, Henny & Wight (1969) calculated the production requirements for a stable Osprey population at 0.95–1.30 fledglings per breeding pair. They assumed that Ospreys started breeding at age 3 and that all individuals of three years and older bred. However, Michigan's Ospreys maintained their numbers during the early 1970s with a mean productivity somewhat less than 0.95 young per breeding pair. Following a rise in mean productivity above 1.0 in the mid-1970s, the statewide population started increasing in 1977 and doubled during the following decade (7 % mean annual growth rate). In the Lower Peninsula the increase was more moderate, about 60 % since 1976, or 5 % per year. From the empirical relationship between previous reproduction and annual changes in the size of the breeding population, and taking into account that Ospreys start breeding at different ages, Spitzer (1980) calculated an "adjusted recruitment productivity" for New England Ospreys of 0.79 young per breeding pair. My data suggest that this break-even value is close to the mark for Michigan too.

Fidelity to nest site (or territory) and mate are high, but "divorce" does occur occasionally, especially among young adults following unsuccessful breeding attempts. Moves to different territories by older individuals are uncommon, but more frequent in females than in males; the latter typically move only short distances (to a nearby territory). Adults establishing new territories are most likely 3–5 years old; if they stay paired in subsequent years, the age difference between pair members will remain small (0–2 years), as found on a small sample by Poole (1982).

However, although new pairings between two older individuals do occur following mate loss, the replacement is more often a young recruit just entering the breeding population. Thus the oldest individuals are often paired with very young birds. For example, in the past two seasons my oldest female (19+ years) was paired with a male at least 15 years her junior.

Assumptions and assessment of data

This analysis of lifetime reproduction in Ospreys rests on four sets of data for known-age (17) and uncertain-age (34) females and known-age (23) and uncertain-age (10) males. The known-age birds were banded as nestlings in the study area and were found breeding there three or more years later. The uncertain-age groups consist of individuals which were unbanded when first found breeding, but were believed to be young adults breeding for the first time. Such birds had either settled on a new territory, or had replaced previous known occupants on established territories. The greater proportion of females of uncertain age, compared to males, arose mainly because more of the females were immigrants to the study area, a finding expected from the greater dispersal distances of females from natal to breeding site (Greenwood 1980, Greenwood & Harvey 1982, Spitzer 1980). Data on known-age Ospreys ($n = 82$) in the study area showed that 55 % of first-time breeders were three years old, 33 % were four years, 10 % were five years, and 2 % were six years old, with no obvious difference between the sexes. Therefore most individuals in the two uncertain age categories were 3–5 years old when first found breeding. I found no evidence that two-year old Ospreys attempted breeding or nest-building.

I assumed that each male had fertilized the eggs and thus had sired the young in "his" nest, and that no "dump laying" by strange females occurred.

For each individual, lifetime production was taken as the total number of young raised to fledging during the period when that individual was present in the breeding population. Each bird was assumed to start breeding the first year it was identified at a nest containing eggs in the study area and to have stopped breeding when it had not been found for at least one year. The identity of some individuals, which could not be trapped in their first breeding year, was established later, if peculiarities of plumage or behaviour, or the colour of single ("hatching year") bands matched those observed in the preceding year(s). This occurred more often with males, which were more difficult to trap than females. Most females were successfully trapped during their first breeding season. A few individuals, both females and males, were apparently absent for one or two years, only to re-appear later. Some birds changed territories between years. Therefore, a few birds which I counted as dead may yet re-appear, or may have moved out of the study area permanently, making the lifetime production totals for some individuals underestimates. Probably this error was low for females and negligible for males.

Lifespan was counted from the year of hatching to the year that the bird was last identified. Many birds handled during the study were excluded from the analysis because they were still alive in the last breeding season (1987), or were already established breeders of unknown age at the start of the study, or because their production record was otherwise incomplete. It was therefore appropriate to evaluate how representative my samples were of the whole population, in terms of longevity.

Estimates of lifetime production were available for 17 known-age females and 23 known-age males. These two samples spanned nearly two-thirds of the known maximum lifespan for American Ospreys. The expected age distribution for my population was calculated from survivorship estimates of 48.5 % for the first year (Henny & Wight 1969), 82.1 % for the second and third years (calculated from data in Henny & Wight 1969), and 85.0 % for each subsequent year (my estimate from return rates of colour-banded individuals). These survival values are also used in subsequent analyses. The age structure of the sample of known age birds whose lifetime reproductive rates were known did not differ significantly from the expected age structure of the whole population (Fig. 18.1).

Results

LIFETIME REPRODUCTIVE OUTPUT

A wide range of lifetime fledgling productions was found for individuals of both sexes (Fig. 18.2), but mean productions (and variance) did not differ between males and females (Table 18.1). Some 22 % of females which laid eggs produced no young during their lifetime, and of those that did raise young, the number ranged from one to 29. Similarly, 12 % of males whose mates laid eggs reared no fledglings during their reproductive lives, and among those which successfully raised young the number ranged from one to 18. Relatively few birds raised only one young. This was because such individuals would have bred successfully only once, and broods of one occurred at lower frequency than broods of two or three. Approximately 16 % of females produced 50 % of the

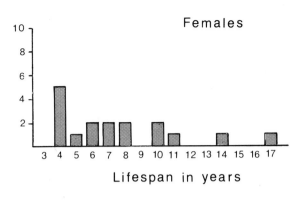

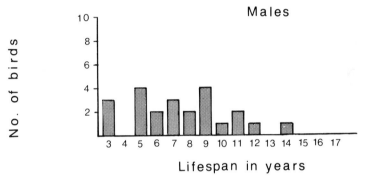

■ Observed   □ Expected

Figure 18.1 Age composition of the samples of 17 known-age females and 23 males for which estimates of lifetime production were obtained. The observed (■) age composition of breeders of both sexes did not differ significantly from the expected (□) age composition of breeders ($\chi^2_{22} = 9.85$, $P < 0.99$), calculated assuming that 55 % start breeding at age 3, 33 % at age 4, and 12 % at age 5, and survival rate 0.327 to 3 years (0.485×0.821^2, Henny & Wight 1969) and 0.85 for each subsequent year (this study).

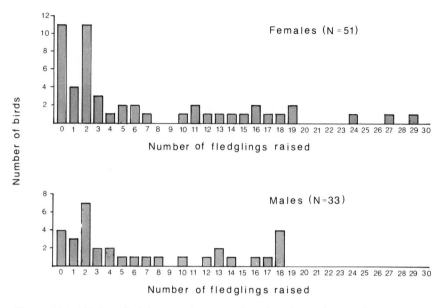

Figure 18.2 Lifetime fledgling productions of 51 females and 33 males.

Table 18.1 Mean number of fledglings produced by individual female and male Ospreys during their lifetimes.

	Total parent birds	Total fledglings	Mean lifetime production (fledglings per female or male)	Standardized variance[a]
Females of known age	17	115	6.8	—
Females of uncertain age	34	224	6.6	—
All females	51	339	6.7	1.099
Males of known age	23	160	7.0	—
Males of uncertain age	10	64	6.4	—
All males	33	224	6.8	0.918

[a] Calculated as $\dfrac{\sigma^2}{\bar{x}^2}$ (Wade & Arnold 1980).

young, and 28 % produced 74 % of young. Among males 24 % raised 51 % of young and 42 % raised 76 % of all young.

In a stable population with equal sex ratio, the female (or male) in each pair must during her (his) lifetime produce two offspring which survive to breed. How many fledglings must be produced to give rise to two recruits at ages 3–6? An estimate is calculated in Table 18.2 from the proportion of breeders starting at different ages and survival rates between fledging and first breeding attempts. The estimated number of 6.79 required fledglings agrees closely with the mean lifetime productions reported in this study (Table 18.1). However, the population was not stable during the study period, but grew from 32–36 pairs during 1970–1976 to 55 pairs in 1987 at an annual rate of about 5 %. This suggests that the mean lifetime production estimates, or the survivorship estimates, or both are slightly low (or there was net immigration). The production estimates may be slightly low for reasons given earlier, that some individuals may have entered or left the study area during their career as breeders, and thus raised some young elsewhere.

As for survival estimates, I believe that the adult rate is fairly accurate, in view of the close correspondence of values obtained by two independent methods: from band recoveries of birds found dead (83.8 %; Henny & Wight 1969) and from returns of individually marked adults (85.0 %; this study). The only available estimates of subadult survival rates are those from band recovery data, which tend to underestimate first-year survival, due to the greater vulnerability of young birds (Newton 1979).

Table 18.2 Estimate of minimum mean lifetime production of fledglings required to maintain a stable Osprey population. Assuming survival rate 0.327 to three years (0.485×0.821^2 from Henny & Wight 1969) and 0.85 for each succeeding year (Postupalsky, unpublished).

Year of first breeding	Proportion of breeders which started at different ages	Survival rate to breeding	Original fledglings which gave rise to breeders starting at different ages
3	0.55	0.327	1.682
4	0.33	0.327×0.85	1.188
5	0.10	0.327×0.85^2	0.423
6	0.02	0.327×0.85^3	0.100
			3.393

Number of fledglings required to give rise to two recruits starting to breed at different ages: $2 \times 3.393 = 6.79$.

CORRELATION WITH LONGEVITY

For both sexes lifetime production was correlated with longevity (Fig. 18.3). Production records to date for several known-age individuals 10 years old or older which were still alive in 1987 are also shown in Fig. 18.3 for comparison, but are not considered in further analysis. Up to three young were produced by females which survived only four years and may have bred only once. Ten or more young were recorded only for individuals which survived for more than eight years, and 20 or more young were recorded only for females which lived for at least 13 years. No males, whose entire lifetime reproductive outputs were known, produced more than 18 young. Two 12-year old males, still alive in 1987, have each produced 19 young to date. Total lifetime production varied widely within age groups, and some long-lived individuals produced no more fledglings than did other birds which lived only 3–7 years.

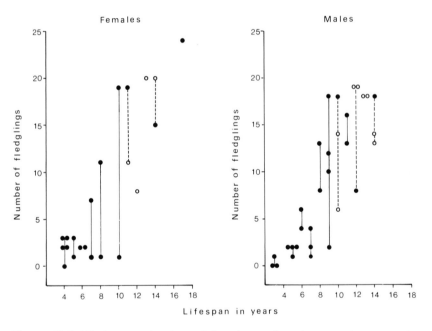

Figure 18.3 Lifetime production of females and males in relation to their longevity. Each spot (●) represents a different individual. Open spots (○) represent older (10+ years) individuals which were still alive in the last breeding season (i.e. have not stopped breeding) and are shown for comparison. Regression relationships of lifetime production (y) against age: females: $b = 1.675$, $r = 0.819$, $n = 17$, $P < 0.001$; males: $b = 1.732$, $r = 0.803$, $n = 23$, $P < 0.001$.

As Ospreys can raise broods of three young (broods of four are extremely rare), an individual which lives for 25 years could in theory produce 69 young in 23 breeding seasons. No age group attained the maximum potential for their age, but a few individuals came close. One female, which survived for 10 years, came closest by raising 19 young (maximum potential 24), and one male, which lived nine years, produced 18 young (maximum potential 21). Both these "high achievers" first bred at age three.

PROPORTION OF ONE GENERATION WHICH CONTRIBUTED TO THE NEXT

To calculate the proportion of any one Osprey generation which contributed genetically to the next requires the following information: (a) age-specific annual survival rates, (b) numbers of birds starting to breed at different ages (Table 18.3), and (c) the proportion of birds starting to breed at different ages which produced fledglings during their lifetime (Table 18.3). From the numbers first breeding at ages three, four, five or six years, I was able to calculate the number of fledglings from which each of the categories was derived, by working back using available survival estimates (Table 18.4 after Newton, 1985).

From fledglings which produced breeders starting at age three, 13.4 % produced young during their lifetimes. Corresponding figures for fledglings which first bred in their fourth, fifth and sixth year were 9.7 %, 2.9 %, and 0.7 % respectively, yielding an overall proportion of 26.7 %. From Table 18.4(b), I also calculated that 70.5 % of all fledglings failed to contribute offspring to the next generation because they died before they could attempt to breed, regardless at what age they would have started. The remaining 2.8 % survived long enough to attempt breeding, but

Table 18.3 Numbers of adults first found breeding at ages 3–6 respectively and numbers which produced fledglings.

Age at first breeding	Numbers (%) which started to breed at each age	Numbers for which estimates of life-time production were obtained	Numbers (%) which produced fledglings
3	45 (55)	18	15 (83)
4	27 (33)	12	12 (100)
5	8 (10)	8	8 (100)
6	2 (2)	2	2 (100)

Table 18.4 Calculations of proportions of non-breeders among different age groups and of proportions of fledglings which first bred at different ages and produced young themselves.

(a) Calculation of the proportion of fledglings which gave rise to birds which first bred at ages 3, 4, 5, and 6, respectively. The numbers of birds first found breeding at these different ages was known (underlined in the table) and all other values were calculated, assuming a 0.327 survival to age 3, and 0.85 in each succeeding year. B = breeding, NB = non-breeding.

Year of first breeding	Fledglings produced (% of total)	Numbers still alive (% breeding) in these years			
		3	4	5	6
3	137.6 (49.4)	<u>45</u> B (49.3)	38.3 B } (84.3)	32.5 B } (96.4)	27.6 B } (100)
4	97.1 (34.9)	31.8 NB	<u>27</u> B	22.9 B	19.5 B
5	33.9 (12.2)	11.1 NB	9.4 NB	<u>8</u> B	6.8 B
6	10.0 (3.6)	3.3 NB	2.8 NB	2.4 NB	<u>2</u> B

(b) Proportion of fledglings which first attempted breeding at 3, 4, 5, and 6 years, respectively, and which produced young.

Year of first breeding	(i) Original fledglings which gave rise to breeders starting at different ages (from (a) above)	(ii) Survival per year to breeding	(iii) % of breeders which started at different ages and produced young (from Table 18.3)	% of original fledglings which produced young themselves (i × ii × iii)
3	49.4	0.327	83 (61)[a]	13.4 (9.9)[a]
4	34.9	0.327×0.85	100	9.7
5	12.2	0.327×0.85^2	100	2.9
6	3.6	0.327×0.85^3	100	0.7
				26.7 (23.2)[a]

[a] Assumes that seven of 18 of breeders, which started at age 3, produced no young during their lifetime (see text).

failed to raise any young. In my study 54.5 % of eggs laid ($n = 1,161$) subsequently resulted in fledglings. Therefore the proportion of eggs which contributed to the next generation of Osprey fledglings was only 14.6 % (26.7×54.5 %).

The data upon which the above calculations are based contain one bothersome anomaly: in the known-age sample ($n = 40$) used here, only three individuals (7.5 %) produced no young in their lifetime, while in the uncertain-age sample ($n = 44$) a much greater proportion—27 %, or 12 individuals failed to produce fledglings. I can offer no compelling biological reason for this difference and regard it as a random occurrence. This means that the proportions of breeders which produced fledglings in the righthand column in Table 18.3 are probably overestimates, and this bias carries over to the two righthand columns in Table 18.4(b) and to some subsequent calculations. Of the 84 individuals (both sexes, known-age and uncertain-age) for which lifetime reproductive output was obtained, 15 (18 %) produced no fledglings. Applying this proportion to the known-age sample of 40 birds yields seven non-producers. In Table 18.4(b) these seven non-producing breeders would have the greatest effect if all were birds starting to breed at age three. This is a "worst case scenario", as it is improbable that all seven birds would have started breeding at the same age. In that case only 61 % of this cohort would have produced fledglings, or 9.9 % of the original fledglings which gave rise to breeders starting at age three, and the summed proportion of all original fledglings starting at ages 3–6 becomes 23.2 %. This means that only about one-quarter (23.2–26.7 %) of the fledglings in any one generation contributed offspring to the next generation. The proportion of fledglings which lived long enough to start breeding, but failed to produce fledglings themselves becomes 6.3 %, and the proportion of eggs laid which contributed to the next generation of Osprey fledglings becomes 12.6 %. The two sets of values obtained should be viewed as maximum and minimum estimates.

RECRUITMENT TO THE BREEDING POPULATION

Evolutionary fitness of an individual can be defined as the proportionate contribution of its offspring to the next generation. Therefore lifetime production should really be measured, not by the number of fledglings, but by the number of recruits which enter the breeding population. Of 339 fledglings produced by 51 female Ospreys in this study, only 38 (13 females, 25 males) were later found breeding in the study area, and of 224 fledglings produced by males, only 24 (11 females, 13 males) were found breeding later. If fledglings produced in recent years, i.e. those which would not have reached breeding age by 1987, are subtracted, the proportions which became breeders were 12.3 % for recruits produced by females and 12.2 % for those produced by males. As the number of recruits found in the study area was well below the number expected to survive to breeding at ages 3–6, it is clear that some recruits will have settled to breed outside the study area. This is supported by a report of one of my colour-banded Ospreys (sex unknown) breeding in Wisconsin, and by my recent trapping of a three-year old banded female in the study area; this bird fledged in Wisconsin, 420 km away. Of the recruits found breeding in the study area, 24 (39 %) were females and 38 (61 %) were males. This ratio in favour of males is probably a result of the longer dispersal distances of females (Greenwood & Harvey 1982, Spitzer 1980).

The number of recruits from individual parent birds (Table 18.5) was correlated with the number of fledglings they had produced (excluding parent birds which raised no fledglings). Some 27 % of females which fledged 1–10 young produced recruits (none more than one), compared with 67 % of females which raised 11–20 fledglings (up to four recruits),

Table 18.5 Relationship between numbers of fledglings produced and numbers of recruits breeding in the study area.

	Females which produced		
	1–10 fledglings	11–20 fledglings	21–29 fledglings
% females which produced recruits	27 ($n = 26$)	67 ($n = 12$)	100 ($n = 3$)
Total number (%) of recruits	7 (18)	20 (53)	11 (29)

	Males which produced	
	1–10 fledglings	11–18 fledglings
% males which produced recruits	26 ($n = 19$)	80 ($n = 10$)
Total number (%) of recruits	7 (29)	17 (71)

Note: Comparing birds in each category which did, or did not, produce recruits: females, $\chi^2_2 = 9.21$, $P < 0.01$; males, $\chi^2_1 = 5.62$, $P < 0.1$. Regression relationship between the number of recruits (y) and number of fledglings (x) produced by individuals: females, $b = 0.113$, $r = 0.674$, $n = 16$, $P < 0.001$; males, $b = 0.148$, $r = 0.668$, $n = 21$, $P < 0.001$. Birds producing no fledglings were excluded.

and all females which reared 21 or more fledglings (up to seven recruits). A similar pattern prevailed among males: 26 % of individuals which had raised 1–10 fledglings produced recruits (1–2), while 80 % of those which fledged 11–18 young produced recruits (up to five).

The correlation noted between parental longevity and lifetime output of fledglings also extended to recruits produced by individual parent birds (Table 18.6). Most recruits were produced by long-lived parents.

Discussion

The principal findings in this study were that (a) 51 female Ospreys produced between 0 and 29 fledglings in their lifetimes, and 33 males produced between 0 and 18 fledglings; (b) the lifetime fledgling productions of 17 known-age females and 23 known-age males were correlated with

Table 18.6 Relationship between parental longevity and number of recruits settling to breed in the study area.

	Longevity of females	
	3–9 years	*10–17 years*
% females which produced recruits	27 (*n* = 11)	100 (*n* = 5)
Total number (%) of recruits	3 (21)	11 (79)
	Longevity of males	
	3–7 years	*8–14 years*
% males which produced recruits	20 (*n* = 10)	73 (*n* = 11)
Total number (%) of recruits	2 (10)	18 (90)

Note: Comparing birds in each category which did, or did not, produce recruits indicated a significant difference in production associated with longevity: Fisher's Exact Test, females, $P = 0.013$; males, $P = 0.021$. Regression relationship between the number of recruits produced by individuals (y) and their longevity (x): females, $b = 0.161$, $r = 0.482$, $n = 17$, $P < 0.05$; males, $b = 0.252$, $r = 0.551$, $n = 23$, $P < 0.01$.

longevity; (c) the numbers of young raised to fledging were correlated with the numbers subsequently recruited to the local breeding population; (d) no more than 13–14 % of eggs laid, and no more than 23–27 % of fledglings, contributed to the next generation of fledgling Ospreys; (e) mortality before attaining breeding age prevented most individuals (70.5 %) from contributing fledglings to the next generation, while failure to raise young after breeding was attempted affected a much smaller proportion of individuals (3–6 %).

These figures are remarkably similar to those previously produced for European Sparrowhawks *Accipiter nisus* by Newton (1985), despite the differences in survival and reproductive rates between the two species. With a maximum life span (25 years) 2.5 times as long as the Sparrowhawk's (10 years; Newton 1985), higher annual survival rate, and delayed reproductive maturity, the Osprey invests relatively less in each breeding attempt. Early breeding attempts in particular tend to be less productive and may be preceded by nestings with no eggs laid. In my lifetime production sample, only one three-year old female raised three fledglings on her first breeding attempt (the same individual which raised 19 young in her lifetime). No three-year old males raised three young, but three four-year old males did. Reproductive success of an individual Osprey is probably influenced by the "quality" and previous experience of its mate, as well as by environmental factors, such as food supply and weather. These other factors are still under investigation.

Acknowledgements

My research on Michigan Ospreys was supported by Conservation for Survival, the National Audubon Society, and the Michigan Nongame Wildlife Fund. Additional support was provided by the Thunder Bay Audubon Society (Alpena), Alpena Power Company, Petoskey Regional Audubon Society, US, Inc., and Chippewa Nature Center, Inc. (Midland). K. Baldwin, T. U. Fraser, T. V. Heatley, J. B. Holt, Jr, J. M. Papp, L. Scheller, S. M. Stackpole, personnel of the Michigan Department of Natural Resources, and others assisted in the field. I am also indebted to I. Newton for encouragement and critical advice which greatly improved this chapter.

References

Clapp, R.B., Klimkiewicz, M.K. & Kennard, J.H. 1982. Longevity records of

North American birds: Gaviidae through Alcidae. *J. Field Ornithol.* **53**: 83–124.

Greenwood, P.J. 1980. Mating systems, philopatry and dispersal in birds and mammals. *Anim. Behav.* **28**: 1,140–1,162.

Greenwood, P.J., & Harvey, P.H. 1982. The natal and breeding dispersal of birds. *Ann. Rev. Ecology & Systematics* **13**: 1–21.

Henny, C.J. & Van Velzen, W.I. 1972. Migration patterns and wintering localities of American Ospreys. *J. Wildl. Manage.* **36**: 1,133–1,141.

Henny, C.J. & Wight, H.M. 1969. An endangered Osprey population: estimates of mortality and production. *Auk* **86**: 188–98.

Levenson, H. 1979. Time and activity budgets of Ospreys nesting in northern California. *Condor* **81**: 364–9.

Newton, I. 1979. *Population Ecology of Raptors*. Berkhamstead: Poyser.

Newton, I. 1985. Lifetime reproductive output of female Sparrowhawks. *J. Anim. Ecol.* **54**: 241–53.

Österlöf, S. 1977. Migration, wintering areas, and site tenacity of the European Osprey *Pandion h. haliaetus* (L.). *Ornis Scand.* **8**: 61–78.

Poole, A. 1982. Courtship feeding and Osprey reproduction. *Auk* **102**: 479–92.

Postupalsky, S. 1977. Status of the Osprey in Michigan. In *Transactions of the North American Osprey Research Conference*, ed. J.C. Ogden, pp. 153–65. Washington, DC: USDI National Park Service.

Postupalsky, S. 1978. Artificial nesting platforms for Ospreys and Bald Eagles. In *Endangered birds: management techniques for preserving threatened species*, ed. S.A. Temple, pp. 34–45. Madison: University of Wisconsin Press.

Postupalsky, S. & Stackpole, S.M. 1974. Artificial nesting platforms for Ospreys in Michigan. In *Management of Raptors*, ed. F.N. Hamerstrom *et al.*, pp. 105–17. *Raptor Research Report* No. 2. Raptor Research Foundation, Inc, Vermillion, South Dakota.

Scharf, W.C. 1968. Improved techniques for trapping harriers. *Inland Bird Banding News* **40**: 163–5.

Spitzer, P.R. 1980. Dynamics of a discrete coastal breeding population of Ospreys (*Pandion haliaetus*) in the northeastern United States during a period of decline and recovery, 1969–1978. Unpublished Ph.D. dissertation, Cornell University, Ithaca, New York.

Wade, M.J. & Arnold, S.J. 1980. The intensity of sexual selection in relation to male behaviour, female choice and sperm precedence. *Anim. Behav.* **28**: 446–61.

19. Screech-Owl

FREDERICK R. GEHLBACH

The Eastern Screech-Owl *Otus asio* L. is a small, cavity-nesting bird of temperate and subtropical woodlands in the eastern half of North America (Johnsgard 1988). It eats small vertebrates of all classes, particularly birds, and many invertebrates, especially crayfish, beetles, crickets, and moths. The owls are monogamous, permanent residents that occupy 6–108 ha home ranges, produce a single brood annually, and live up to 13 years. During the breeding season the male provides most of the food, while the female incubates and broods the young. If the first nest fails at the egg stage, re-nesting usually occurs.

In central Texas, where I have studied *O. a. hasbroucki* (Ridgway) since 1967, egg laying starts about 23 March, and clutches contain 2–6 (usually 4) eggs. Incubation requires 30 days, while the nestling period is 27 days. The young begin to fledge about 20 May; but remain nearby, fed by their parents, until July–October. They then disperse an average of 1.7 km. Of those that survive, about 85 % attempt to breed at 10–12 months of age. The owls have a survival rate of 36 % in their first year, improving to an average of 69 % per year thereafter. Mean generation time is 2.8 years.

During 1976–1987 I concentrated on a population in suburban Waco, Texas, and a rural population 3 km away, using 20 nest boxes in 135 ha plots of culturally modified, deciduous woodland at each locale. The suburban nestboxes were used by the owls in essentially the same average annual proportion (68 %) as rural boxes (57 %, Wilcoxon signed-ranks, $P = 0.21$), although the suburban breeding population was numerically more stable from year to year because of greater environmental stability (Gehlbach 1988). There was no correspondence between annual densities in the two areas ($r_s = 0.18$, $P = 0.55$) and only minimal interchange of individuals (two of 219 fledglings changed areas, no adults).

Predation caused most rural nest failures (76 %), and desertion most of the suburban ones (65 %, Fisher's exact test, $P < 0.001$). In fact,

LIFETIME REPRODUCTION IN BIRDS
ISBN 0-12-517370-9

natural nest failures were so much higher in the rural area (71 % versus 35 % in suburbia, Wilcoxon two-sample test, $P = 0.001$), and vandalism of my nestboxes was so great there, that I obtained much more information on the suburban owls (Table 19.1). Moreover, males were caught less readily than females, so the present account focuses mostly on females.

Seventy-six females, 1–7 years old, furnished lifetime reproductive data. These birds enabled me to examine the hypothesis that a relatively few, long-lived females contributed the majority of fledglings and recruits to the breeding population (cf. Newton 1985, 1986). The ageing of unbanded birds, first encountered while nesting, depended upon the narrower, worn, fault-barred, primary feathers of yearlings as opposed to broader, fresher, non-faulted primaries of older birds. Since individuals first moult primaries about a year after hatching, older owls were aged only if they had been banded as nestlings or yearlings. All owls were presumed dead or emigrated, if they were missing from nestboxes or natural cavities, including those within about 2 km of the study plots, for three successive years.

One-third of each year's nesting females were yearlings that attempted to nest once and then disappeared. These I called one-timers in order to compare their productivity with that of yearlings which survived to nest two or more seasons. The latter were called long-lived. Disappearance of the one-timers was 2.2 times greater annually than the loss of all long-lived females. Among these survivors a few seemed to be especially productive in that they fledged more young per egg, regardless of their lifespan. Because the loss of eggs and nestlings is wasted energy, I termed the fledglings per egg value "production efficiency", and postulated that it contributes to lifetime reproduction independently of lifespan.

Results

SUBURBAN AND RURAL COMPARISONS

In suburbia, long-lived females were more successful in their individual nesting attempts than were the one-timers (Table 19.1). The number of nests found each year did not differ between the two groups ($F = 2.6$, $P = 0.16$), but the long-lived females produced more total fledglings than one-timers ($F = 7.8$, $P = 0.01$) and more per clutch ($F = 6.1$, $P = 0.03$) on an annual basis. In these respects there were no differences among years ($F < 1.1$, $P > 0.05$, in two-way ANOVAS). Conversely, in the rural area long-lived females were not significantly more successful than one-timers, possibly because the high rate of nest predation negated any

Table 19.1 Frequencies of successful (at least one fledgling) and unsuccessful nests of one-time (yearling) and long-lived (1–7 years old) female Eastern Screech-Owls in suburban and rural environments (column percentages in parentheses).[a]

Females	*Suburban nests (n = 90)*		*Rural nests (n = 28)*	
	Successful	*Unsuccessful*	*Successful*	*Unsuccessful*
One-time	18 (29)	19 (68)	6 (43)	8 (57)
Long-lived	44 (71)	9 (32)	8 (57)	6 (43)
Fisher's exact test,	$P < 0.001$		$= 0.35$	

[a] Also note the higher proportion of successful nests of long-lived females in suburbia (Fisher's, $P = 0.04$) and the higher proportion of successful nests there overall (Fisher's, $P = 0.05$), but no significant difference in the frequency of one-timer nests between the two areas (Fisher's, $P = 0.48$).

advantage of experience linked to longevity (Table 19.1). There was no difference between areas in the lifetime output of one-timers, and in both areas long-lived females produced the majority of fledglings (Table 19.2). Lifespan contributed significantly to fledgling production in both populations (Fig. 19.1).

The lifespan of 20 long-lived females in suburbia was 3.2 ± 1.6 years, while five rural counterparts averaged 2.9 ± 1.3 years ($\bar{x} \pm 1$ SD here and below). This difference was not significant (Wilcoxon two-sample test, $P > 0.50$). Furthermore, production efficiency did not differ significantly in the two populations (0.61 ± 0.33 in 49 successful suburban nests versus

Table 19.2 Lifetime fledgling production of one-time (yearling) and long-lived (1–7 years old) female eastern Screech-Owls in suburban and rural environments (column percentages in parentheses).[a]

	Suburban nests (n = 90)			*Rural nests (n = 28)*		
	Females	*Fledglings*	*Fledglings/ female*	*Females*	*Fledglings*	*Fledglings/ female*
One-time	37 (65)	52 (30)	1.4	14 (74)	19 (44)	1.4
Long-lived	20 (35)	124 (79)	6.2	5 (26)	24 (56)	4.8
Fisher's exact test,	$P < 0.001$			$= 0.02$		

[a] Also note no significant difference in the proportion of long-lived females in suburban versus rural areas (Fisher's, $P = 0.34$) and the greater number of fledglings produced per female in suburbia (Fisher's, $P = 0.05$).

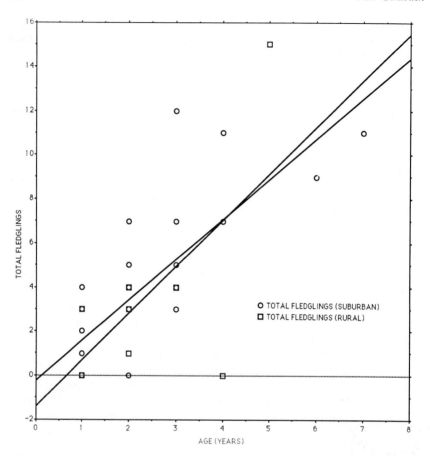

Figure 19.1 Fledgling production related to lifespan in female Eastern Screech-Owls. The equation for 57 suburban individuals is y = 1.82x − 0.20 (r^2 = 0.65, P < 0.001); that for 19 rural individuals is y = 2.10x − 1.38 (r^2 = 0.49, P < 0.001). The two regression coefficients are not significantly different (t = 1.1, P = 0.26). Duplicate symbols are omitted from the graph.

0.43 ± 0.35 in 14 rural ones, Wilcoxon, P = 0.09). Thus, the generally higher nest success in suburbia explains the greater lifetime productions of long-lived females there, compared to the rural area.

AGE-RELATED REPRODUCTION IN SUBURBIA

Long-lived females showed greater nest success in their first year than did the one-timers (Table 19.3), but other reproductive attributes of the

Table 19.3 Successful and unsuccessful nests of one-time yearling and surviving, female eastern Screech-Owls in suburbia (column percentages in parentheses).

Nests	One-time yearlings	Survivors as yearlings	Survivors ages 2–7
Successful	18 (49)	16 (80)	26 (84)
Unsuccessful	19 (51)	4 (20)	5 (16)
Fisher's exact test,		$P = 0.02$	0.50

two groups did not differ significantly (Table 19.4). In birds which survived beyond one year, there seemed to be some improvement in breeding success with age. Yet this was significant only for brood sizes and for the number of fledglings per egg, and only when comparing individuals in their second or third year with the one-timers. Body weights during incubation and brooding were also higher in second- and third-year birds compared to the one-timers (Table 19.4).

There was some indication that breeding performance began to deteriorate by age 4 (Table 19.4). Seven 5–7-year old females were too few for the multivariate analyses, but their average clutch size (3.6 ± 1.1) and brood size (2.4 ± 1.2) were lower than those of 2–4 year-olds. Furthermore, lifetime numbers of eggs and fledglings were more strongly influenced by inputs at ages two and three than by inputs earlier and later in life ($r^2 = 0.67$ for eggs, 0.60 for fledglings $P < 0.02$, in separate, multiple stepwise regressions).

I found no significant correlations between successive annual clutch sizes or brood sizes (r values between -0.31 and 0.21, $P > 0.05$) and no negative relationships between these parameters and longevity. This suggests that early breeding attempts were not "costly" to later ones and that productivity did not affect survival of the long-lived females (Reznick 1985). However, coefficients of variation of all the reproductive features in Table 19.4, except laying date, seemed to decline with age ($r_s \geq -0.80$, $P \leq 0.20$).

During the 12-year study, an average of 2.0 ± 1.6 fledglings were produced by 90 nest attempts, little different from the 1.9 needed for population stability as calculated by the method of Henny *et al.* (1970). All yearlings together averaged 1.6 ± 1.5 fledglings in 57 attempts, whereas older females averaged 2.7 ± 1.4 in 33 attempts ($F = 9.9$, $P = 0.002$). The significantly higher value of the older females was probably necessary for population stability, as the total number of fledglings they produced each year offset the low annual output of the yearlings ($r_s = -0.62$, $P = 0.03$).

Table 19.4 Annual reproductive features of one-time (yearling) and long-lived, female, Eastern Screech-Owls in suburbia, 1976–1987. Data are mean ±1 standard deviation of complete, initial clutches and include those that failed (sample sizes in parentheses). [a]

Features (univariate F, P)	One-time females (27)	Long-lived females at ages			
		One (20)	Two (17)	Three (9)	Four (8)
First egg date in March ($F = 1.7$, $P = 0.16$)	26.2 ± 9.0	25.9 ± 8.8	24.1 ± 5.9	19.0 ± 7.9	20.8 ± 5.5
Clutch size ($F = 1.8$, $P = 0.14$)	3.6 ± 0.9	3.7 ± 0.9	4.1 ± 0.7	4.3 ± 0.8	4.1 ± 0.4
Fledgling number ($F = 2.9$, $P = 0.03$)	1.4 ± 1.6	1.9 ± 1.3	2.6 ± 1.5[b]	2.7 ± 1.8[b]	2.5 ± 1.3
Fledglings/eggs (%) ($F = 2.8$, $P = 0.03$)	34.3 ± 39.7	53.4 ± 34.0	63.6 ± 36.9[b]	62.1 ± 34.5	60.6 ± 32.3
Incubation–brooding weight (g) ($F = 2.6$, $P = 0.04$)	169.2 ± 11.3	174.6 ± 9.9	176.0 ± 8.5[b]	179.1 ± 6.7[b]	175.4 ± 6.1[b]

[a] From a MANOVA, Wilk's lambda = 0.59, $F = 2.0$, $P = 0.009$ (data transformed).
[b] Significantly larger values compared only to one-timers (Least significant difference tests, $P < 0.05$; another MANOVA of just the four age-classes of long-lived females gives a Wilk's lambda of 0.64, $F = 1.4$, $P = 0.16$.

NEST SITES, MATES, AND FEMALE SIZE

Long-lived females not only retained the same mate from year to year, they also tended to re-use successful nest sites (including boxes) in suburbia. Perhaps a superior male provider and a desirable location promoted the success of the few females that were major fledgling producers. Also, lifetime reproductive success may have been linked to female weight, as heavier females laid larger clutches and fledged more offspring (Table 19.4). When these three potential determinants of reproduction were analysed simultaneously, I found that continued use of an initial, successful, nest site was most important to lifetime success; number of mates and weight were insignificant by comparison (Table 19.5).

Nestboxes near human dwellings, and with reduced shrub cover around them, had the lowest predation–desertion rates and were chosen most often by experienced owls (McCallum & Gehlbach 1988). The age of these long-lived females and their box-site fidelity was correlated (Table 19.5), so particular sites may be an important reason for the high frequency of successful nests (Table 19.1). Yet neither lifetime fledgling number nor production efficiency were related to distance from a house or shrub cover ($r < 0.32$, $P > 0.05$).

Heavier females were perhaps larger than lighter ones, since weights were not related to site fidelity (Table 19.5) and hence to site-specific food supplies (most prey were caught within 100 m of the nest). Heavy females could have laid more eggs and fledged more owlets (Table 19.4) because of more stored energy and greater incubation–brooding ability. Indeed, lifetime egg and fledgling outputs were positively related to

Table 19.5 Influence of nest site, mate, and body weight on lifetime fledgling production by 20, long-lived suburban, female, eastern Screech-Owls; results of a multiple stepwise regression analysis.

Determinants by step (all positive, x̄ ± 1 SD)	r^2 added	F	P
1. Nests at initial nestbox (2.6 ± 0.8)[a]	0.51	18.7	< 0.001
2. Incubation–brooding weight (174.2 ± 7.7 g)	0.06	2.2	0.155
3. Number of different mates (1.3 ± 0.5)	0.01	0.5	0.512
$R^2 = 0.58$		11.1	< 0.001

[a] This factor and weight are uncorrelated ($r = 0.37$, $P = 0.11$), but site fidelity is correlated with female age ($r = 0.78$, $P < 0.001$) and number of mates ($r = 0.46$, $P = 0.03$).

Table 19.6 Comparative importance of lifespan and mean incubation–brooding weight in determining the lifetime egg and fledgling productions of 20, long-lived, suburban, female, Eastern Screech-Owls; results of separate multiple stepwise regression analyses.[a]

Determinants by step (all positive, x̄ ± 1 SD)	r^2 added	F	P
Eggs (10.5 ± 4.2)			
Age	0.59	25.8	< 0.001
Weight	0.11	5.9	0.03
R^2 = 0.70		19.4	< 0.001
Fledglings (6.2 ± 3.3)			
Age	0.49	17.2	0.001
Weight	0.14	6.1	0.02
R^2 = 0.63		14.1	< 0.001

[a] Age and weight are uncorrelated ($r = 0.15$, $P = 0.53$).

weight, but much more strongly to lifespan (Table 19.6); while production efficiency was independent of both weight and lifespan ($r < 0.10$, $P > 0.42$). Thus, long-lived females in suburbia produced the most eggs and fledglings mainly because they were long-lived.

Females were not long-lived because they selected safe nest sites, for 27 one-timers used the same 16 nest boxes as the 20 long-lived females and had relatively fewer successful nests (Fisher's exact test, $P = 0.003$) and fledglings ($P < 0.001$) at these sites (cf. Table 19.2). The one-timers did not add significantly to explanations of variation in fledgling production among the nestboxes, although they and the long-lived owls nested in almost equal numbers. Only long-lived females explained the inter-box differences (Table 19.7). This suggests that nest sites had less influence on reproductive output than did the particular females involved, and that a small group of long-lived females was superior in reproductive performance.

POPULATION RECRUITMENT

Among the total of 57 suburban females, only 12 (21 %), all long-lived, fledged the majority of suburban offspring (55 %, Fisher's exact test, $P < 0.001$) and from just 40 % of all eggs laid (Fisher's $P = 0.004$). Their production efficiencies and lifespans exceeded those of the other eight, long-lived females (72.8 ± 17.3 % versus 39.1 ± 19.9 %, Wilcoxon two-sample test, $P = 0.007$; 3.9 ± 1.8 years versus 2.3 ± 0.5 years,

Table 19.7 Comparative importance of 20, long-lived and 27, one-time (yearling) female Eastern Screech-Owls to nest success and total fledgling production at 16 nestboxes used by both groups in 1976–1987; results of separate multiple stepwise regression analyses.[a]

Determinants by step (all positive, $\bar{x} \pm 1$ SD)		r^2 added	F	P
Successful nests (3.8 ± 1.6)				
Long-lived females		0.52	15.2	0.002
One-time females		0.11	4.1	0.06
	$R^2 =$	0.68	11.3	0.001
Fledglings (10.6 ± 5.1)				
Long-lived females		0.33	6.8	0.02
One-time females		0.06	1.2	0.30
	$R^2 =$	0.39	4.0	0.04

[a] Mean numbers of long-lived and one-time users are essentially the same (1.4 ± 0.6 and 1.7 ± 1.6 per box, respectively; Wilcoxon signed-ranks $P > 0.50$).

Wilcoxon, $P = 0.02$). Interestingly, three of four females, whose owlets became breeding members of the suburban population, were among the top 12; but their longevity did not differ significantly from that of the other survivors (4.0 ± 2.2 years versus 3.2 ± 1.6 years, Wilcoxon, $P = 0.46$).

Of five recruits in suburbia (2 % of 219 fledglings), four were from lineages with comparatively high lifetime productivities. The first was a male, fledged by a three-year old female in her second year. He paired with another yearling and produced a female recruit in his first of three seasons. She was killed by a car after a successful first year, but her production efficiency, averaged with that of her father and grandmother (71 %), exceeded the 58 % average of 18, possibly unrelated females.

The third recruit was another female, the offspring of a seven-year old female in her third year. This lineage's average production efficiency (70 %) was also higher than that of 19, possibly unrelated females (59 %). Similarly, the fourth recruit was a four-year old female that fledged a male recruit during her third year. In his first breeding season, paired with another yearling, he helped to fledge all of their clutch but was lost to a predator shortly thereafter. Taken together, the lineage's production efficiency was 87 % compared to 59 % in 19, possibly unrelated females.

From the three recruit-producing lineages, the seven owls were grouped for a comparison of production efficiency with 16, possibly-unrelated, long-lived females. The difference, though striking, only approaches statistical significance (75.2 ± 11.7 % versus 56.9 ± 27.0 %, Wilcoxon two-sample test, $P = 0.08$), probably because of the small sample. Among the 20, long-lived suburban females, production efficiency explained significant variation in the number of lifetime fledglings ($r^2 = 0.12$, $P = 0.03$), in addition to that simultaneously determined by longevity ($r^2 = 0.49$, $P = 0.001$, in a multiple stepwise regression), but only fledgling numbers explained significant variation in recruitment ($r^2 = 0.19$, $P = 0.04$).

The final record of recruitment involved a two-year old, rural female, whose female fledgling moved 3.3 km into suburbia and nested unsuccessfully one year before disappearing. Thus, 25 long-lived rural plus suburban females fledged all five recruits, while 51 one-timers from both areas left no descendants in either area. This difference is significant (Fisher's exact test, $P = 0.005$), although there is no significant difference when recruits and fledglings of the two groups of females are compared ($P = 0.14$, cf. Table 19.2).

Discussion

Five major findings emerged from this study: (a) lifespan contributed significantly to the number of fledglings per lifetime; (b) a few, long-lived females experienced the greatest nest success, even as yearlings, and produced the most offspring; (c) surviving females tended to improve in reproductive performance up to age 3 but were more productive at ages 2 and 3 only by comparison with one-time breeders; (d) particular nest sites, selected in the first year, characterized the high productivity of certain owls, but had less influence on lifetime production than did the females themselves; and (e) recruits to the suburban population were mostly members of a few lineages, in which survival from egg to fledging was high.

My first two results substantiate Newton's (1985, 1986) for *Accipiter nisus*, in which less than a quarter of the females produced the majority of fledglings, and longevity was the principal correlate of lifetime production. The third finding, concerning the age-related increase in reproductive performance, characterizes many birds (e.g. Curio 1983), including *A. nisus* (Newton 1986) and another cavity-nesting owl (*Aegolius funereus*; Korpimaki 1988a). An age-related decline in productivity is less well known generally, but has been found in both *A. nisus* and *A. funereus* and may obtain in *O. asio*.

First-year productivity was not low among all breeding females of *O. asio*, only among those that disappeared after their first breeding attempt.

The few survivors were uniformly heavier, and showed less variation in their annual reproductive outputs. They showed some tendency to improve in breeding performance with age, but not significantly so. Perhaps these long-lived individuals were a special subset of the suburban population. If so, they resemble *Parus major* (van Noordwijk *et al.* 1981), one of only two birds previously studied in a manner that distinguished among several hypotheses about age-related reproduction (Hamann & Cooke 1987).

My fourth finding about productive nest sites also applies to *A. nisus* (Newton 1986) and *A. funereus* (Korpimaki 1988b); but my fifth, about recruit-producing lineages, has not been investigated previously in birds of prey. The number of lifetime fledglings ascribed to the suburban females of *O. asio* was attributable to their longevity, but also to a greater production efficiency independently of lifespan (perhaps true of *A. nisus*; Newton 1986). Although only fledgling numbers explained recruitment, recruits were not produced by all long-lived *O. asio*, but by lineages with high production efficiencies. This suggests that recruitment is not wholly a function of lifetime fledgling production, and that lifespan and production efficiency provide different measures of fitness.

Acknowledgements

K. J. Gutzwiller, E. Korpimaki, D. A. McCallum, I. Newton and B. A. Pierce made constructive comments on this chapter, which is based on a life-history monograph currently being written.

References

Curio, E. 1983. Why do young birds reproduce less well? *Ibis* **125**: 400–4.
Gehlbach, F.R. 1988. Population and environmental features that promote adaptation to urban ecosystems: the case of eastern Screech-Owls (*Otus asio*) in central Texas. *Proc. Int. Orn. Congr.* **19**: 1809–1813.
Hamann, J. & Cooke, F. 1987. Age effects on clutch size and laying dates of individual female Lesser Snow Geese *Anser caerulescens*. *Ibis* **129**: 529–32.
Henny, C.J., Overton, W.S. & Wight, H.M. 1970. Determining parameters for populations by using structural models. *J. Wildl. Manage.* **36**: 690–703.
Johnsgard, P.A. 1988. *North American Owls*. Washington and London: Smithsonian Inst. Press.
Korpimaki, E. 1988a. Effects of age on breeding performance of Tengmalm's Owl *Aegolius funereus* in western Finland. *Ornis Scand.* **19**: 21–6.
Korpimaki, E. 1988b. Effects of territory quality on occupancy, breeding

performance, and breeding dispersal in Tengmalm's Owl. *J. Anim. Ecol.* **57**: 97–108.

McCallum, D.A. & Gehlbach, F.R. 1988. Nest-site preferences of flammulated owls in western New Mexico. *Condor* **90**: 653–61.

Newton, I. 1985. Lifetime reproductive output of female Sparrowhawks. *J. Anim. Ecol.* **54**: 241–53.

Newton, I. 1986. *The Sparrowhawk.* Calton: T. and A.D. Poyser Ltd.

Noordwijk, A.J. van, Balen, J.H. & Scharloo, W. 1981. Genetic and environmental variation in clutch size of the Great Tit. *Neth. J. Zool.* **31**: 342–72.

Reznick, D. 1985. Costs of reproduction: an evaluation of the empirical evidence. *Oikos* **44**: 257–67.

20. Ural Owl

PERTTI SAUROLA

The Ural Owl *Strix uralensis* is a greyish, medium-sized and relatively long-tailed bird of prey, which breeds in northern taiga forests of the Palearctic, ranging from Norway to Japan. South of this extensive main breeding range, some small isolated populations occur in mountain forests. The species breeds in coniferous, mixed or deciduous areas provided with suitable nest sites and foraging opportunities. However, ideal nest sites, big trees with cavities or chimney-like stubs, are lacking in forests in commercial use in wide areas across the potential breeding range. In the absence of tree cavities, Ural Owls may breed in disused twig-nests built by large raptors or Ravens *Corvus corax*, or even on the ground at the base of a tree, but the breeding success in these alternative nest-sites is poor.

The Ural Owl is a generalist feeder, eating a wide variety of vertebrates, from frogs to birds and mammals up to several hundred grams. However, for breeding the Ural Owl is a specialist, highly dependent on fluctuating microtine populations: in Fennoscandia mainly on the Field Vole *Microtus agrestis*, Bank Vole *Clethrionomys glareolus* and Water Vole *Arvicola terrestris* (Linkola & Myllymäki 1969, Korpimäki & Sulkava 1987).

In southern Finland, where my study was made, the median laying date is 6 April, when the ground is normally covered by snow, and the average clutch size is 3.0 (range 1–8, annual means varying from 2.0 in the worst year to 4.0 in the best). In years of low microtine populations, most pairs (up to 90 %) do not lay and the median laying date is up to four weeks later than in peak microtine years (e.g. Saurola 1989). Incubation and brooding together take roughly two months but, after leaving the nest, the young are fed by their parents for a further two and a half months, into August or September. On average 17 % of breeding attempts fail. Repeat clutches are seldom laid, and only if the failure occurs at the very start of the breeding season.

LIFETIME REPRODUCTION IN BIRDS
ISBN 0-12-517370-9

Both sexes can breed successfully at one year old (in their second calendar year), but do so only in ideal circumstances (see Saurola 1989). On average, about half the females must postpone their first breeding attempt until the age of four or more.

At the moment, the longevity record of a ringed Ural Owl is 20 years (three individuals), but as practically no Ural Owls were ringed 30 years ago, this record may soon be exceeded. Preliminary estimates for the average annual mortality, based on ring recoveries and retraps, indicate 40 % during the first year, 30 % during the second year, 20 % during the third and 15 % during the fourth and later years of life. The most important natural causes of death include starvation, which is highly dependent on winter food availability, and predation. Full-grown Ural Owls are eaten by Eagle owls *Bubo bubo*, and sometimes by Goshawks *Accipiter gentilis*, while eggs and nestlings are taken by Pine Martins *Martes martes*, whose depredations form one of the most important causes of nesting failure, apart from food shortage and harsh weather.

Ural Owls defend their nests aggressively against predators. During incubation and brooding, less than one half of all females fly away from the nest on disturbance. The others stay with the eggs or young and try to frighten the intruder away by bill snapping or by striking with their feet. From the time the young are two weeks old, the female stays near the nest and defends the young (including diving at the intruder, see below) with an intensity, which varies with the individual and with the external circumstances.

During the breeding season, as in most other raptors, there is a clear division of roles between the sexes: the male is smaller (weight 720 g) and provides all the food for the female and young from courtship to the third week of the nestling period, while the female (945 g) incubates, broods and feeds the young (see Lundberg 1986). Both sexes defend the nest, but the female is usually the more aggressive.

The Ural Owl is a year-round resident: nearly all males and 90–95 % of females stay near their previous nest sites even during the least favourable years (Saurola 1987). Ring recoveries show, however, that during very harsh winters, some females have moved more than 100 km away from their territories. If such emigrants survive, they may totally change their breeding area (two cases known).

Associated with the high nest site fidelity, members of a pair are highly likely to stay together in successive years: the divorce rate in the Ural Owl is less than 3 % (Saurola 1987). Only one case of polygyny is known, where a male had two females and broods 1.5 km apart.

Material

In the 1960s, Finnish conservationists, mainly amateur bird ringers, started to provide nestboxes for hole-nesting owls in compensation for the old trees and stubs cut down by foresters. Now, more than 12,000 nestboxes are available for owls, and are checked annually by bird ringers (Haapala & Saurola 1986).

This "nestboxes for owls" scheme started in southern Finland, around the city of Hämeenlinna (61° 00′N, 24° 30′E). In 1965, I joined these "Strix-fans" with my own nestbox area along the northern boundary of the initial one. The whole study area (including mine) covers about 3000 km², two-thirds of which is forest, mainly coniferous at different stages of succession, 10–15 % is lakes and 15–20 % is fields. At the start, only nestlings and a few brooding adults were ringed each year. In the late 1960s the original chimney-like nestboxes were changed to big "tit-type" ones, and practically all breeding females have been ringed (or retrapped) since. In 1973, catching of males was started, but during most years males were checked extensively only in my own part of the study area.

This chapter is based on all data obtained so far from the whole study area. Up to 1988, in total 1,643 active Ural Owl nests were recorded and almost 3,000 nestlings were ringed; in 1,200 cases (443 individuals) the female, and in 440 cases (219 individuals) the male was checked. The population increased during the 1960s, at least partly due to the provision of nest-sites, but from the early 1970s it remained at the same level, although the number of pairs which started to breed fluctuated greatly between years (Fig. 20.1). For further details see Linkola & Myllymäki (1969), Saurola (1980, 1989) and Pietiäinen et al. (1984, 1986).

Assessment of data for LRS analysis

For unbiased analysis of lifetime reproductive success, a representative sample of completely known life-histories is required. This means that only those cohorts from which all individuals have died can be included. For a long-lived species this requirement is seldom fulfilled, and could not be met here. For males our data were wholly inadequate, and for females the following criteria were used in accepting individuals for inclusion:

1) Females ringed as nestlings: all individuals hatched up to 1977 were included, providing that information on life-history was otherwise complete enough.

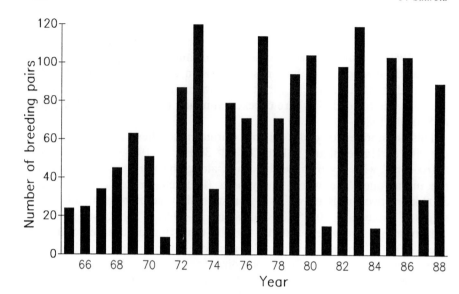

Figure 20.1 Number of breeding pairs 1965–1988.

2)   Females ringed as breeding adults: all individuals breeding for the first time up to 1980 were included, if the information was otherwise sufficiently complete.

Most of the females start to breed before their sixth birthday (Table 20.1). However, two females did not appear as breeders until the age of eight years and two others at nine years. These birds may have been genuine late-starters, or they may have bred elsewhere previously, so that their lifetime success would have been underestimated. Likewise, a small percentage of females may have left the study area after breeding there, which would have reduced their recorded lifetime success too.

Results

LIFETIME REPRODUCTIVE SUCCESS

In total 203 out of 443 females met the criteria given above, and 82 of them were ringed as nestlings, so were of precisely known age. The ages of nine females still alive (but included) were: 20 +, 3 × 15, 14, 12 +, 11, 10 + and 9 + years. Inevitably the lifetime productions of at least some of these birds will have been underestimated. On average, for 82

Table 20.1 Age distribution of females breeding for the first time (all birds ringed as nestlings).

Age	Number
1	2
2	8
3	24
4	24
5	9
6	9
7	2
8	2
9	2

females of known age, the life span was eight years and age at first breeding was four years, which implies a mean breeding lifespan of four years (Table 20.2). For all 203 females, however, the breeding lifespan was almost five years, which perhaps indicates that some of the females which apparently had started at the age of six or later (cf. above) had in fact bred elsewhere first.

The average lifetime production for all females included was 8.2 fledglings (range 0–35) or 0.7 known recruits (range 0–7; Table 20.2). For females which produced one or more fledglings, the correlation between number of fledglings and number of recruits was highly significant

Table 20.2 Mean breeding parameters for 203 Ural Owl females of known lifetime output.

	Mean	SD	n	Min.	Max.
Lifespan	7.9	4.0	82	1	17[a]
Age at first breeding	4.1	1.6	82	1	9
Breeding lifespan	4.9	3.8	203	1	20[a]
Breeding attempts[b]	3.9	2.9	203	1	14
Clutch size	2.9	0.6	203	1	5
Brood size	2.3	0.7	203	0	4
Number of fledglings	8.2	6.7	203	0	35
Number of (known) recruits	0.7	1.2	203	0	7
Laying date	6 Apr	7.9	197	5 Mar	3 May

[a] These figures differ because the breeding lifespan is based on a larger sample of individuals some of which were first caught as breeding adults.
[b] Egg-laying taken as a breeding attempt.

($r = 0.50$, $P < 0.001$), although there were some females which produced many fledglings but no known recruits.

The individual variation was wide: 23 % of breeding females produced 50 % of fledglings, and the top 5 % of females produced 16 % of fledglings (Figs 20.2 and 20.3). Using the average mortality estimates given above and the distribution of ages at first breeding in the 82 known-age females, I estimated, using the method of Newton (1985), that 72 % of all Ural Owl fledglings died before they could attempt to breed. This meant that 6.4 % of fledglings of one generation produced 50 % of fledglings in the next.

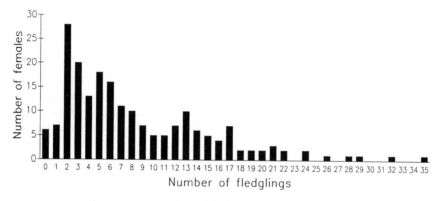

Figure 20.2 Number of fledglings raised per lifetime.

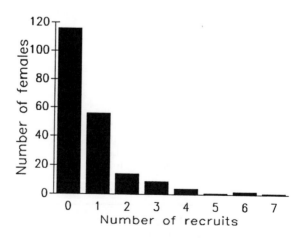

Figure 20.3 Number of known recruits produced per lifetime.

COMPONENTS OF VARIANCE IN LRS

For evaluating the contribution of different components to variance in LRS I used the method of Brown (1988), in which LRS is viewed as a product of lifespan, breeding frequency, fecundity and survival of offspring to various stages.

I made four analyses based on the following models:

(1) LRS (fl)   = LS × CY × EC × FE
(2) LRS (fl)   = BL × CY × EC × FE
(3) LRS (rec) = LS × CY × EC × FE × RF
(4) LRS (rec) = BL × CY × EC × FE × RF

where LRS (fl) = lifetime output of fledglings; LRS (rec) = lifetime output of recruits (known); LS = lifespan; BL = breeding lifespan; CY = clutches per year (related either to LS or BL); EC = eggs per clutch; FE = fledglings per egg; and RF = recruits per fledgling.

Analyses (1) and (3) included only those females which were ringed as nestlings and thus of known age. In case (1) at least one fledgling, and in case (3) at least one known recruit, was produced during a lifetime by those females included. Analyses (2) and (4) were similar, but also included females of unknown ages.

The main results from these four analyses (Tables 20.2–5) can be summarized as follows (for comparison, see Clutton-Brock 1988a):

(1) Lifetime Fledgling Production (> 0) of Known-age Females (Table 20.3a).

The most variable single components were LS (39 % of the variance of LRS (fl); see also Fig. 20.4) and CY (26 %). These components were also positively correlated (contribution of covariation 27 %), because the influence on CY of non-breeding years preceding the first breeding attempt was smaller the longer the bird lived.

(2) Lifetime Fledgling Production (> 0) of All Females (Table 20.3b).

Because in this data set real lifespan was substituted with breeding lifespan, the variance in LRS caused by the differences in age at first breeding was eliminated and BL remained as the only impдrtant source of variation (117 %) in LRS (fl). (The value of a component higher than 100 % means in Brown's method that this component is more variable in relation to its average than the product of all components, LRS itself.

Table 20.3 Variance components of lifetime production (see text).

a) Fledgling production of known-age females

Component	Original data			Upward partition		
	Mean	Variance	Standardized variance	Total	Independent component	Covariation component
LS	8.05	15.77	0.24	39		
CY	0.47	0.04	0.16	26		
EC	2.89	0.40	0.05	8		
FE	0.78	0.05	0.07	12		
LSCY	4.01	8.81	0.61	33	6	27
LSEC	23.60	164.21	0.30	2	2	
LSFE	6.21	11.50	0.29	−4	3	−7
CYEC	1.37	0.41	0.22	2	1	1
CYFE	0.36	0.02	0.18	−9	2	−10
ECFE	2.24	0.54	0.11	−3	1	−3
LRS(fl)	8.74	45.63	0.62			

b) Fledgling production of all females

Component	Original data				Upward partition	
	Mean	Variance	Standardized variance	Total	Independent component	Covariation component
BL	5.04	14.59	0.57	117		
CY	0.85	0.03	0.04	8		
EC	2.87	0.39	0.05	10		
FE	0.78	0.04	0.07	13		
BLCY	3.96	8.70	0.48	−27	4	−32
BLEC	14.55	132.13	0.63	2	6	−4
BLFE	3.82	8.89	0.57	−13	8	−21
CYEC	2.43	0.47	0.08	−1	1	−2
CYFE	0.66	0.05	0.11	2	1	1
ECFE	2.23	0.51	0.10	−3	1	−3
LRS (fl)	8.59	45.19	0.49			

c) Recruit production of known-age females

| Component | Original data | | | | Upward partition | |
	Mean	Variance	Standardized variance	Total	Independent component	Covariation component
LS	10.03	16.08	0.16	50		
CY	0.52	0.03	0.10	30		
EC	2.88	0.28	0.03	10		
FE	0.81	0.04	0.06	20		
RF	0.19	0.02	0.57	178		
LSCY	5.50	9.88	0.36	33	5	28
LSEC	29.43	173.23	0.21	4	2	3
LSFE	7.77	11.09	0.17	−17	3	−20
LSRF	1.58	1.33	0.38	−110	28	−139
CYEC	1.49	0.25	0.11	−6	1	−7
CYFE	0.41	0.02	0.12	−11	2	−13
CYRF	0.09	0.00	0.50	−50	17	−67
ECFE	2.32	0.49	0.09	−2	1	−3
ECRF	0.51	0.11	0.38	−69	6	−75
FERF	0.16	0.02	0.94	97	11	86
LRS (rec)	1.69	1.67	0.32			

d) Recruit production of all females

Component	Original data		Standardized variance	Total	Upward partition	
	Mean	Variance			Independent component	Covariation component
BL	6.79	15.58	0.34	131		
CY	0.81	0.02	0.03	13		
EC	2.91	0.29	0.03	13		
FE	0.79	0.03	0.06	21		
RF	0.20	0.02	0.48	188		
BLCY	5.25	9.24	0.30	−27	4	−31
BLEC	19.72	140.02	0.36	−5	5	−10
BLFE	5.19	9.83	0.34	−21	7	−28
BLRF	1.03	0.63	0.35	−184	63	−247
CYEC	2.35	0.32	0.06	−4		−4
CYFE	0.65	0.04	0.11	7	1	6
CYRF	0.17	0.02	0.78	103	6	97
ECFE	2.29	0.44	0.08	−3	1	−3
ECRF	0.55	0.13	0.40	−47	7	−53
FERF	0.16	0.02	0.71	67	10	56
LRS (rec)	1.75	1.63	0.26			

Table 20.4 Characteristics of females which produced 0, 1 and 2–7 known recruits.

	Number of recruits		
	0	1	2–7
Average wing-chord	361.1	364.3	365.7
Average level of nest defence	4.6	4.8	4.8
Median laying date	April 7	April 6	April 3

Table 20.5 Classification of female nest defence behaviour.

Rank	Number	Explanation
1	0	no reaction
2	6	hoots and bill-snaps from a distance of some metres
3	9	makes pseudo-attacks very near the intruder, but does not strike him
4	18	attacks and strikes once, but not more
5	64	attacks and strikes many times
6	17	attacks and strikes continuously, often before the intruder starts to climb; often also attacks during incubation

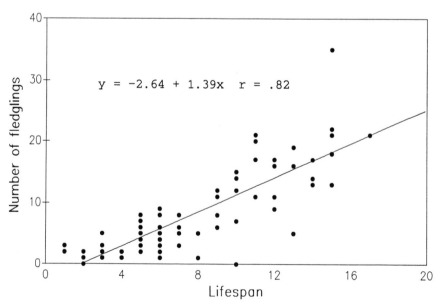

$y = -2.64 + 1.39x \quad r = .82$

Figure 20.4 Relationship between lifespan and number of fledglings produced.

This is possible if some other component(s) counteract the component concerned.) Negative covariation between BL and CY (− 32 %) was mainly caused by females which bred only once or in two successive good years and no more. The other negative covariation, between BL and FE (− 21 %) resulted from the probability of nesting failure becoming higher as the number of nesting attempts increased.

(3) Lifetime Recruit Production (> 0) of Known-age Females (Table 20.3c).

The contributions of LS (50 %) and CY (30 %) to variance in LRS (rec) were as significant as they were to variance in LRS (fl), but much less than the contribution of RF (178 %!). Of the two-way contributions, the most remarkable was the negative covariation (−139 %) between LS and RF. It could have been caused mainly by the fact that, because individuals without any recruits were excluded, a few lucky females, in spite of their short lifespan, produced one recruit with an artificially high probability.

(4) Lifetime Recruit Production (> 0) of All Females (Table 20.3d).

RF (188 %), BL (131 %) and their negative covariation (− 247 %) were the most important contributors to variance in LRS (rec) in this data set.

It was also possible to estimate the contribution to variance in LRS of those fledglings which totally failed to produce fledglings themselves: 72 % of fledglings which died before they could breed and those six females which attempted to breed but failed (see Brown 1988). Their contribution was a little bit more than half of the total variance in LRS (e.g. 54 % if calculated on the basis of Table 20.3b).

FACTORS AFFECTING LRS

Lifetime reproductive success of an individual may depend on several morphological, behavioural and environmental factors (Clutton-Brock 1988b). In Ural Owls, such factors may include body-size, nest defence behaviour, breeding phenology, territory quality and conditions in the year of first breeding (and/or birth). As yet, little can be said about the roles that these different factors play in lifetime production in the Ural Owl. However, I ranked the females in three classes according to their production of (known) recruits and compared these classes on the basis of the factors mentioned above (Table 20.4).

Body Size

Wing-chord measurements were available only for 66 of the 203 females included in this analysis. These birds revealed a significant difference ($P = 0.03$) in wing-chord between classes 0 and 2, so successful females seemed to be slightly bigger than unsuccessful ones (Table 20.4).

When females were divided into groups, small and big, the lifespan of smaller birds was found to be significantly shorter ($P < 0.05$). Further, offspring survival from fledgling to recruitment was lower if the female was small ($P < 0.05$). No significant differences were found between small and big birds in average clutch size, breeding phenology or age at first breeding.

Nest Defence

The behaviour of 114 females during chick ringing was ranked (either based on field notes or on memory) in six categories (Table 20.5). More than 50 % of females fell in the second-most aggressive category, and perhaps partly for this reason no connection was found between the level of nest defence and lifetime production (Table 20.4).

Breeding Phenology

In order to eliminate the effect of those females which bred only in one or two (often early and productive) years, only females which had attempted to breed at least three times were selected for this analysis. They were grouped in three categories (early, middle and late) on the basis of the average deviation of their laying dates from respective annual medians of the whole population (Table 20.4). In addition, a significant difference between early and late females was found in production of fledglings and in production of recruits.

Annual Differences in Recruitment

Recruitment of fledglings varied greatly from year to year, but not in simple relation to the quality of year (Fig. 20.5, cf. Fig. 20.1). High recruitment rates were observed for cohorts hatched before (e.g. 72, 79, 82, 85) and during (66, 69, 73) a microtine peak, but also for small cohorts hatched during a microtine low (74, 84).

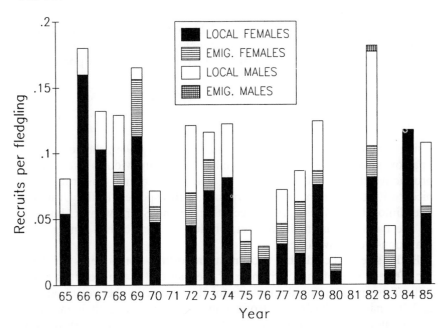

Figure 20.5 Number of known recruits per fledgling produced in different years. Recruits to local population (sexes separately) are separated from emigrants.

Discussion

COMPONENTS OF LIFETIME REPRODUCTIVE SUCCESS

The most variable components of lifetime production in Ural Owl females were recruitment rate (RF), lifespan (LS) and the number of clutches per year (CY). The contributions of clutch size and egg/nestling survival were small in all data sets studied. In the analysis which stopped at the fledgling-stage, most of the variation in lifetime production could be attributed to the LS and CY, or simply breeding lifespan in the case of birds not known by age. In general, the contributions of different single components were similar to those given for the Great Tit *Parus major* by van Balen *et al.* (1987).

Is it appropriate to incorporate the recruitment rate of fledglings in the model, if the information on recruits is incomplete? I assumed that the Ural Owl population has stayed around the same level, and concluded that one-third of recruits are known to me (0.7 versus 2.0 recruits per female in a stable population). I believe that this sample of one-third of all recruited offspring is random and big enough to give an unbiased estimate of the variation in this component, and further, that the most revealing models were those in which recruitment was incorporated.

FACTORS AFFECTING LIFETIME REPRODUCTIVE SUCCESS

Body Size

Using wing-chord as an indicator of size, Newton (1989) showed that, in Sparrowhawk *Accipiter nisus* females, large individuals were more productive than small ones, while in males the reverse was true, fitting nicely with the extreme size dimorphism of this species. In the Ural Owl, size dimorphism is less pronounced than in the Sparrowhawk, but still, on the basis of this small data set, productive females had longer wings than non-productive ones.

Body size affected the two most important components of LRS, recruitment rate of the offspring and lifespan of the female. It is not so difficult to speculate on the reasons for the connection between body size and lifespan: big birds are stronger in competition and their metabolic rate is more efficient. But this does not explain the better survival after independence of the offspring of big females, unless there is an inherited component to body size, giving offspring the same competitive advantage as their mothers.

Nest Defence Behaviour

Most Ural Owl females defend their nestlings vigorously. In Finland, at least five Ural Owl females have lost their lives through hitting a ringer wearing a hard helmet. Such risky defence behaviour should be adaptive: "aggressive" females should be more successful than "cowards". But no differences in LRS were found which could be attributed to the different defence levels of females. Maybe the classification of defence behaviour was not the best possible, and a more detailed and accurate division of aggressive behaviour might give a different result. However, the female which produced most fledglings (35) and was the second best in producing recruits (6) showed only moderate nest defence, diving at the intruder but never touching him. This example shows that, if there is a positive relationship between nest defence and LRS, there is at least one exception to the rule.

Breeding Phenology

In the Ural Owl, as in many other bird species, there is a significant negative correlation between the laying date and clutch size and, consequently breeding success, both between and within years (e.g. Saurola 1989). From this it follows that the LRS of average early breeders should be higher than the LRS of the late ones, as was found above, at least if we stop the analysis at the fledging stage. But even the survival of fledglings to recruits was higher if the females were early breeders. The breeding season of the Ural Owl is long, almost five months from egg-laying to independence of offspring, which means that the difficult conditions of winter soon become a reality for a young independent Ural Owl. It would not be astonishing if those young owls, which fledge early in the season and have more experience before the onset of harsh conditions, survive better than those which fledge late.

Quality of Territory

In the Sparrowhawk, quality of territory was an important factor affecting LRS (Newton, Chapter 17). Because the Ural Owl lives much longer and stays, if possible, in the same territory all its life, I could not separate the effect of territory from the effect of female. The reason for the high site fidelity is probably the importance of keeping a nest hole once found (Lundberg 1979).

In addition to hunting and nesting opportunities, predation pressure may also influence territory quality. The low recruitment rate of some females which produced many fledglings could have been due to heavy predation of the young in the post-fledging period by Eagle Owls or Goshawks.

Conditions in the Year of First Breeding

Breeding performance in the Ural Owl is highly dependent on voles, but not so much on their numbers as on their availability. The Ural Owl is much less adept at catching voles by diving through a deep snow layer than is the Great Grey Owl *Strix nebulosa*. Thus feeding conditions in a given year do not depend solely on vole numbers but on the duration, depth and hardness of snow cover.

In the Great Tit *Parus major*, recruitment varied greatly from year to year in relation to the beech mast crop, and since most individuals breed in only one year, LRS is highly dependent on the year of first reproduction (or birth) (van Balen *et al.*, 1987). Probably it would be advantageous to a Ural Owl to be hatched one year before a peak year for voles, but not to start to breed as a one year old (Saurola 1989). This is because the extent of annual moult in the Ural Owl seems to depend on the level of reproductive effort (Pietiäinen *et al.* 1984); and it must be important for a one-year old bird to change its low quality juvenile flight feathers before the second winter.

Also, the proportion of fledgling Ural Owls which were recruited varied greatly between cohorts from different years. However, the preliminary examination showed no simple correlation between recruitment and number of voles. Nor was LRS greater in females which started to breed one year before a peak in voles than in those which started during a peak and thus before a decline. High producers of a long-lived species such as the Ural Owl experience anyway many different kinds of breeding seasons and for this reason the quality of the first season is not very important for lifetime production.

Summary

Lifetime reproductive output of Ural Owl females was highly variable: 0–35 fledglings in different females; 50 % of fledglings were produced by 23 % of breeding females or by 6 % of fledglings from the previous generation. Assuming a stable population, one third of all recruits was found, and 50 % of these were produced by 11 % of females.

The recruitment rate of fledglings contributed most strongly to variance in lifetime reproductive rate, followed by lifespan and number of breeding attempts per unit length of life.

Larger females lived longer, and produced more recruits per fledgling, and as a result had higher lifetime reproductive output, than did small ones. Females which bred, on average, early in the season were more productive than those which bred later.

Acknowledgements

In addition to my own data, I used results from long-term field-work by Juhani Koivu, Timo Larm, Pentti Linkola and Väinö Valkeila and their assistants. Heikki Lokki kindly wrote the computer programs needed for data analysis, and Ian Newton edited my manuscript to its present form.

References

van Balen, J.H., van Noordwijk, A.J. & Visser, J. 1987. Lifetime reproductive success and recruitment in two Great Tit populations. *Ardea* **75**: 1–11.
Brown, D. 1988. Components of lifetime reproductive success. In *Reproductive*

Success, ed. Clutton-Brock, T.H., pp. 439–453. Chicago: University Press.

Clutton-Brock, T.H. (ed.) 1988a. *Reproductive Success*. Chicago: University Press.

Clutton-Brock, T.H. 1988b. Reproductive success. In *Reproductive Success*, ed. Clutton-Brock, T.H. pp. 472–485. Chicago: University Press.

Haapala, J. & Saurola, P. 1986. Breeding of raptors and owls in Finland in 1986 (in Finnish with English summary). *Lintumies* **21**: 258–67.

Korpimäki, E. & Sulkava, S. 1987. Diet and breeding performance of Ural Owls under fluctuating food conditions. *Ornis Fennica* **64**: 57–66.

Linkola, P. & Myllymäki, A. 1969. Der Einfluss der Kleinsäugerfluktuationen auf das Bruten einiger kleinsäugerfressender Vögel in südlichen Häme, Mittelfinnland 1952–1966. *Ornis Fennica* **46**: 45–78.

Lundberg, A. 1979. Residency, migration and compromise: adaptations to nest-site scarcity and food specialization in three Fennoscandian owl species. *Oecologia* (Berl.) **41**: 273–81.

Lundberg, A. 1986. Adaptive advantages of reversed sexual size dimorphism in European owls. *Ornis Scand.* **17**: 133–40.

Newton, I. 1985. Lifetime reproductive output of female Sparrowhawks. *J. Anim. Ecol.* **54**: 241–53.

Newton, I. 1989. Individual performance in Sparrowhawks: the ecology of two sexes. *Proc. Int. Orn. Congr.* **19**: 125–154.

Newton, I. 1989. Sparrowhawk. In *Lifetime Reproduction in Birds*, ed. I. Newton. London: Academic Press.

Pietiäinen, H., Saurola, P. & Kolunen, H. 1984. The reproductive constraints on moult in the Ural Owl *Strix uralensis*. *Ornis Fennica* **17**: 309–25.

Pietiäinen, H., Saurola, P. & Väisänen, R. 1986. Parental investment in clutch size and egg size in the Ural Owl *Strix uralensis*. *Ornis Scandinavica* **17**: 309–25.

Saurola, P. 1980. *Strix uralensis* D-21037 in memoriam (in Finnish with English summary). *Lintumies* **15**: 121–8.

Saurola, P. 1987. Mate and nest-site fidelity in Ural and Tawny Owls. In *Biology and Conservation of Northern Forest Owls: Symposium Proceedings*, February 3–7, 1987, Winnipeg, Manitoba, eds Nero, R.W., Clark, R.J., Knapton, R. J. & Hamre, R.H. USDA Forest Service. Gen. Tech. Rep. RM-142.

Saurola, P. 1989. Breeding strategy of the Ural Owl. In *Raptors in the modern world*, ed. B.U. Meyburg & R.D. Chancellor, pp. 235–240. Berlin: World Working Group on birds of prey and owls.

Part V. Long-lived Waterfowl and Seabirds

Two waterfowl species are represented in this section, the Barnacle Goose *Branta leucopsis* studied chiefly on its wintering grounds in Britain, and the Mute Swan *Cygnus olor*, in which both dispersed (territorial) and colonial populations were studied in Denmark. Both species are long-lived, with some individuals reaching more than 20 years of age, and both are monogamous, with long-term pair bonds. Moreover, in both species LRS is greatly affected by the incidence of severe weather.

In the Barnacle Goose many individuals are prevented from breeding in some years by harsh spring conditions on the arctic breeding grounds; and despite a long potential lifespan, few (if any) birds breed successfully more than four times in their lives. Success is helped by a stable pair bond and, after a mate change (resulting from a death or divorce), breeding is often depressed for one or more years in otherwise experienced individuals. The findings from this study, by Owen & Black, show many parallels with those from arctic-nesting Snow Geese *Anser caerulescens* (Cooke & Rockwell 1988) and Bewick's Swans *Cygnus columbianus* (Scott 1988).

In the Mute Swan, Bacon & Andersen-Harild show how LRS is affected by the severity of winter weather. This is reflected in several aspects of performance, including the likelihood of nesting, first-year survival and age of first breeding. In any one year success also differs between nesting situations, being much higher in territorial pairs than in colonial ones. The latter bred so poorly that their numbers could be maintained only by the continued immigration of young from the more successful territorial sector.

The two seabird species represented in this section include the inshore-feeding Red-billed Gull *Larus novaehollandiae* of New Zealand the pelagic Short-tailed Shearwater *Puffinus teniurostris* of Australia. These studies complement two previous studies of LRS in long-lived seabirds, on the Kittiwake *Rissa tridactyla* (Thomas & Coulson 1988) and Fulmar *Fulmarus glacialis* (Ollason & Dunnet 1988) respectively.

The Red-billed Gull, studied by Mills, is unusual in having a marked surplus of females among birds of breeding age. This leads not to widespread polygyny, but to the formation of many female–female pairs. Not unexpectedly, same-sex-pairing lowers the fitness of the individuals concerned, though some such pairs manage to produce young through extra-pair copulations. Despite a long-life and high potential breeding success, lifetime production is generally low, even in male–female pairs, owing to high chick loss. Nonetheless, the population continues to grow.

The Short-tailed Shearwater has been studied on Fisher Island near Tasmania for more than 40 years, so even though individuals reach more than 30 years of age, many complete lifetimes are represented in this study by Wooller, Bradley, Skira & Serventy. The species shows extreme delay in first breeding, up to 15 years in some individuals, and frequent non-breeding years among established breeders. In its general life history, it shows many parallels with the Fulmar and other long-lived seabirds.

References

Cooke, F. & Rockwell, R.F. 1988. Reproductive success in a Lesser Snow Goose population. In *Reproductive Success*, ed. T.H. Clutton-Brock, pp. 237–50. Chicago: University Press.

Ollason, J.C. & Dunnet, G.M. 1988. Variation in breeding success in Fulmars. In *Reproductive Success*, ed. T.H. Clutton-Brock, pp. 263–78. Chicago: University Press.

Scott, D.K. 1988. Reproductive success in Bewick's Swans. In *Reproductive Success*, ed. T.H. Clutton-Brock, pp. 220–36. Chicago: University Press.

Thomas, C.S. & Coulson, J.C. 1988. Reproductive success of Kittiwake Gulls, *Rissa tridactyla*. In *Reproductive Success*, ed. T.H. Clutton-Brock, pp. 251–62. Chicago: University Press.

21. Barnacle Goose

MYRFYN OWEN & JEFFREY M. BLACK

The Barnacle Goose *Branta leucopsis* breeds in colonies on small islands or cliff faces between 70° and 80° N in east Greenland, Svalbard and western Siberia. There are three discrete populations. That breeding in Greenland winters in western Scotland and Ireland; that from Svalbard winters on the Solway Firth in northern Britain, whereas the largest population from Siberia migrates through the Baltic States to the Netherlands. This account relates to the Svalbard–Solway population which has increased from a few hundred birds in the late 1940s to 11,400 individuals in 1987. The geese have been studied each year on their wintering grounds since 1970 and periodically on their breeding and migration staging areas. Males are larger and heavier than females (1,800 g compared with 1,600 g when lean during moult), but the sexes are similar in appearance, with a striking black, grey and white plumage pattern. The birds are highly gregarious, occurring in large flocks, and are protected from shooting throughout their range.

Geese compete with other flock members for food (vegetation), mates and territories (Owen & Wells 1979, Black & Owen 1988, 1989a). Young remain with their parents as a family unit for 4–11 months. The adult males are involved in most agonistic encounters, but females and offspring can also participate. Single birds do not establish nesting territories and within the flocks are ranked lowest in the dominance hierarchy (Black & Owen 1984, 1989a).

Like most other geese, the Barnacle Goose shows life-long monogamy; pairing usually takes place in the second year of life and the earliest breeders nest when two years old (Owen 1984: Owen, *et al.* 1988). The species is a determinate layer, with a single clutch of 3–5 eggs. The eggs are laid largely from body reserves and there is little chance of re-nesting if the first nest is lost. During the study period the number of pairs

LIFETIME REPRODUCTION IN BIRDS
ISBN 0-12-517370-9

breeding successfully in the entire population has varied between 85 (3 %
of potential breeders) in 1977, and 1,000 (40 %) in 1978.

Study population and methods

It has been known for some time that this population is virtually closed
(Boyd 1961): only 23 of the 5,264 ringed birds (0.4 %) have been seen
in either of the other populations in the last 15 years and seven of these
subsequently returned (Owen & Black 1989). The annual resighting rate
of marked (readable colour ringed) geese is around 95 %. If a bird was
not seen for two successive seasons it was assumed to have died soon
after its last sighting date; in other words, the bird did not live to return
the next season. The chance of a live bird being missed two years in a
row is only 0.14 % (Owen 1982). However, there will be a slight
underestimate of survival in the last year of this study (birds not seen
recorded as dead), although the resighting rate in that year was
exceptionally high, perhaps 98 %, as it was in 1986–1987 (Owen & Black
1989). Ring loss is a potential problem, but over at least the first 10 years
of the study was as low as 0.6 % per annum. Worn or lost rings were
replaced on recaptured birds (which were also fitted with a metal ring),
so that actual loss was minimal.

The criterion of successful breeding was identification on the wintering
grounds accompanied by young. There are several potential problems
with this method:

(a) Birds may not be seen before their young have split from the
     family. If a bird was not recorded until after January (except in
     very poor breeding years when all families were identified), it was
     recorded as breeding uncertain, so that birds with incomplete
     records could later be excluded from analysis.
(b) Some pairs may rear goslings which are not their own, although
     extra-pair copulation has not been reported. Others may lose their
     goslings to other adults through brood amalgamation. This is
     common in some goose species (see Owen 1980), but is unusual
     in wild Barnacles reared at rather low density. Broods of unnatural
     size (more than five) are almost unknown.
(c) Young sometimes become separated from their parents before
     they arrive on the wintering grounds; exceptionally up to 15 %
     of goslings may be orphaned when they arrive. However, many

families regroup during the autumn and any remaining orphans probably have poorer chances of survival than family goslings (Black & Owen 1984, 1989a).

Difficulties with correctly classifying birds as breeders and with allocating them the correct number of young were not thought to have had a substantial effect on the data presented here. We could not follow the fate of the progeny further than the first winter, as they were not ringed, but young which survive to arrive on the wintering grounds have an 83 % chance of surviving to two years (based on large samples of yearlings) when they become potential breeders (Owen 1982, 1984).

The oldest sample of birds was a group of 45 males and 44 females ringed as yearlings in 1973 (born in 1972). Of these only 27 males and 32 females had complete reliable breeding histories. We decided to add to this sample birds which had uncertain status one year in five or less (maximum three uncertain years in 15). Effectively, we recorded the geese as not breeding in these years. This is not unreasonable for the following reasons. The chance of a goose breeding was, on average, 15.8 % ($n = 17,660$ bird years). Geese with families were much more easily seen than non-breeders, since they were usually found at flock edges more accessible for viewing. The vast majority of uncertain breeding records occurred in poor breeding years when virtually all birds failed (Table 21.1). By adding these birds, the sample increased to 37 males and 41 females.

Results

LIFETIME PRODUCTION

Geese are long-lived; eight males (22 %) and four females (10 %) born in 1972 were still alive in 1987–1988. Longevity and lifetime success for these birds were estimated from the performance of a sample of 1962-ringed geese which were monitored from their 11th year onwards. The sample was small so data for males and females were combined. The oldest of the 37 birds in the sample (a female) died at 23 and three others (8 %) survived to more than 20 years. The longevity of the 1972 cohort is shown in Fig. 21.1.

Using the figures for both sexes, we calculated the mean expectation of further life from 1987–1988, as 4.0 years. We calculated each bird's future breeding success using the following assumptions. Birds with proven

Table 21.1 Population parameters during the study period. Mortality rate is based on ringed birds that did and did not return each autumn to Scotland. To reduce potential problems due to ring loss, birds whose rings were older than five years were not included. Sample sizes are in parentheses.

Year	Numbers	% Juveniles (October)	% Annual mortality (n)	
1973	5,100	21.0	4.4	(340)
1974	5,200	15.0	7.7	(325)
1975	6,050	20.6	11.1	(369)
1976	7,200	28.0	15.5	(444)
1977	6,800	2.4	8.7	(1,764)
1978	8,800	26.0	10.5	(1,626)
1979	7,700	3.7	9.8	(1,747)
1980	9,050	23.6	12.9	(1,613)
1981	8,300	2.3	9.7	(1,089)
1982	8,500	13.5	8.4	(1,054)
1983	8,400	8.0	8.7	(667)
1984	10,500	26.0	8.1	(517)
1985	10,400	9.6	12.4	(394)
1986	10,500	11.9	—	

reproductive success have a 20 % chance of breeding in each year, or a 80 % chance of breeding once in their four years of further life (the 15.8 % annual breeding success for all birds given above included young geese, which have less chance of success than proven breeders). Therefore, the geese that were still alive in 1987 were attributed two additional offspring, given that the mean brood size on arrival in Scotland for all birds over all years is 1.96. Birds that did not breed successfully in their first 16 years were assumed to continue to fail.

Among the 1972 sample, only one female and two males bred successfully more than three times during their lifetimes (Fig. 21.2). This by no means implies that they did not nest or hatch young. In the poor breeding year of 1977, in an area holding a fifth of the population, 158 females had brood patches, indicating that they had nested, but only 30 broods were present when the young were 3–4 weeks of age (Owen et al. 1978). Moreover, in 1986 35 % of young that survived to the age of 3–5 weeks (n = 485) failed to arrive on the wintering grounds (Owen & Black 1989).

Only two females and six males raised more than five young in their lives despite the fact that some pairs raised that number in a single season. The mean lifetime production for the whole sample was 2.6 ± 0.44 (SE) young for males and 1.5 ± 0.29 young for females. The difference

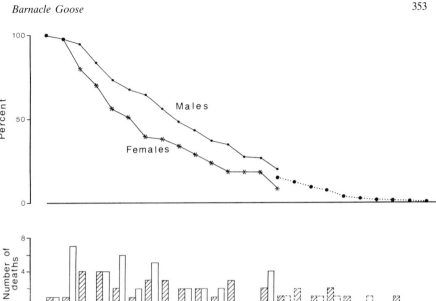

Figure 21.1 The proportion of the 1972 cohort alive in subsequent autumns and the distribution of lifespans included in the sample. Sample sizes 37 males and 41 females. The dotted line is an estimate based on the longevity of 37 older birds from a different cohort. Males (●) are left and females (*) right of each pair of histograms. Columns outlined by dashed lines are projections for birds still alive.

between the sexes was not significant. The most successful goose in the 1972 sample was a leucistic (all white-feathered) male which did not pair until its fifth year (and with a normal female), but subsequently bred successfully four times and reared 14 young. The only other leucistic bird to survive to breeding age was its female sibling, which survived to 10 years of age and reared only one young.

FACTORS AFFECTING PRODUCTION

Longevity

Females' annual mortality rate is consistently higher than males' (Owen 1982); in this sample the median lifespan was 10 years for males and eight for females ($t = 1.79$, $P < 0.05$, $df = 76$, one-tailed test). This leads to a disparity in the adult sex ratio (Owen *et al.* 1978); yet all birds eventually find partners. Unattached adult males often pair with younger

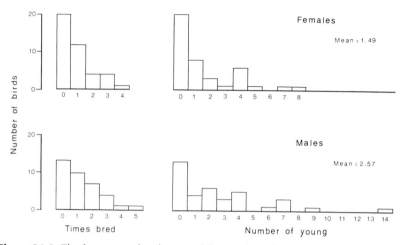

Figure 21.2 The frequency distribution of the number of seasons individuals brought young to the wintering grounds and the total number of young reared (goslings are four months old upon reaching Scotland). For females (upper), *n* = 41 and for males (lower), *n* = 37. The minimum age at first breeding is two years. Includes extrapolation for birds still alive.

females, so that females on average pair and breed at an earlier age than males (see below).

The relationship between lifetime production and years of life was significant for both sexes (Fig. 21.3). The slope was linear and significantly different from zero for both sexes ($P < 0.001$). The proportion of variation accounted for by the regression was, however, only 29 % for males and 30 % for females. Thirteen of the males (35 %) and 20 females (49 %) from the 1972 sample died without producing any young. The difference between the sexes was not statistically significant, but the productivity for a given age class was significantly different for the two sexes (comparison of slope, $t = 2.46$, $P < 0.02$).

It appears that both female and male reproductive rates peak at nine years of age (Fig. 21.4) which is strikingly similar to the situation in Lesser Snow Geese *Anser caerulescens caerulescens* (Radcliffe *et al.* 1988). However, one of the females and three of the males that were still alive in the spring of 1988 (16 years old) had still not returned to Scotland with any young. A close look at the older 1962 sample revealed that when these birds were 17 plus years, they recruited young at a similar level as the 1972 birds did at 13 years of age (Fig. 21.4) (see Discussion).

Age at First Breeding

The range in the years of first successful breeding was between the second and 10th summer of life. Females bred successfully for the first time at

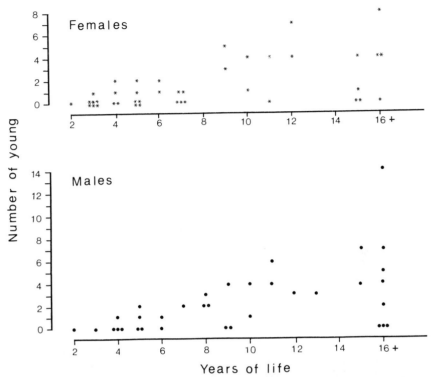

Figure 21.3 The relationship between longevity and the number of young reared in the lifetime for females (upper, slope 0.245 $P < 0.001$) and males (lower, slope 0.249 $P < 0.001$).

a slightly younger age (mean $= 6.0 \pm 0.50$ (SE), $n = 21$) than males (mean $= 6.7 \pm 0.52, n = 24$), although this difference was not significant.

There was no significant relationship betweeen the age of first breeding and the number of young produced in a lifetime for either sex. However, this statement takes no account of the fact that many birds, which might have started to breed later in life, would have died before they could do so. Taking this mortality into account, the lifetime productions of the early starters may well have been greater.

Effect of Mate Change

The proportion of birds which bred successfully in the first (8 %) and second year (16 %) after re-pairing with a new mate was lower than in longer established pairs (21 %), although only the first year difference was significant ($P < 0.05$, first year $n = 357$, second year $n = 410$ and three-plus years $n = 11,214$; from Owen *et al.* 1988). Therefore, each

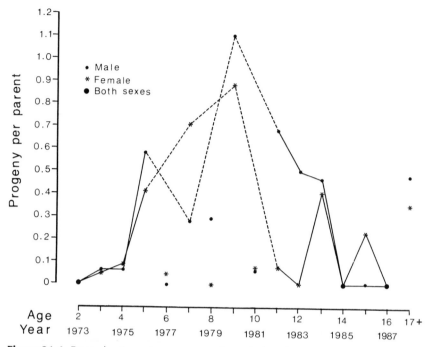

Figure 21.4 Reproductive rate per parent through their lives. Females are depicted with asterisks and males with dots. The dashed line signifies an omission of poor breeding years (1977, 1979, 1981) from the trend line. The points above the 17+ age axis denotes the reproductive rate for birds caught in 1962 (see text). The respective number of parents in each year are as follows: females 41, 40, 33, 29, 23, 21, 16, 16, 14, 12, 10, 10, 10, 6, 4, 20; males 37, 36, 35, 31, 27, 25, 24, 21, 18, 16, 14, 13, 10, 10, 8, 17.

time a bird re-paired a substantial reduction in breeding output occurred; the penalty for re-pairing once in a 4.5 year period was calculated to be 14 % of the average production in the period (Owen *et al.* 1988). The effect of mate loss is therefore likely to be significant for an individual's lifetime production.

Breeding Area

In order to test whether the breeding area affected lifetime reproductive success, the first nine years of reproductive life of birds from two different areas were compared: the Hornsund cohort considered above and a cohort from Nordenskioldkysten born in 1976. Hornsund is blessed with several large nesting islands but poor feeding areas for families, whereas Nordenskioldkysten (a stretch of coastline about 150 km to the north) has fewer islands but ample feeding places. Since these cohorts lived in different periods, and hence experienced different weather conditions and goose densities (the population continually increased, see Table 21.1), we must first show that the productivity in two sets of years are similar. To do this we compared the production of young in the whole population (lumping all colony areas) during the two nine-year periods: 1974–1982 and 1978–1986.

The overall expectation of rearing young in these two sets of years was only slightly different, 1.74 young per pair per year for the earlier period and 1.67 for the later period (Mann-Whitney U test for difference in expected annual reproductive success, $U = 40.5$, $n1$ and $n2 =$ nine years, $P > 0.10$). In other words, the different situations the cohorts experienced were not different enough to invalidate the comparison between the breeding areas; the difference, if anything, favoured the earlier (Hornsund) cohort.

For the following comparison we assumed that the majority of the females showed a high degree of natal and breeding philopatry. This seems to be true for both Barnacle Geese (Prop *et al.* 1984, Owen & Black, unpub.) and Snow Geese females (Cooke *et al.* 1975, Rockwell & Cooke 1977). Females from Nordenskioldkysten produced significantly more offspring in their first nine years of life (2.13 ± 0.19 (SE) per female, $n = 203$) than did Hornsund females (1.10 ± 0.23 per female, $n = 41$); ($t = 2.35$, $P < 0.02$).

Additional evidence for a difference in breeding success between geese in these two areas was obtained in the 1986 breeding season. On the Nordenskioldkysten 50 % of the nests produced young (207 broods/410 nests) compared to only 21 % on a portion of the Hornsund coast (21 broods/100 nests) (Chi Square $= 11.7$, $df = 1$, $P < 0.001$). This result may have been due to a cohort effect rather than a breeding area effect, but if so, the Hornsund males should also have shown reduced success, even though (presumably) scattered across the breeding range (males show less philopatry than females). This was not the case. Offspring production per male was similar for the two areas (1.78 ± 0.21 (SE) per Nordenskioldkysten male, $n = 181$, and 1.89 ± 0.37 per Hornsund male, $n = 37$) ($t = 0.22$, ns).

Seasonal Effects

Comparing years, the proportion of young in the autumn population is negatively correlated with the extent of snow cover on the breeding grounds in the preceding spring (Owen & Norderhaug 1977). Snow cover delays nesting and causes females to deplete body reserves before laying;

this in turn causes nest desertion in the later stages of incubation (Prop *et al.* 1984). The 1972 and 1976 cohorts each lived through three disastrous breeding seasons that occurred close together: 1977, 1979 and 1981 (see Table 21.1). In these years there were only 2.3–3.7 % young in the autumn population. In addition to nest failures and poorer survival, significantly more adults failed to establish nesting territories in these late seasons (late seasons 26 % non-breeding, $n = 231$; earlier seasons 11 % non-breeding, $n = 205$; Chi square $= 14.4$, $df = 1$, $P < 0.001$); (data in Prop *et al.* 1984). Presumably birds that do not suffer such a run of poor breeding seasons would have markedly higher lifetime success than those that do.

To examine weather effects further, we predicted the average lifetime production for cohorts of geese living through different time periods. Figure 21.5 gives the estimated lifetime production of 100 geese of each sex surviving for the median lifespan (eight years for females and 10 for males) and starting to breed in their third year. The chances of breeding in any year is taken as the proportion of all ringed geese of each sex that succeed in bringing young to the wintering grounds in that season.

As may be seen, expected lifetime production varies greatly depending on year of birth. For both sexes the most fortunate cohort produced

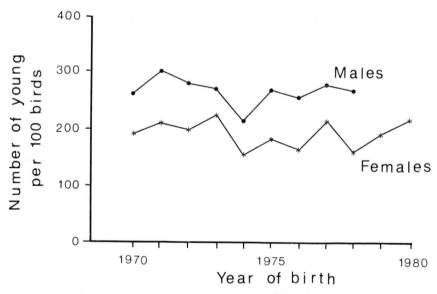

Figure 21.5 Hypothetical lifetime production of 100 male and 100 female geese of median lifespan (males = 10 and females = 8 years) alive during different periods to illustrate effects of different years/seasons on lifetime production. The year on the x axis is the year of hatching: breeding is assumed to begin in year 3. The mean sample size for each point = 576 ± 53 (SE).

more than 40 % more young than the least fortunate. Moreover, in periods of poor breeding conditions, the variability in lifetime production between individuals would be expected to be higher, as competition for food and nesting sites intensifies (see Discussion). The annual mortality from natural causes has increased over the years as numbers in the population increased, although this has been counterbalanced by a decline in the mortality from illegal shooting (unpublished data). In addition, the number of successful breeders is inversely related to the number of potential breeders (Owen 1984).

Body Size

Total body mass was the only measure of size available for the whole of the 1972 sample. However, all the birds were caught within a few days of one another at a time (the flightless moult) when they were lean and when daily mass changes were small (Owen & Ogilvie 1979; Owen 1980). At this time body mass is likely to be a good approximator of body size at least for non-breeders (this sample was caught as non-breeders when one year old).

Since the sample was small, we compared the longevity and lifetime production of those birds above and below the mean body mass. In no case was there a significant difference, but in both sexes the relationship appeared negative, contrary to that in swans (Scott 1988).

Discussion

Lifetime production in long lived species is generally highly variable; some individuals contribute disproportionately to the future population (Coulson & Thomas 1985) and others produce no young despite living for several years (Newton 1985, Scott 1988). Although this study has demonstrated similar variability, it is not as great as might have been expected. For example, a goose which has a productive life of 18 years, as did some of the geese in the 1972 cohort, has the potential to raise 90 young given a clutch size of five. The largest number raised by any individual in this cohort, however, was 14 and only three birds reared more than seven young. The main reasons for the shortfall are the high rate of non-breeding, nest desertion and high gosling mortality following nesting, either through predation from foxes or gulls, severe weather and/ or competition for food in the brood rearing areas (see Prop *et al.* 1984). For example, in 1986 only 8 % of the potential eggs (if every female had nested) resulted in young on the wintering grounds, even though the summer was favourable (Owen 1987).

Preliminary examination of the cohort from the Nordenskiold coast (above) revealed that these birds will end up with a larger number of young; after only nine years seven of the Nordenskiold geese have already produced more offspring than the most productive Hornsund bird produced in its lifetime. This asymmetry in reproductive success could be related to many density dependent or density independent factors acting on the breeding grounds. These include feeding quality and opportunity, terrain of the colonies and distance to brood rearing areas, predator pressure, nesting opportunities and competition for territories and food (see Owen & Wells 1979, Prop et al. 1984, Owen & Black 1989).

If different situations influence a bird's chance to survive and reproduce, then it may be equally possible for resource quality to affect a gosling's development, perhaps to the extent that it affects its lifetime production (see Albon et al. 1987). We have already shown that gosling growth rate and body mass varies between breeding areas and that larger goslings survive their first migration better (Owen & Black 1989). Similarly, offspring quality may be affected by the type of parental care that is received. In semi-captive geese, parental effort appears to affect gosling survival until fledging and may determine how long offspring remain in families, where they have large benefits in terms of putting on weight and learning social and feeding skills (Black & Owen 1987, 1989b). Possibly, therefore, goslings reared under different conditions and/or experiences will be expected to have different lifetime reproductive performance. Detecting these differences, however, will be difficult (see Findlay et al. 1985).

The marked difference in mean productivity between the sexes was explained by the difference in longevity. The shorter mean lifespan of females has been attibuted to the energetic stress of breeding, which results in higher annual mortality (Owen 1982, Owen & Black 1989). The fact that females breed at a younger age than males may reduce female longevity.

Perhaps breeding successfully at an early age is more costly than breeding later in life because of the reduced experience and competitive ability when young. For example, young birds increase their aggressive effort against other flock members during daily foraging prior to their first successful breeding season (Black & Owen 1989b). The age of first breeding has been shown to be important for survival in Great Tits Parus major; those breeding for the first time in years of limited food have lower survival (Balen et al. 1987, McCleery & Perrins 1988). Moreover, Goldeneye ducks Bucephala clangula that lay earlier and produce larger clutches than normal in their first and second breeding attempts have higher mortality than older birds (Dow & Fredga 1984). This aspect will have priority in future analyses on larger samples.

Based on the 1972 cohort and the information from the breeding grounds (Prop *et al.* 1984), the following statements can be made in summary. About half the goslings that hatch survive to migrate to the wintering grounds and 83 % of these live to reproductive age. In some seasons, however, as few as 17 % of the pairs that attempt to breed succeed in hatching eggs. Each year, between 11 % and 26 % of the adult birds do not establish nesting territories. Forty-nine per cent of the adult females and 35 % of the adult males died without recruiting young into the next generation. The best 10 % of the female and male breeders produced 34 % and 39 % respectively of all the young and it only took 15 % of the potential breeders to produce half of the next generation's recruits.

Acknowledgements

Over the years this project has relied on contributions from many students, volunteer observers and expedition participants too numerous to mention individually. We record our grateful thanks for their contribution and to those who have given logistical and financial support. We thank in particular the staff of the Wildfowl Trust, both at the refuge at Caerlaverock and at Headquarters.

References

Albon. S.D., Clutton-Brock, T.H. & Guinness, F.E. 1987. Early development and population dynamics in red deer. II. Density-independent effects and cohort variation. *J. Anim. Ecol.* **56**: 69–82.

Balen. J.H., van Noordwijk, A.J. & Visser, J. 1987. Lifetime reproductive success and recruitment in two Great Tit populations. *Ardea* **75**: 1–11.

Black, J.M. & Owen, M. 1984. On the importance of the family unit to Barnacle Goose offspring—a progress report. *Norsk Polarinstitutt Skrifter* **181**: 79–85.

Black, J.M. & Owen, M. 1987. Determinants of social rank in goose flocks: acquisition of social rank in young geese. *Behaviour* **102**: 129–46.

Black, J.M. & Owen, M. 1988. Variation in pairbond and agonistic behaviours in Barnacle Geese on the wintering grounds. In *Waterfowl in Winter*, Ed. M. Weller, pp. 39–57. Minneapolis: University Press.

Black, J.M. & Owen, M. 1989a. Agonistic behaviour in goose flocks; assessment, investment and reproductive success. *Anim. Behav.* **36**: 199–209.

Black, J.M. & Owen, M. 1989b. Parent-offspring relationships in wintering Barnacle Geese. *Anim. Behav.* **36**: 187–198.

Boyd, H. 1961. The number of Barnacle Geese in Europe in 1959–60. *Wildfowl Trust Annual Report* **12**: 116–24.

Cooke, F., MacInnes C.D. & Prevett, J.P. 1975. Gene flow between breeding populations of Lesser Snow Geese. *Auk* **92**: 493–510.

Coulson, J.C. & Thomas, C. 1985. Differences in the breeding performance of individual Kittiwake Gulls, *Rissa tridactyla* (L). In *Behavioural Ecology: Ecological Consequences of Adaptive Behaviour*, ed. Sibly, R.M. & Smith, R.H., pp. 489–503. Oxford: Blackwell.

Dow, H. & Fredga, S. 1984. Factors affecting reproductive output of the Goldeneye duck *Bucephala clangula*. *J. Anim. Ecol.* **53**: 679–92.

Findlay, C.S., Rockwell, R.F. & Cooke, F. 1985. Does clutch size vary with cohort in Lesser Snow Geese? *J. Wildl. Manage.* **49**: 417–20.

McCleery, R.H. & Perrins, C.M. 1988. Life-time reproductive success of the Great Tit, *Parus major*. In *Reproductive Success*, ed. T.H. Clutton-Brock, pp. 136–153. Chicago: University Press.

Newton, I. 1985. Lifetime reproductive output of female Sparrow-hawks. *J. Anim. Ecol.* **54**: 241–53.

Owen, M. 1980. *Wild Geese of the World*. London: Batsford.

Owen, M. 1982. Population dynamics of Svalbard Barnacle Geese 1970–1980. The rate, pattern and causes of mortality as determined by individual marking. *Aquila* **89**: 229–47.

Owen, M. 1984. Dynamics and age structure of an increasing goose population— the Svalbard Barnacle Goose. *Norsk Polarinstitutt Skrifter* **181**: 37–47.

Owen, M. 1987. Barnacle Goose project. 1986 Report. Unpubl. Rep. Wildfowl Trust.

Owen. M. & Black, J.M. 1989. Factors affecting the survival of Barnacle Geese on migration from the breeding grounds. *J. Anim. Ecol.* **58**: 603–617.

Owen. M. & Norderhaug, M. 1977. Population dynamics of the Barnacle Goose *Branta leucopsis* breeding in Svalbard 1948–1976. *Ornis Scandinavica* **8**: 161–74.

Owen, M. & Ogilvie, M.A. 1979. Wing molt and weights of Barnacle Geese in Spitsbergen. *Condor* **81**: 45–52.

Owen, M. & Wells, R. 1979. Territorial behaviour in breeding geese—a re-examination of Ryder's hypothesis. *Wildfowl* **30**: 20–6.

Owen, M., Drent, R.H., Ogilvie, M.A. & Van Spanje, T.M. 1978. Numbers, distribution and catching of Barnacle Geese *Branta leucopsis* on the Nordenskioldkysten Svalbard in 1977. *Norsk Polarinstitutt Aarbok* **1977**: 247–58.

Owen, M., Black, J.M. & Liber, H. 1988. The duration of the pair bond and the timing of its formation in Barnacle Geese. In *Waterfowl in Winter*, ed. M. Weller, pp. 25–38. Minneapolis: University Press.

Prop. J., Van Eerden, M.R. & Drent, R.H. 1984. Reproductive success of the Barnacle Goose *Branta leucopsis* in relation to food exploitation on the breeding grounds western Spitsbergen. *Norsk Polarinstitutt Skrifter* **181**: 87–117.

Ratcliffe, L., Rockwell, R.F. & Cooke, F. 1988. Recruitment and maternal age in Lesser Snow Geese *Chen caerulescens caerulescens*. *J. Anim. Ecol.* **57**: 553–63.

Rockwell, R.F. & Cooke, F. 1977. Gene flow and local adaptation in a colonially nesting dimorphic bird: the Lesser Snow Goose. *Amer. Nat.* **111**: 91–7.

Scott, D.K. 1988. Breeding success in Bewick's Swans. In *Reproductive Success*. ed. T. Clutton-Brock. pp. 220–236. Chicago: University Press.

22. Mute Swan

PHILIP J. BACON & PELLE ANDERSEN-HARILD

Mute Swans *Cygnus olor* are common in lowland lakes, marshes, slow rivers and coastal shallows. They are vegetarian, eating mainly soft water-plants, but sometimes grazing in fields and marshes. The species occurs naturally in the western Palearctic, with a few small populations in eastern Asia. It has been introduced to Australia, New Zealand, southern Africa and the eastern sea-board of North America.

The Mute Swan is one of the heaviest flying birds, and is sexually dimorphic in size. Typical weights of adults are 8 kg for females and 11 kg for males (Bacon & Coleman 1986). Adults accumulate body reserves both to enhance over-winter survival and to permit early breeding within a season (Reynolds 1972, Andersen-Harild 1981, Birkhead *et al.* 1983, Beekman in press). Young birds are grey-brown, with slate grey bills and feet. During their first year the plumage moults to all white. As birds develop to sexual maturity, at around 3–5 years old, the bill changes through reddish-grey to orange, and a fleshy black knob develops above the bill.

Non-breeding swans comprise about half the population, and live in flocks away from territorial pairs. Mute Swans typically pair at 2–3 years old, and breed a year or so later. Once swans have bred, they usually breed every subsequent year. They are strictly monogamous and almost invariably keep the same mate for as long as that mate survives.

Breeding normally takes place within vigorously defended territories, comprising some 2–3 km of river or 4–5 ha of lake (Cramp & Simmons 1977). Territorial birds sometimes kill intruding swans that are unable to escape (up to 3 % of deaths, Ogilvie 1967). But in unusual circumstances of superabundant food coupled with limited and concentrated nesting sites, Mute Swans may nest colonially. In some colonies aggression is rare and birds nest only 10–30 m from one another (Bloch 1970, Perrins & Ogilvie 1981): in other colonies aggression is intense, and causes considerable loss of eggs.

LIFETIME REPRODUCTION IN BIRDS
ISBN 0-12-517370-9

The single annual brood (repeat clutches only replace those lost early in the season) typically hatches from 5–8 eggs and comprises 3–6 cygnets (young) at fledging. About 60 % of nests hatch young and some 50 % of hatched young survive to fledge (for regional differences see: Bacon 1980b, Coleman & Minton 1980, Perrins & Ogilvie 1981). Survival is around 70 % per year for immatures, but rather higher, 80–85 %, for breeding adults. The longest recorded lifespan for a wild Mute Swan is 26 years, in southern Britain. Only some 50 % of fledged cygnets survive until they are first likely to pair (2–3 years), and only around 6 % until they are likely to fledge cygnets themselves (one generation is about eight years, Bacon 1980a).

The Danish Mute Swan population

Our study was made in Denmark where, due to excessive hunting, the breeding population reached a low of three or four pairs in 1924. The introduction of legal protection in 1926 permitted a rapid increase in numbers, which persists to the present. The increase came partly from the small wild population and partly from cygnets of "ornamental" pairs that were allowed to fly free. The contribution of immigrants from other countries is unknown, but could have been considerable in later years.

Recently migrants to Denmark have become common, both for the summer moult (ca. 40,000 swans) and for winter feeding (ca. 70,000). These migrants include swans bred in Germany, Poland, western USSR and Scandinavia (Andersen-Harild & Preuss 1978). In severe winters Danish Mute Swans migrate southwest, to the Netherlands for example. We are unable to discuss here the frequency of this behaviour, or its possible effects on lifetime reproductive success. Despite the influx of migrants, most Mute Swans bred in Denmark return to nest close to their place of birth; females return closer than males (Coleman & Minton 1980).

Colonial breeding was unknown in Denmark before 1900, and occurred in only a few localities between 1943 and 1957, accounting for only 1–2 % of all pairs. By 1966 about 22 % (650 pairs) bred in some 25 colonies (Bloch 1971), and by 1978 this had risen to 38 % (1,500 pairs). The percentage is probably even higher today.

In 1966 the average production of cygnets to fledging was 1.9 for colonial pairs but 3.5 for solitary pairs on fresh water lakes. In the 1970s similar calculations for North Sjælland gave 1.1 for colonial pairs but 2.5 for solitary pairs. Recent studies of ringed birds showed that the production of cygnets by colonial breeders was too low to maintain numbers. But the colonies persisted because cygnets hatched on territories bred in the colonies. It was estimated in the 1980s that half the colonial breeders had themselves been hatched on territories, although the converse of a colonial raised cygnet breeding territorially was unknown.

Despite the high influx to the colonies of swans that had been raised on territories, adjacent colonial and territorial populations showed significantly different genotype frequencies at two biochemical "marker" loci. The gene frequency differences probably arose because some genotypes were more likely than others to attempt nesting in the colonies (Bacon & Andersen-Harild 1987).

STUDY AREAS AND PROCEDURES

We studied two adjacent populations in northeast Sjælland, separated by only 25 km. The Kobenhavn pairs were territorial, whereas the Roskilde birds bred colonially on several islands in the fjord.

The numbers of breeding pairs each year, together with the numbers of cygnets fledged and ringed, are given in Table 22.1 and Fig. 22.1. Around 70 % of cygnets raised each year were caught and marked, and those missed were unlikely to have been a biased sample. Note the rapid increase in the Roskilde population during the study.

As the study started in 1966, we are unable to cover complete lifetimes, by 1986, for birds that can live for 26 years. Moreover, as numbers breeding in the Roskilde colonies were very low until the early 1970s, we had an even shorter period for the majority of birds in the population. We chose to work with three sets of data because we wished to assess both individual lifetime reproductive successes and to compare the success of territorial and colonial breeders.

The first set, *Ko*Benhavn-*born*-*E*arly (KBbE), comprised cygnets raised on territories between 1966 and 1975. The last of these cohorts had a 10-year period (to 1986) for recoveries, by which time some 94 % would have died. This set provided a close approximation to true LRS values. The second set, KBbL, comprised a later series of Kobenhavn cohorts (born 1972–1981) that was directly comparable to the third set of colonial cohorts from Roskilde. This set, *Ros*Kilde-*born*-*L*ate (RKbL), comprised cygnets bred at the Roskilde colonies between 1972 and 1981. Details of all sets are given in Table 22.1.

The last samples from the KBbL and RKbL cohorts would have had only a five-year recovery period to 1986; by that time only some 80 % of that last cohort would have died. In addition, as no comprehensive survey of Roskilde was made in 1978, we omitted the 1978 cohort from *both* these later series to allow fair comparison. The KBbL and RKbL

Table 22.1 Winter severity, number of nests and cygnet production.
For each year of the study the numbers of ice-days, nesting pairs and numbers of cygnets both fledged and ringed in each study area are shown. Ice-days refer to the winter ending that year, i.e. 99 ice-days for 1962/1963. The KB sample is split into two categories, to allow comparison with data for RK (see text).

Year	Winter ice-days	Nesting pairs		Ringed cygnets included in sample			
		KB	RK	KBbE		KBbL	RKbL
1962	21	*	*	*		*	*
1963	99	*	29	*		*	*
1964	27	*	42	*		*	*
1965	21	*	101	*		*	*
1966	54	60	91	88		*	*
1967	6	59	*	130		*	*
1968	24	62	*	129		*	*
1969	53	54	*	155		*	*
1970	95	24	*	41		*	*
1971	22	58	*	87		*	*
1972	39	67	*	134	→	134	86
1973	4	73	*	118	→	118	59
1974	3	70	*	133	→	133	47
1975	1	57	*	140	→	140	95
1976	17	55	*			127	68
1977	26	59	*			128	100
1978	18	55	300			−103−	not-known
1979	83	44	207			88	69
1980	43	60	338			145	149
1981	11	49	425			115	93
1982	77	45	292				
1983	4	49	431				

samples were appreciably truncated in time and hence underestimated lifetime reproductive success, but were directly comparable with each other. We therefore defined a *Life-period* to be the lifespan of a swan or, occasionally, its apparent lifespan as truncated by our cut-off date, depending on the individual.

Swans caught in the study areas were given a unique metal ring which lasted "for life", plus an engraved plastic ring which could be read at a distance (Ogilvie 1972). These plastic rings normally lasted several years, longer than most of our Mute Swans lived. Swans breeding at the colonies were also given engraved plastic neck-collars to aid identification at long distances (Sladen 1974).

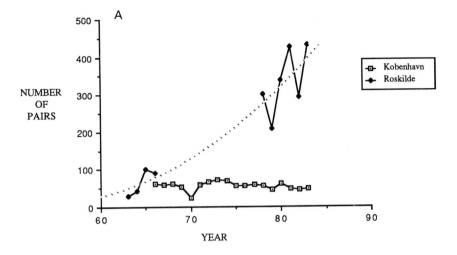

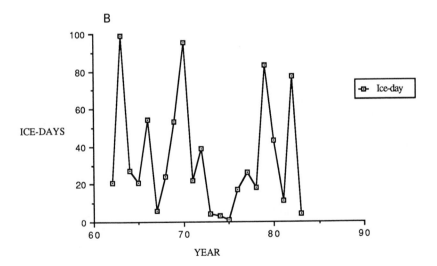

Figure 22.1 A. Changes in numbers of nesting pairs at two sites against time in years. Year 60 is 1960. The dotted line through the points for the Roskilde colonial site is the quadratic regression with time. B: Number of Ice-days on inner Danish coastal waters in the winter of the preceeding year (hence the value for 61 is the number of Ice-days in the 1960/1961 winter). Note the decreases in numbers of nesting pairs, in A above, coinciding with high numbers of Ice-days.

Very few successful breeding attempts would not have been correctly ascribed to known birds. Even if breeders had lost their plastic rings, they could be re-caught at nests or with cygnets, identified from their metal rings and fitted with new plastic rings and collars. A few unsuccessful attempts by known birds may have gone undetected, if the birds failed early in the nesting cycle, but unsuccessful attempts added nothing to lifetime cygnet production anyway.

For mortality estimates we used only data on swans reported dead through the national ringing scheme. Re-sighting probabilities might have varied considerably over the regions to which our non-breeders dispersed. Although Mute Swans show high tendencies both to return and nest close to their natal site, and to re-occupy the same territory in successive years, we chose to exclude breeding data about swans of unknown background, raised outside our study areas.

Ringed swans were sexed by cloacal examination. This technique was reliable for adults, but less so for cygnets. Apart from the first few years of the study, when cygnets were not routinely sexed, the sexes of over 95 % of swans were reliably known. Body weight and a size measurement (the width of the foot web) were recorded for most swans on capture.

Territorial and colonial pairs were not equally easy to study. Differences in accessibility restricted our choices of life history and breeding success measures to those that were comparable between the areas. Ages of first pairing were not comparable between areas, as pair formation was harder to record in the colonies. The job of finding nests and identifying breeders was comparably effective in both situations, but the time taken to visit some solitary nests in extensive reed-beds prevented accurate recording of all clutch sizes. The success of pairs that hatched and fledged cygnets could be recorded with equal confidence in both areas; similarly, most young from every brood raised to fledging (some four months from hatching, about 1 October) could be caught in both areas.

About a third of fledged broods could not be observed closely enough to be certain of final brood size. Those brood sizes were estimated from numbers hatched and mean survival rate, or were taken as the average brood size, for that area and year. This degree of correction will have reduced the variance of lifetime fledging estimates but, as both brood sizes and mortality of large cygnets are low in Mute Swans, it will have had little effect on total numbers fledged per lifetime.

We used the following parameters to describe life histories and breeding success: (a) age at first nesting; (b) lifetime sum of years the swan nested; (c) lifetime sum of years the swan successfully hatched cygnets (Mute Swans have only one successful clutch per year); (d) lifetime sum of estimated number of cygnets fledged.

SOME ANALYTICAL DIFFICULTIES

It is unfortunate that the data normally collected in population studies refer to cohorts of individuals, because classical statistical theory requires an independent random sample of birds. Cohorts clearly do not fit this requirement: individuals within cohorts are born in similar areas and conditions, and may experience similar habitats and weather in their early lives. Within cohorts, even if we could get large enough samples, we are unsure of the number of degrees of freedom because of the inter-dependence. But we need data on different cohorts to produce information on the effects of weather on survival and on breeding success (especially if these may change with age).

As our Mute Swans show high mortality in severe winters and also have considerably delayed maturity, we believe that the inter-dependence of individuals within cohorts may well be too high for existing methods of partitioning "LRS components" to be robust (e.g. Brown 1988). Accordingly, we have chosen to use the results of regression analyses to identify the major factors, but have avoided trying to quantify their relative contributions.

Results

WINTER WEATHER AND SURVIVAL

All swans were categorized as to whether they Died or Survived their first winter, the remainder being classed as Unknown if they were never seen again. There were no survival differences between the sexes for KBbE or KBbL swans, but RKbL females survived better than their male counterparts (Table 22.2). It was therefore necessary to compare survival rates between the areas for the sexes separately. First winter survivals did not differ between KBbL and RKbL swans for either sex: the significant differences found when the "Unknown" class was included (Table 22.2) were presumably due to subsequent resighting differences between the areas (including probability of returning).

The first winter survivals of different cohorts varied from 86 % subsequently recorded alive for one cohort to only 30% for another. Survival decreased with increasing winter severity. Scatter plots of known survival and known mortality against the numbers of days with ice-cover on inner Danish coastal waters (Ice-days) showed the effects to be similar for the two areas. The combined data gave the following relationships:

% Known to have died   = 10.0 + 0.120 * Ice-days $P < 0.01$, $r^2 = 0.25$

% Known to have survived = 65.0 − 0.350 * Ice-days $P < 0.01$, $r^2 = 0.24$

Survival after the first winter did not vary, either between sexes within areas or between KBbL and RKbL samples ($0.99 > P > 0.15$ for all comparisons). Annual survivals for the combined data were approximately:

Year of life	0–1	1–2	2–3	3–4	4–5	5–6	6+....
Per cent survival	55	74	78	79	80	82	84

WINTER WEATHER AND NUMBERS NESTING

The number of pairs with nests in the Kobenhavn area remained approximately constant over the years of study (Fig. 22.1), but decreased temporarily after each cold winter. A quadratic regression with the number of Ice-days in the preceding winter explained 59 % of the variance in numbers nesting per year. There were no other significant effects of particular years, nor of Ice-days two winters previously.

KB-nests = 63.45 − 0.2570 * Ice-days $P < 0.001$, $r^2 = 0.46$
KB-nests = 57.87 + 0.2430 * Ice-days
− 0.00567 * Ice-days^2 $P < 0.05$, $r^2 = 0.59$

The Roskilde population increased dramatically during the study (Fig. 22.2): 81 % of the variance in annual nesting numbers per year at Roskilde was explained by a linear increase with time (taking 1966 as Year = 66), but a quadratic regression with time and the number of Ice-days in the two preceding winters explained 86% of the variance.

RK-nests = −1248 + 20.16 * Year $P < 0.001$, $r^2 = 0.81$
RK-nests = 8509 − 246.00 * Year
    + 1.808 * Year^2 $P < 0.05$, $\left.\right\}$ $r^2 = 0.86$
    − 1.484 * Ice-days − $P < 0.05$
    1.214 * Prev-ice days

Table 22.2 Differences in first winter survival by area and sex.

First winter survivals of both sexes for KBbL and RKbL samples. As survival differed between sexes for RKbL swans it was compared between RKbL and KBbL for each sex separately. The first test with 2 df compares all three categories Died, Survived and Unknown; the second test with 1 df compares Died versus (Survived + Unknown).

Fate during 1st Winter	Females				Males			
	KBbL		RKbL		KBbL		RKbL	
Died	62	13 %	32	9 %	80	13 %	64	16 %
Survived	282	57 %	176	50 %	348	58 %	169	45 %
Unknown	147	30 %	145	41 %	168	28 %	155	39 %
Sum	491	100 %	353	100 %	596	100 %	398	100 %

Comparison by sex for RKbL:

$\chi^2_2 = 8.4$, $P < 0.02$ $\qquad$ $\chi^2_1 = 7.6$, $P < 0.01$

Comparison between RKbL and KBbL

Females

$\chi^2_2 = 11.9$, $P < 0.005$ $\qquad$ $\chi^2_1 = 2.3$, $P \simeq 0.1$ ns

Males

$\chi^2_2 = 17.8$, $P < 0.001***$ $\qquad$ $\chi^2_1 = 1.5$, $P \simeq 0.2$ ns

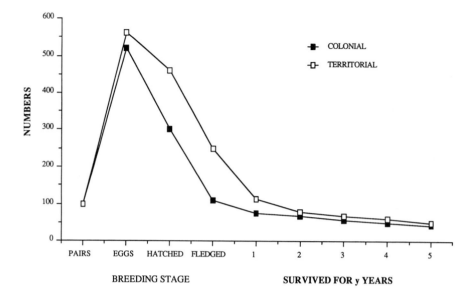

Figure 22.2 Relative production and survival of Mute Swans breeding in Territorial and Colonial sites from data collected during the early phases of colony establishment, following Andersen-Harild (1978). Production and survival are shown relative to a constant starting number of 100 pairs for each breeding type.

WINTER WEATHER, FIRST-TIME NESTING AND BREEDING SUCCESS

The severity of the preceding winter had no significant effect on the number of young raised to fledging. The main factor that influenced the success of annual attempts was whether these were made in territories or colonies. There were some small differences between the years and a small effect of "first-time" breeding. The variance of annual brood size explained increased only slightly when year/site interactions were included (from 20 % to 23 %). The effect of "first-time" breeding was evident only for territorial pairs. It was slightly larger for males than for females, but as both males and females breeding for their first time tended to have partners who were also first-time breeders, we were unable to show separate effects for each sex.

AGES AT FIRST NESTING

Ages at first nesting ranged from two to 15 years, with modes of four or five years for all samples. Less than 1 % of swans first nested before age

3. The modal ages of first nesting given in Table 22.3 were a little high compared with studies elsewhere (Bacon 1980b, Cramp & Simmons 1977), probably because of more severe winters in Denmark.

LIFETIME NUMBER OF NESTING ATTEMPTS

Significantly different proportions of males and females were recorded as *returning to the study areas* and nesting (KBbL, Females 13 %, Males 8 %, $P < 0.005$; RKbL, Females 15 %, Males 6 %, $P < 0.001$). This is perhaps not surprising, as more males than females died in their first winter (Table 22.2), and females return closer to their natal sites than males (Coleman & Minton 1980). However, amongst the swans that did return to nest, there were no significant differences in ages at first nesting, either between sexes within areas, or between areas.

The numbers of years in which individual swans nested per lifetime are given in Fig. 22.3 and Table 22.4. In round figures, each swan that survived and nested within the study areas had means and standard deviations of two attempts, ranging from one to nine attempts, with a mode of one attempt in all areas.

The number of years in which nesting was attempted was correlated both to the birds' life-periods (YEARS LIVED) and, particularly for KB swans, the period from first nesting to death (NEST PERIOD). The KB swans were significantly more likely to nest again in subsequent years of their lives (66 % ± 4 %) than were RK swans (24 % ± 10%). These re-nesting probabilities were rather low compared to other studies, where subsequent attempts are almost always annual. Likely explanations are the harder winters in Denmark and, perhaps, the poorer hatching success in the colonies subsequently leading to greater disruption of pair bonds in future years.

Table 22.3 Ages at first nesting, by area and sex.

Area	Sex	Nesting Number	(%)	Mean age at first nesting	Standard deviation	Range
KBbE	Female	70	(55)	6.0	2.4	2–15
KBbE	Male	57	(45)	5.9	2.1	2–11
KBbL	Female	65	(59)	5.8	1.9	3–10
KBbL	Male	46	(41)	5.4	2.0	2–11
RKbL	Female	54	(68)	5.3	1.8	2–11
RKbL	Male	25	(32)	5.9	2.1	2–11

YEARS   NESTING   YEARS HATCHING

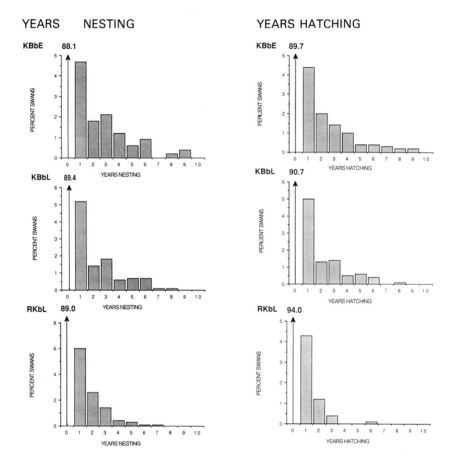

Figure 22.3 Comparison of life-period breeding performance between three different areas, shown as rows of histograms, for each of three measures of life-period production, shown as columns of histograms. Values in bold type at the left of each histogram show the percentage of swans that were never recorded to nest, hatch or fledge young within the study areas. *Left-column*: Percentages of swans, from each area, known to have nested within the study areas in different numbers of years. *Centre-column*: Percentages of swans, from each area, known to have hatched young within the study areas in different numbers of years. *Right-column*: Percentages of swans, from each area, having different life-period productions of cygnets of their own within the study areas.

YEARS   FLEDGING

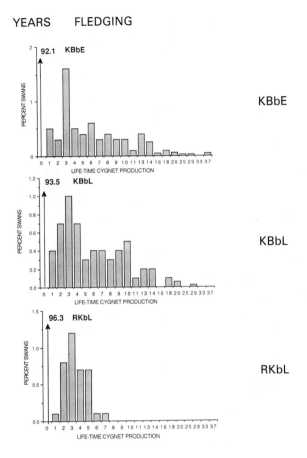

KBbE

KBbL

RKbL

Table 22.4 Numbers of nesting attempts per life-period for individual swans. Only birds known to have nested are included: there were no differences in numbers of nesting attempts between the sexes within areas.

Area	Number of swans	Mean	Standard deviation	Range
KBbE	138	2.8	2.1	1–9
KBbL	120	2.4	1.7	1–8
RKbL	84	1.8	1.2	1–7

For the Kobenhavn territorial area (df = 49):
Years-nested = −0.85 + 0.36 * Years-lived $P < 0.001, r^2 = 0.24$
Years-nested =   0.0 + 0.66 * Nest-period $P < 0.001, r^2 = 0.62$

For the Roskilde colonial area (df = 14):
Years-nested = −0.18 + 0.22 * Years-lived $P < 0.01, \ r^2 = 0.32$
Years-nested =   0.96 + 0.24 * Nest-period $P < 0.02, \ r^2 = 0.24$

LIFETIME NUMBER OF BROODS HATCHED

Only 12 % of territorial swans failed to hatch any eggs (KBbE = 13 %, KBbL = 11 %), whereas 46 % of colonial nesters failed. This difference between the KBbL and RKbL birds was highly significant ($P < 0.001$). There were no differences in hatching success between sexes within any sample.

The number of successful hatchings per life-period was also significantly different between KBbL and RKbL samples ($P < 0.01$, Table 22.5 and Fig. 22.3 centre). Some 17 % of KBbL swans hatched eggs in four or more years, compared to only 2 % of RKbL birds. Compared to KBbL swans, RKbL birds that nested were only 61 % as likely ever to succeed in hatching, and those that did hatch achieved only 67 % as many successes.

LIFETIME NUMBER OF CYGNETS FLEDGED

Ignoring birds that failed to fledge any cygnets, life-period cygnet productions ranged from 1 to 37, with a mode of 3 in all areas. There were no differences between the sexes (Table 22.6 and Fig. 22.3 right).

Table 22.5 Numbers of successful hatchings per life-period for individual swans. Only birds known to have hatched young are included: there were no differences in numbers of successful hatchings between the sexes within areas.

Area	Number hatching	Mean	Standard deviation	Range
KBbE	119	2.5	1.9	1–9
KBbL	105	2.1	1.6	1–8
RKbL	46	1.4	0.9	1–6

However, compared to KBbL born swans, RKbL born birds had lower proportions fledging cygnets (0.065 versus 0.037, $P < 0.001$), and much lower numbers fledged per life-period for those that did succeed (means of 7.8 and 3.5 respectively, distributions significantly dissimilar at $P < 0.001$). No RKbL swan fledged more than seven cygnets, whereas 42 % of KBbL swans fledged more than seven cygnets.

The average territorial bred cygnet (that also returned to nest within the study areas) raised 0.704 ± 3.235 SD cygnets; the average territorial cygnet that bred successfully raised 8.961 ± 7.730 cygnets. Colonial swans raised only 25 % as many young as territorial ones when failed breeders were included, and 45 % as many when failed breeders were excluded.

Only 2 % of KB swans produced 10 % of KB cygnets, while about a fifth of them produced half the total cygnets (Table 22.6B). This disproportionate production by a few birds was less evident in RK swans, which had much lower chances of nesting repeatedly and of hatching young.

Life-period cygnet productions were bimodal at 0 and 3 for all areas and sexes, while the means and variances of both the RK males and females were lower than for the KB swans (Table 22.6C).

TERRITORIAL OR COLONIAL NESTING

The results given above classified swans as KB or RK according to their natal origin, irrespective of where they subsequently bred. The majority of swans returned to nest in the area of their birth and we have no records of any swans breeding outside the study areas. However, the fledging successes of some 10 % of KB-hatched cygnets which subsequently nested in the RK colonies were more similar to the RKbL sample ($P \backsimeq 0.80$) than to the KBbE or KBbL samples ($P \backsimeq 0.08$). This result confirms that colonial breeding entailed low success, even for birds that had themselves been raised on territories. Nor was there any evidence that the sizes or weights of adults that produced cygnets differed between KB and RK breeders (but see discussion below).

PREDICTING LIFE-PERIOD CYGNET PRODUCTION

Although many of the life history features considered above were inter-correlated some useful conclusions could be drawn. Life-history parameters relating to how many times a swan bred, and where it bred, were highly predictive, explaining 49–85 % of overall variance of life-period cygnet production. Conversely, the swans's sexes, and the ages at which they

Table 22.6A Estimated lifetime cygnet production by area and period.

Estimated life-period cygnet productions by individual swans which were themselves raised in each area, both including (Incl.0) and excluding (Excl.0) birds that failed to produce young.

	KBbE		KBbL		RKbL	
	Incl.0	*Excl.0*	*Incl.0*	*Excl.0*	*Incl.0*	*Excl.0*
Total swans	1159	91	1130	73	766	28
Mean	0.7	9.0	0.5	7.8	0.1	3.5
SD	3.2	7.7	2.5	6.2	0.7	1.4

Table 22.6B Proportions of total cygnet production by different percentages of successful adults.

Area	Number of adults	Total cygnets produced	Percentage of adults which produced the column's percentage of cygnets		
			50 %	*10 %*	*5 %*
KBbE	91	802	22	2	1
KBbL	73	566	23	3	1
RKbL	28	98	40	7	4

Table 22.6C Life-period cygnet productions (to fledging), for swans known to have nested, whether or not they produced cygnets, by area and sex.

Area	Sex	Numbers of swans	Modes		Mean	Standard deviation	Variance	Range
KBbE	Female	70	0,	3	7.05	8.04	64.69	0–37
	Male	58	0,	3	5.17	7.36	54.25	0–37
KBbL	Female	65	0,	3	5.28	6.48	42.00	0–32
	Male	47	0,	2	4.67	6.12	37.45	0–23
RKbL	Female	54	0,	3	0.72	1.43	2.05	0–6
	Male	25	0,	3	2.04	2.24	5.04	0–7

first attained differing degrees of breeding success, had no significant effects on lifetime productions. Area of birth was just significant (explaining 8 % of variance), largely because it was associated with the subsequent nesting situation.

In a series of regression models, various life-history features predicted 68–85 % of the variance in life-period cygnet productions. The predictor variables were not independent, but the explanatory power of the simplest was high, and the degree of improvement given by the others was low: whatever the relative contributions of the inter-correlated parameters, the biological interpretations were clear. In each model the equations referred only to swans that succeeded in fledging one or more cygnets, as these were the only birds for which complete information was available.

Model 1 explained 68 % of variance in life-period cygnet production: each swan fledged 1.7 cygnets, plus 2.7 for each year it bred on a territory but only plus 0.4 for each year it bred in a colony. *Model 2*, added number of years hatching as a further predictive factor and explained 71 % of variance: then, each swan fledged 1.0 cygnets, plus 1.5 per year for each year it hatched its eggs, plus 1.3 cygnets for each year it bred on a territory but minus 0.3 per year for each year it bred in a colony; the colony factor became non-significant, but the equation makes little sense if it is excluded. *Model 3* added number of years cygnets were fledged to the equation and explained 85 % of variance: this then became the only significant factor, 3.9 cygnets per year-fledging; but the effects of territorial and colonial breeding were still positive and negative respectively. *Model 4* added nesting attempts per mate (a possible index of the effects of changing a mate on breeding success) and achieved no further explanation (85% variance): nor could any additional significance be attributed to "nests-per-mate" in several other models.

Major factors in determining the life-period cygnet productions of our Mute Swans were clearly both their numbers of nesting attempts and whether those attempts were made on territories or in colonies. The proportions of variance of life-period cygnet production explained by these two factors were high (Model 1, 68 %; Model 2, 71 %), and similar to the proportion explained by the life-period totals of cygnets hatched (74 %).

Discussion

Winter severity, as indicated by the number of ice-days on Danish coastal waters, affected both survival and size of subsequent breeding populations (see also Andersen-Harild 1981). Although this effect was most clearly established for the Kobenhavn area, it seemed inconceivable that it would not also apply to the adjacent population at Roskilde. We attributed the lower significance of the effect of ice-days on subsequent breeding numbers at the Roskilde colonies to the over-riding influence of rapid population increase there.

The small differences in first winter survival between the sexes in the RKbL sample were consistent with the lower autumn weights of Roskilde cygnets compared to Kobenhavn cygnets. The Roskilde males, being larger than their female siblings and requiring more energy, might have been harder hit by severe weather (in very severe weather there was virtually no food, so the social dominance of larger males might not have helped them). The differences between KBbL and RKbL swans in the proportions of cygnets whose fates were known could probably be explained by higher probabilities of re-sighting the territorial birds if they survived to maturity and attempted to nest.

The number of nesting attempts per life-period (Table 22.4) was slightly higher for KBbE (as might be expected from the less truncated data) and lowest for RKbL. Although none of the distributions were significantly different, regression analysis showed that RK swans had lower prospects than KB swans of nesting in years subsequent to their first attempt (24 % versus 66 % per year of further life).

The difference between the territorial and colonial nesting birds was manifest chiefly in hatching success (Table 22.5). Compared to KBbL swans, RKbL birds were 0.61 times as likely ever to hatch, and hatched only 0.67 times as often if they did hatch; the difference was largely due to the high probabilities of clutch desertion or egg breakage resulting from fights over nesting space in the colonies (Bacon & Andersen-Harild 1987). This gave RK swans a relative disadvantage of 0.41 (= 0.67 * 0.61) for the hatching stage of the breeding cycle alone. No individuals breeding within the colony did consistently well; colonial conditions seemed bad for all swans.

Table 22.6 presents the critical data on production of fledged cygnets per life-period. There were no significant differences between the sexes, but RKbL swans succeeded less often than KBbL birds, and also fledged fewer cygnets per life period if they did succeed. The relative disadvantages per life-period were 0.25 including birds which never produced cygnets and 0.45 if the comparison was confined to those that did produce cygnets.

The number of times a swan was able to start nesting explained some two-thirds of the variance of its life-period production of cygnets to fledging. In view of the effects of severe winter weather on survival and breeding prospects, life-time production must depend heavily on the weather conditions the swans experienced during their lives. Severe winters often occur at long intervals, so a "lucky" bird would fledge the year following a severe winter, develop quickly to maturity before the next, survive, and breed several times before another; conversely, a less fortunate bird would be severely set back by a hard winter during immaturity, and reach breeding condition late, if at all; a really "unlucky" bird would fledge just before a severe winter, and die during it. This is a clear example of the "silver spoon" effect, *sensu* Grafen (1988, page 459), and we outline below why it would be difficult to allow for this effect in our analysis of lifetime reproductive success.

Our Mute Swan data derived from cohorts of broods of individuals, bred in the same sites and years, and were thus not independent. Cold winters would have decreased the survival and condition of particular cohorts and other swans in the population. Cold winters also decreased the size of the following year's breeding population, and subsequently, when the survivors did breed, may have reduced density-dependent competition. Such a spate of deaths, followed by reduced competition, could have increased the chances that new pairs would form for the first time and that survivors would take new mates to replace dead partners. Hence the spring following hard winters, when survivors might also be in poor condition, may have had higher than usual proportions of pairs containing one or two first-time-breeders. All these interesting effects would not have been independent from the direct consequences of the severe winters; nor would analyses confined to within cohort-data remove the worst of the inter-dependencies.

The surest way a swan can promote its reproductive success is to start breeding as early in life as possible (Sibly & Calow 1986), thereby increasing its potential nesting life-span, as there is always a risk of dying if the next winter happens to be severe. As many of our swans only nested once (Fig. 22.3 left), any hypothetical advantages of breeding success increasing with age after deferrment would have to be enormous to outweigh the risks of missing the only chance to breed the bird might get.

We will finally try, with caution, to set our results in a population context. The estimates of life-period production in Table 22.6A are minimal, as our procedures have always been conservative, and some birds were still alive and breeding at the end of the study. Hence, it is possible that the absolute figures for mean production (including zero fledgings) should be closer to 0.8 for KB territorial birds and 0.25 for RK colonial birds. However, neither of these figures took account of cygnets bred within the study areas that survived and, unknown to us, bred successfully elsewhere.

Taking the figure of 0.8 life-time cygnets per swan for the Kobenhavn area, we would only need another fifth of our fledged cygnets (e.g. an extra 20 swans for the KBbE sample) to have survived and bred equally successfully elsewhere to have a stable population (1.0 cygnets fledged per fledging-cygnet). If one and a half times as many survived and bred we would have a production of 2.0 cygnets per fledging-cygnet, a rate sufficient to produce the observed increase in the Danish population over the past 60 years without immigration.

In contrast, even optimistic upper limits for production from the Roskilde colony would need to be increased four-fold just to achieve replacement of dead breeders. As we had no evidence of Roskilde-bred cygnets managing to breed in the Kobenhavn area (but several cases of the converse), it seemed, in the circumstances pertaining at Roskilde, most unlikely that colonial breeding *per se* was a viable reproductive strategy.

Breeding at Roskilde was probably a consequence of desperation: if population density was such that the prime habitats required for territorial breeding were full, then less successful breeding in colonies would have been better than not breeding at all; especially if the prospects for survival and attaining territories were low. However, to account for our lack of observations of colonial bred cygnets breeding in territories, we must additionally hypothesize that colonial bred cygnets are unable to acquire territories; behavioural imprinting on natal sites, or inability to reach adequate physical condition, would be plausible explanations for this failure.

COMPARISON WITH BEWICK'S SWANS

It is not easy to compare our results with those of Scott (1988) for Bewick's Swan *Cygnus columbianus bewickii*, because of the differing ways in which the data were collected. Scott relied on winter observations of pairs with families, so the birds included in her samples had already successfully completed migration. Thus her mortality estimates, which were lower than ours at all ages, were taken post-migration, whereas ours were post-fledging, and *C.olor* in our study sometimes had very high first-winter mortalities. These higher mortalities gave us small numbers of birds for which the effects of age and mate changes could be assessed, while our shorter runs of data per bird showed considerable stochastic variation, depending on which winters the birds experienced.

Our *C.olor* first nested at ages between two and 15 years: Scott recorded ages at "first producing cygnets that had survived migration" of

2–10 years in *C.c.bewickii*. In addition, she was obliged to assume that breeding life-spans for *C.c.bewickii* started at two years of age in assessing contributing factors to lifetime reproductive success.

Scott reported that in *C.c.bewickii* breeding success increased with age and with the duration of pair bonds; it is unlikely that many of our *C.olor* survived long enough for such effects to be important, and those effects would have been hard to separate from the consequences of "severe-winters". In *C.olor* we could only demonstrate a reduced annual success for "first-time" breeders. However, given the differing ecologies of the two species, differences in the advantages of previous experience might be expected. *C.c.bewickii* lives long enough to benefit from previous experience (Scott 1988), and pairs have such a short time-span for breeding in the arctic tundra that they have to return very early and start incubating in unthawed nest mounds remaining from previous years; also, unlike *C.olor* and most other swans, male *C.c.bewickii* participate heavily (20–50 %) in incubation (Krechmar & Kondratyev 1986). With such premiums on early return to precisely known pre-existing nests and on a high degree of cooperation during incubation, previous experience would be expected to enhance success in *C.c.bewickii*.

Scott was able to demonstrate correlations between life-time fledging success of *C.c.bewickii* and skull length, weight and social dominance in winter, especially for males. One might expect similar relationships for *C.olor*, which has a similar social structure. However, we had no data on dominance hierarchies, and our weights referred to adults in summer, and were not strictly comparable because of variations in timing of breeding and available food supply at different sites. The lack of correlation between adult size of *C.olor* (as indicated by width of foot-web) and breeding success was surprising, but, at least within the colonies, competition and fighting were so intense that the smaller birds may have failed to produce young and so would not have been represented in our samples of adults caught with large cygnets.

Despite these differences, both *C.olor* and *C.c.bewickii* seem to have lifetime fledgling productions dependent on their lifetime number of breeding attempts, rather than on brood sizes in their years of success.

In conclusion, lifetime cygnet production of *C.olor* in Denmark was heavily influenced by winter weather, which affected both survival through the first winter and age at first breeding. Once swans had begun to breed, their lifetime productions were strongly affected by their attained numbers of nesting attempts, which were also affected by winter weather. Annual breeding success was heavily dependent on whether the swans were nesting in large territories (85 % hatching success per lifetime) or in a colony (55 % hatching success). Colony swans that succeeded in producing cygnets achieved only 45 % of the fledgling production of territorial swans. But, as many colonial swans failed to fledge any cygnets at all, their relative success was even lower, 25 %, assuming that all swans fledged from each area were equally likely to be found when breeding.

Acknowledgements

The Danish Mute Swan study was supported by the Bird Ringing Centre, Zoological Museum, Kobenhavn. We thank Eric Hansen, Eddie Fritze, Niels Preuss and numerous other bird ringers for assistance with censuses and ringing, and Eric Hansen for setting up and maintaining the initial computer files of swan records. Frances Mitchelmore, Drs Jane Sears and Ian Newton and an unknown referee made constructive comments about earlier drafts. Finally we thank our wives, not only for their enthusiastic support of our swan studies, but for their frequent understanding when we let these take precedence over urgent domestic matters.

References

Andersen-Harild, P. 1978. *Knopsvanen.* Holte, Denmark: Skarv Nature Publications.
Andersen-Harild, P. 1981. Weight changes in *Cygnus olor.* In *Proceedings of Second International Swan Symposium, Sapporo, Japan,* ed. G.V.T. Matthews & M. Smart, pp. 359–78. Slimbridge GL2 7BX, UK: International Waterfowl Research Bureau.
Andersen-Harild, P. & Preuss, N.O. 1978. Optaelling of Ynglende Knopsvaner i 1970. *Feldornitologen* **20**: 34–5.
Bacon, P.J. 1980a. Population genetics of the Mute Swan (*Cygnus olor*). D.Phil. thesis, University of Oxford. D321193/80 (BLLD F).
Bacon, P.J. 1980b. Status and dynamics of a Mute Swan population near Oxford between 1976 and 1978. *Wildfowl* **31**: 37–50.
Bacon, P.J. & Andersen-Harild, P. 1987. Colonial breeding in Mute Swans (*Cygnus olor*) associated with an allozyme of lactate dehydrogenase. *Biol. J. Linn. Soc.* **30**: 193–228.
Bacon, P.J. & Coleman, A.E. 1986. An analysis of weight changes in the Mute Swan, *Cygnus olor. Bird Study* **33**: 145–58.
Beekman, J.H. *et al.* 1990. Laying date and clutch size in relation to body condition in the Mute Swan (*Cygnus olor). Wildfowl,* in press.
Birkhead, M.E., Bacon, P.J., & Walter, P. 1983. Factors affecting the breeding success of the Mute Swan (*Cygnus olor*). *J. Anim. Ecology* **52**: 727–41.
Bloch, D. 1970. Knopsvaner som kolonifugl i Danmark. *Dansk Ornithologiisk Forenings Tidsskrift* **64**: 152–62.
Bloch, D. 1971. Ynglebestanden af Knopsvane (*Cygnus olor*) i. Danmark i 1966. *Vildtundersogelser* **16**: 1–47 (Vildbiologisk Station, 1971).

Brown, D. 1988. Components of Lifetime Reproductive Success. In *Reproductive Success*, ed. T.H. Clutton-Brock. pp. 439–453. Chicago: University Press.

Coleman, A.E. & Minton, C.D.T. 1980. Mortality of Mute Swan progeny in an area of South Staffordshire, UK. *Wildfowl* **31**: 22–8.

Cramp, S. & Simmons, K.E.L. 1977. *Handbook of Birds of Europe, the Middle East and North Africa; the Birds of the Western Palearctic*: I. Oxford: University Press.

Grafen, A. 1988. On the Uses of Data on Lifetime Reproductive Success. In *Reproductive Success*, ed. T.H. Clutton-Brock. pp. 454–471. Chicago: University Press.

Krechmar, A.V. & Kondratyev, A.Ya. 1986. Comparative ecological analysis of *Cygnus bewickii* and *Cygnus cygnus* breeding. In *Experimental Methods and Results of their Applications in Northern Bird Studies*, eds. A.V. Andreev & A.V. Krechmar, Vladivostock: Academy of Sciences of USSR Far Eastern Centre.

Munroe, R.E., Smith, L.T. & Kupa, J.J. 1968. The genetic basis of color differences in the Mute Swan (*Cygnus olor*). *Auk* **85**: 504–5.

Perrins, C.M. & Ogilvie, M.A. 1981. A study of the Abbotsbury Mute Swans. *Wildfowl* **32**: 35–47.

Perrins, C.M. & Reynolds, C.M. 1967. A preliminary study of the Mute Swan (*Cygnus olor*). *Wildfowl* **18**: 74–84.

Ogilvie, M.A. 1967. Population changes and mortality of the Mute Swan in Britain. *Wildfowl* **18**: 64–73.

Ogilvie, M.A. 1972. Large numbered leg bands for individual identification of swans. *J. Wild. Mgmt.* **36**: 1261–5.

Reynolds, C.M. 1972. Mute Swan weights in relation to breeding. *Wildfowl* **23**: 111–18.

Scott, D.K. 1988. Breeding success in Bewick's swans. In *Reproductive Success*, ed. T.H. Clutton-Brock, pp. 220–236. Chicago: University Press.

Sibly, R.M. & Calow, P. 1986. Why breeding earlier is always worthwhile. In *Proc. 20th Population Genetics Group meeting*. Nottingham Univ., UK.

Sladen, W.J.L. 1974. International colour marking codes for swan and goose studies. In: M Smart (ed) pp. 310–317. *International conference on the conservation of wetlands and waterfowl*, Heilignhafen, FRG, 2–6 Dec 1974. pp. 310–17. Slimbridge, GL2 7BX, UK: International Waterfowl Research Bureau.

23. Red-billed Gull

J.A. MILLS

This chapter examines the number of fledglings and recruits to the breeding population produced by individual Red-billed Gulls *Larus novaehollandiae scopulinus* and identifies some factors which influence lifetime production for each sex. The relationship between the productivity of the parents and their offspring is also examined. The data for the study were collected over a 24 year investigation of the Red-billed Gull at the Kaikoura Peninsula, New Zealand.

Characteristics of the Red-billed Gull and the Kaikoura population

The Red-billed Gull is common on the New Zealand coast, and generally breeds in large colonies at high densities. Recognized as subspecifically different from the geographic races of the Australian Silver Gull (Dwight 1925), it has close affinities to *L. hartlaubii* from south-western Africa (Johnstone 1982). The sexes are alike but males (mean 299.5 g ± SD 27.6) tend to be slightly larger than females (259.7 g ± 23.0). Adult plumage, which is almost completely white except for the pale grey mantle, back and wing coverts, and black primaries, is attained together with the scarlet bill, eyelids and feet in the second year.

At Kaikoura, the third largest breeding colony in New Zealand, the bird feeds predominantly in the inshore region on surface swarms of the planktonic euphausiid *Nyctiphanes australis*.

The Kaikoura population is virtually closed. Searches at nearby colonies have revealed that less than 1 % of Kaikoura birds emigrate and there is little immigration from other colonies (Mills 1973).

The Kaikoura breeding population increased from 4,380 pairs in 1964–1965 to 5,678 in 1968–1969, but stabilized in the 1970s and 1980s

LIFETIME REPRODUCTION IN BIRDS
ISBN 0-12-517370-9

at between 5,400 and 6,400 pairs. There is a large non-breeding component. For example, in 1983–1984, when 5,888 pairs nested, the total adult population was estimated by capture—recapture analysis to be 23,192 ± SE 1,674 individuals. The non-breeders therefore numbered between 9,742 and 13,090 individuals, forming roughly half the population. Most non-breeders were females (Table 23.1), the imbalance arising because females had a higher annual survival (89.4 % ± SE 2.6) than males (84.4 % ± SE 2.9) with the sex ratio widening in favour of females with age (unpublished data), and through differential survival of some cohorts prior to breeding (Table 23.7).

Red-billed Gulls are monogamous, with 82 % of pair bonds being retained from the previous season (Mills 1973). The age of first breeding is variable (Table 23.2). In heterosexual pairings males commence breeding on average 1.05 years younger than females, and by the fourth year some 96 % of males have bred compared with only 63 % of the females. No female paired with a male commenced breeding at two years of age, but 11 % of females in female/female associations bred as two year olds (Table 23.2). In three of the 24 seasons studied males were found breeding as yearlings. Only one brood is raised per year, but re-laying occurs frequently after the loss of a clutch or a brood. The usual clutch is two eggs (76 %), but clutches of one (17 %) and three (7 %) are reasonably common (Mills 1973). Seldom, however, are three chicks fledged. Nests which contain more than three eggs are invariably owned by female–female pairs. Predation is a major cause of nest mortality, accounting for about 25 % of eggs and 17 % of chicks. Introduced Ferrets *Mustela putorius furo*, Stoats *M. erminea* and cats prey on adults, chicks

Table 23.1 Proportion of colour-marked female and male Red-billed Gulls of known age (colour-marked prior to 1983) breeding at Kaikoura in 1983–1984.

	Females		Males	
Age	No. alive	% bred	No. alive	% bred
5– 9 years	19	74	38	84
10–14 years	46	74	40	90
15–19 years	111	55	62	85
20–25 years	58	43	22	91
Unknown age	261	46	145	86
Total	495	51	307	86

Table 23.2 Age of first breeding of Red-billed Gulls colour-marked as chicks in 1980–1981.

Sex	Number recovered breeding	Age of first breeding (years)						Mean	SD
		1	2	3	4	5	6		
Male	28	0 %	18 %	35 %	43 %	4 %	0 %	3.32	0.82
Female	28	0 %	4 %	18 %	46 %	21 %	11 %	4.18	0.98
Female paired with male	19	0 %	0 %	11 %	52 %	26 %	11 %	4.37	0.83
Female paired with female	9	0 %	11 %	33 %	33 %	12 %	11 %	3.78	1.20

and eggs; Black-backed Gulls *Larus dominicanus* take small and medium-sized chicks and some Red-billed Gulls specialize in egg-robbery.

Following breeding the majority of birds disperse and are usually found within 380 km of Kaikoura.

Collection, assessment and analysis of data

The banding of nestling Red-billed Gulls at Kaikoura has been undertaken annually for 29 years (1958–1986) and more than 70,000 have been marked. Up until 1986, 4,408 adults and 1,078 chicks were colour-banded for identification.

Each season all nests of colour-marked gulls were numbered and the fate of each determined. Chicks were banded with numbered metal bands within four days of hatching. Very few colour-marked gulls bred without being observed, and efforts were made to identify those which were non-breeders. Over the 20 years since colour-banding began, fledging success was monitored in all but four seasons (1969–1970, 1972–1973, 1973–1974 and 1974–1975). Because of the four missing seasons, lifetime analysis could be applied only to the period 1975–1976 to 1986–1987.

Two measures of lifetime productivity were calculated for both sexes: the total number of young raised to fledging, and the total number of progeny recruited into the local and nearby breeding populations. The numbers of recruits are minimum estimates because not all birds which were marked only with metal bands and which survived to breed were captured. Nevertheless, efforts were made to identify as many banded breeders as possible, and some 500–900 were captured annually. An individual was considered to have bred when it or its partner laid an egg.

Only a small proportion of the colour-marked individuals studied was eligible for inclusion in the lifetime analysis. The age of first breeding had to be known and uncertainty over lifespan and fledging success eliminated many individuals.

Only 66 females and 81 males (Fig. 23.1) satisfied the selection criteria. The oldest individual was 16 years old, but because Red-billed Gulls can live for 28 years, the measures of lifetime reproductive success will be underestimated. Nevertheless, the data in this paper cover a large range of the lifespans and so provide a useful approximation of lifetime production of the population.

Lifetime production

Fledging Success

Both sexes showed considerable variation in the number of young fledged per lifetime, with the variability being greater in females (mean 3.4 with

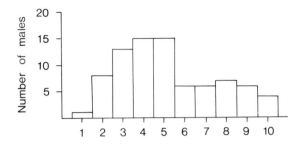

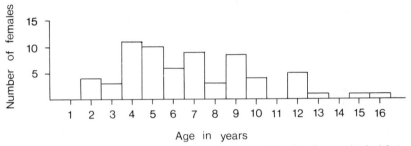

Figure 23.1 Age composition of 81 males and 66 females from which lifetime production was calculated.

variance 8.9) than in males (mean 3.0, variance 4.7) (Fig. 23.2). For such a long-lived species, the maximum number fledged was surprisingly small, six for males and nine for females. Of those which attempted to breed, 36 % of the males and 39 % of the females produced no fledglings in their lives. Among those which fledged young, 42 % of the males and 48 % of the females fledged only one or two young. Overall, 20 % of the males in the population produced 58 % of all fledglings and 15 % of females produced 52 %.

Recruitment to the Breeding Population

Only 17 % of males and 24 % of females which bred produced young which survived to breed. The probability of a fledgling being recruited into the breeding population was independent of the number of fledglings produced by either parent in their lifetimes (Fig. 23.3). In other words, post-fledgling survival was not related to the quality of the parent as measured by the total number of young they fledged. Birds which produced a large number of fledged young tended, simply by weight of numbers, to have more progeny recruited into the population than less productive individuals; this relationship was strongly correlated for females

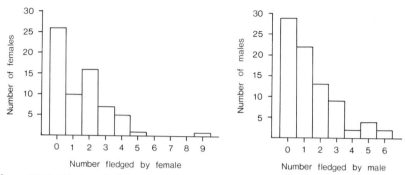

Figure 23.2 Lifetime fledging success of 66 females and 81 males.

($r_s = 0.533$, $n = 40$, $P < 0.01$), but less so for males ($r_s = 0.295$, $n = 52$, $P < 0.05$).

FACTORS INFLUENCING LIFETIME REPRODUCTIVE RATE

Longevity

Male and female longevity accounted for approximately 30 % of the variance in lifetime production of fledged young (Fig. 23.4), and about 5 % of the variance in production of breeders (males, $r = 0.232$, $n = 81$, $P = 0.04$; females, $r = 0.244$, $n = 66$, $P < 0.05$).

Considering that the average clutch size was 1.8 eggs (Mills 1973), the actual lifetime production of fledged young was well short of the maximum possible relative to the individual's age. Within each age group there was variation in number fledged, a disparity which increased with age (Fig. 23.4).

The fledging of four or more young was achieved only by parents which lived at least six years. Some 31 % of the females which lived for 7–12 years were unproductive, compared with only 9 % of males.

Frequency of Breeding

For birds with the same lifespan, males fledged on average more young than females (Fig. 23.4), largely because males began breeding at a younger age and bred more often than females. These differences resulted from the excess of females in the population, and the difficulty which some females had in obtaining or retaining a mate. The 66 females in the sample bred once for every 3.2 (± 1.9 SD) years lived, whereas the 81 males bred once every 2.6 (± 1.2 SD) years lived. Males living 9–10

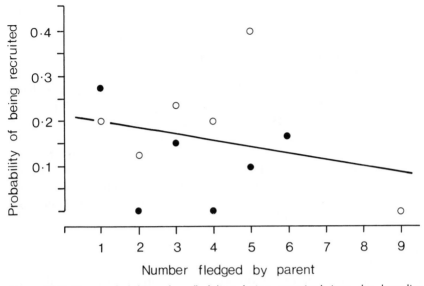

Figure 23.3 The probability of a fledgling being recruited into the breeding population in relation to the number of fledglings male (●) or female (○) parents raised in their lifetime. Females $y = -0.013x + 0.202$; $r = -0.244$, $n = 12$, P = ns.

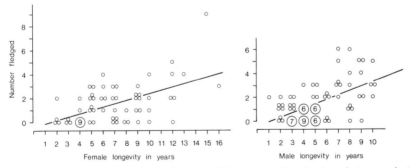

Figure 23.4 Lifetime fledgling production of females and males in relation to their lifespan. Females $y = 0.28x - 0.456$, $r = 0.543$, $n = 66$ $P = < 0.05$; males $y = 0.37x - 0.515$, $r = 0.56$, $n = 81$, $P = 0.04$.

years bred 5.4 times, whereas females of the same age bred only 3.8 times.

In a typical year at Kaikoura only 51 % of the adult females attempted to breed, compared with 86 % of the males (Table 23.1). Amongst males the proportion which bred did not change significantly with age, but for females the proportion breeding decreased with age. For birds which survived 10 years after their first known breeding attempt, 49 % of males bred in all seasons, and only 4 % missed three or more seasons, whereas only 16 % of females bred every year and 51 % missed three or more seasons (Table 23.3).

The number of seasons a bird bred in its lifetime accounted for almost half the variance in fledgling production among both males (47.1 %) and females (46.3 %) (Fig. 23.5). The number of offspring surviving to two years of age was also correlated with the number of seasons the parents bred (females $r = 0.219$, $n = 66$, $P = 0.07$; males $r = 0.260$, $n = 81$, $P < 0.02$); however, the number of offspring recruited into the breeding population showed no such relationship among the females ($r = 0.182$, $n = 66$, $P = 0.14$) and accounted for only 5 % of the variability amongst males ($r = 0.233$, $n = 81$, $P < 0.04$).

Body Weight

Body weight was the only morphological feature recorded and this was taken during the breeding season to the nearest 5 g. Individuals showed little variation in weight from one season to the next.

Table 23.3 Frequency of breeding by male and female Red-billed Gulls which lived 10 years from their first known breeding attempt.

Sex	No. gulls considered	Percentage of birds which bred in different numbers of years over a 10-year period									
		1	2	3	4	5	6	7	8	9	10
Male	45						2	2	11	36	49
Female	45		2	7	4	7	22	9	20	13	16

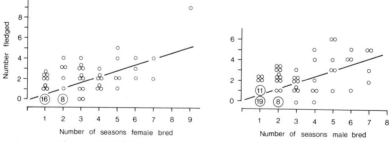

Figure 23.5 Lifetime fledgling production of females and males in relation to the frequency of breeding. Females $y = 0.610x - 0.167$, $r = 0.680$, $n = 66$, $P = 0.0001$; males $y = 0.605x - 0.066$, $r = 0.686$, $n = 81$, $P = 0.0001$.

During the course of their reproductive lives, heavier females hatched more eggs than lighter females ($r = 0.252$, $n = 58$, $P = 0.05$). There was also a tendency for heavier females to fledge more young in their lifetime, but the relationship was not significant ($r = 0.223$ $n = 58$, $P = 0.09$) and accounted for only 5 % of the variance (Fig. 23.6). There was no relationship between the weight of the female and the number of her offspring which survived to breed ($r = -0.020$, $n = 58$, $P = 0.88$).

The weight of the male had no effect on the number of eggs hatched in his lifetime ($r = 0.153$, $n = 70$, $P = 0.21$), but the number of young fledged was higher for heavier males ($r = 0.233$, $n = 70$, $P = 0.05$) (Fig. 23.6). This was probably because heavier males were better foragers, and so were able to provide more food for their young. However, this effect did not seem to ensure better post-fledging survival, for there was no association between the number of young recruited into the breeding population and male weight ($r = -0.015$, $n = 70$, $P = 0.90$).

Heavier females did not live longer than their lighter brethren ($r = 0.124$, $n = 58$, $P = 0.36$), or breed more frequently ($r = 0.222$, $n = 58$, $P = 0.09$), but they had more partners in their lifetime ($r = 0.3603$, $n = 58$, $P = 0.005$). Males showed no relationship between weight and longevity ($r = -0.003$, $n = 70$, $P = 0.97$), number of seasons bred ($r = -0.005$, $n = 70$, $P = 0.97$) or the number of partners per lifetime ($r = 0.014$, $n = 70$, $P = 0.91$).

Laying Date

The Red-billed Gull has an extremely long egg-laying season, extending from late September to the end of December. Laying becomes progressively earlier with increasing age, with an average difference of about 25

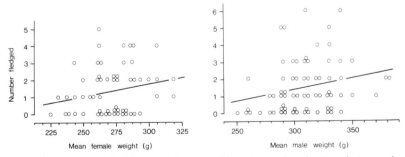

Figure 23.6 Lifetime fledgling production of females and males in relation to body weight. Mean weight was calculated for birds weighed on more than one occasion. Females $y = 0.015x - 2.707$, $r = 0.223$, $n = 58$, $P = 0.09$; males $y = 0.013x - 2.584$, $r = 0.233$, $n = 70$, $P = 0.05$.

days between three-year and 10-year old females (Mills 1973). Within this trend, however, individual birds generally remain early or late breeders.

Productivity was related to the timing of laying. The relationship between mean lifetime laying date of the female and the number of young she fledged over her lifespan explained 32 % of the variability in fledging success, with the earliest layers producing the most surviving young ($r = -0.566$, $n = 65$, $P = 0.0001$) (Fig. 23.7). For males, the relationship was highly significant, but explained only 15 % of the variability ($r = -0.393$, $n = 76$, $P = 0.0004$).

An indicator of the future lifetime production of a bird, particularly a female, was the date of laying in the first breeding season. Birds which were late nesters in their first season generally fledged fewer young overall than did early nesters. First laying date accounted for 20 % of the variance in female lifetime fledgling production ($r = -0.444$, $n = 63$, $P = 0.003$), but only 6 % amongst males ($r = -0.254$, $n = 74$, $P = 0.03$). The number of offspring recruited into the breeding population was also highly correlated with the laying date on first breeding for females ($r = -0.377$, $n = 63$, $P = 0.002$), but not for males ($r = -0.122$, $n = 74$, $P = 0.30$).

Longevity of the female was also correlated with the first laying date ($r = -0.364$, $n = 63$, $P < 0.01$), with early nesters living longer. No such relationship was apparent amongst males ($r = -0.042$, $n = 74$, $P = $ ns). Birds which laid late in the season tended to be those which could not compete successfully with others for food and hence could not achieve breeding condition earlier (Mills 1979). Evidently this inefficiency, or lack of competitive ability, remained throughout life.

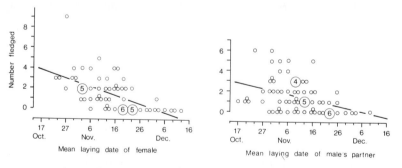

Figure 23.7 Lifetime fledgling production of females and males in relation to laying date. For birds which bred over a number of seasons, mean laying date was calculated. Females $y = -0.083x + 27.97$, $r = -0.57$, $n = 65$, $P = 0.0001$; males $y = -0.057x + 19.26$, $r = -0.39$, $n = 76$, $P = 0.0004$. (For calculating equations the calendar date was converted to sequential days of the year.)

Female/Female Pairings

Due to the difficulty that many females experienced in obtaining a male partner, some resorted to female–female pair bonds. Although no male was usually associated with the nest, approximately a third of the eggs hatched. Such females must either have solicited a male or been forcefully inseminated. Attempted extra-pair copulations commonly occurred.

Among the 66 females from which estimates of lifetime production were calculated, 15 (23 %) had been involved in female/female pairings at least once. As a group, the 66 females bred for a total of 177 breeding seasons and in 12 % of these occasions the pair bond was female/female. Of the 15 individuals involved, nine bred solely in such associations (60 %), whilst the remaining six bred in male/female pairs in some years. Females which paired with other females for at least part of their reproductive lives had lifetime fledging and recruitment rates that were only half those of females that mated exclusively with male partners (Table 23.4).

The fledging success and the proportion of offspring recruited into the breeding population were higher for bisexual females when they bred with males than when they bred with females but the differences were not significant (Table 23.5). Furthermore, the fledging success of bisexual females when breeding with males was lower than for females that bred exclusively with males, but again not significantly so (Table 23.5). These results highlight two features. Firstly, female/female pairings had low fledging and recruitment success, and secondly, the females which had a propensity for breeding in female/female pairs tended to be less successful breeders even with male partners.

COMPARISON OF PRODUCTIVITY, FREQUENCY OF BREEDING AND LAYING DATE IN PARENTS AND OFFSPRING

Productivity

I could not examine total lifetime reproduction of parents and offspring because insufficient time has elapsed to accumulate representative data for two generations. However, I could compare the productivity of parents and offspring over five seasons, starting from their first known breeding attempt (not necessarily the first actual attempt).

On this basis no significant correlations emerged between the number of chicks fledged by a male ($r = -0.024$, $n = 87$, $P = 0.82$) or a female ($r = 0.078$, $n = 119$, $P = 0.40$) and the number of chicks that their offspring fledged. There was also no significant correlation in fledging

Table 23.4 The lifetime productivity of females which were involved in a female–female pairing at some stage of their reproductive lives compared to females which were involved in only male–female pairings.

Pair-bond status	No. females	No. chicks fledged in lifetime	No. fledged per bird	No. recruited into population	No. recruited per bird
Female–female and male–female pairings	15	12	0.80	2	0.13
Male–female pairings only	51	85	1.70	15	0.29

Table 23.5 Comparison of the productivity of females which were: (a) involved exclusively with male–female pairings throughout their reproductive lives; (b) involved exclusively with female–female pairings; (c) involved with male–female and female–female pairings; (c(ii)) when bisexual females were breeding with a female; (c(ii)) when bisexual females were breeding with a male.

Category of pair bond	No. females	Seasons bred	No. fledged	No. fledged per season bred	No. recruited into breeding population	No. recruits produced per season bred
(a) exclusively female–male	51	138	85	0.62	15	0.11
(b) exclusively female–female	9	12	0	0	0	0
(c) bisexual females	6	27	11	0.41	2	0.08
(i) bisexual females when breeding with female	6	11	3	0.27	1	0.09
(ii) bisexual females when breeding with male	6	16	8	0.50	1	0.06

Note: Fledging success between bisexual females paired with males and females, $\chi^2_c = 0.61$, P = ns.
Fledging success between bisexual females breeding with males and exclusively heterosexual females, $\chi^2_c = 0.39$, P = ns.

success between father and daughter, father and son, mother and daughter and mother and son for each generation over the five seasons (Fig. 23.8). The same was true for the number of young recruited into the population (Table 23.6).

There was no relationship between the number of seasons the parents bred over five breeding seasons and the number that their offspring bred over a similar period of time (mother and daughter $r = -0.206$, $n = 45$, $P = 0.17$; mother and son $r = 0.107$, $n = 74$, $P = 0.36$; father and daughter $r = -0.131$, $n = 34$, $P = 0.461$; father and son $r = 0.015$, $n = 53$, $P = 0.92$).

Similarly, there was no significant correlation in mean laying date of mother and daughter ($r = 0.219$, $n = 45$, $P = 0.15$), mother and son ($r = -0.008$, $n = 74$, $P = 0.94$), father and daughter ($r = 0.25$, $n = 34$, $P = 0.15$), and father and son ($r = 0.018$, $n = 53$, $P = 0.90$). Nevertheless, the two correlation coefficients involving daughters were higher than those involving sons, perhaps indicating some heritability of laying date by the daughters.

SURVIVAL BETWEEN FLEDGING AND BREEDING

The analysis so far has examined the lifetime reproductive rate of individuals which survived to breed. Yet large numbers of birds died before they could breed.

Survival between fledging and breeding was calculated from samples of chicks colour-marked from 1979–1980 to 1983–1984 (Table 23.7). The sex of the chicks was not known at banding, but it was assumed that equal numbers of males and females were involved. Survivorship varied markedly between seasons and between sexes. From the small number marked in 1979–1980 none survived to breed. Equal numbers of males

Table 23.6 Correlations between the number of young recruited into the breeding population by parents and by their offspring over five breeding seasons.

	Recruits produced by female offspring over five seasons			Recruits produced by male offspring over five seasons		
Sex of parent	No.	Correlation coefficient	P	No.	Correlation coefficient	P
Females	45	0.0928	ns	74	0.0652	ns
Males	34	−0.0603	ns	53	−0.0743	ns

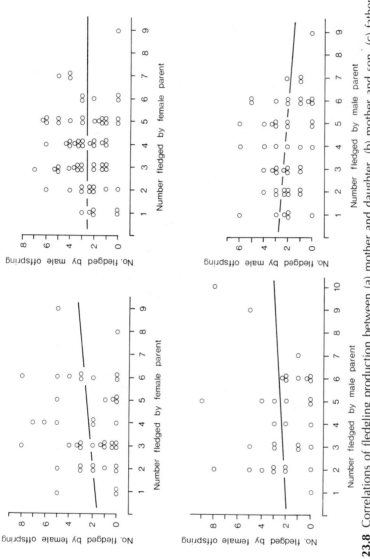

Figure 23.8 Correlations of fledgling production between (a) mother and daughter, (b) mother and son, (c) father and daughter, (d) father and son. Each generation was followed for five breeding seasons, starting from their first known breeding attempt. Each point represents the number of chicks fledged by the parent and its offspring over the five seasons. Mother and daughter $y = 0.18x + 1.647$, $n = 45$, $r = 0.141$, $r = 0.141$, $p = 0.36$; mother and son $y = 0.025x + 2.485$, $r = 0.023$, $n = 74$, $P = 0.85$; father and daughter $y = 0.126x + 1.932$, $r = 0.110$, $n = 34$, $P = 0.53$; father and son $y = -0.143x + 2.79$, $r = -0.176$, $n = 53$, $P = 0.21$.

Table 23.7 Survivorship from fledging to breeding. The number recovered breeding up to and including the 1986–1987 season. It is assumed that equal numbers of each sex were colour-banded as chicks.

Season hatched	No. colour-marked chicks of each sex which were fledged	No. colour-marked males survived to breed	No. colour-marked females survived to breed
1979–1980	14	0 (0 %)	0 (0 %)
1980–1981	82	28 (34 %)	28 (34 %)
1981–1982	32	0 (0 %)	6 (19 %)
1982–1983	50	28 (56 %)	11 (22 %)[a]

[a]More are expected to be recovered breeding as two more seasons are needed before all females alive breed.

and females survived to breed from 1980–1981, but, from the 1981–1982 breeding season, 19 % of the fledged females survived to breed but none of the males. Post-fledging survival of chicks from 1982–1983 was high for both sexes, with 56 % of the males and 22 % of females managing to breed. The last figure is a minimum estimate because insufficient time has elapsed for all females surviving to have commenced breeding.

Discussion

The lifetime analyses of 81 male and 66 female Red-billed Gulls which survived to breed showed considerable variation in the total number of young they raised to fledging. Frequency of breeding, longevity and laying date were the most important sources of variation in lifetime fledging success. As in the Sparrowhawk *Accipiter nisus* (Newton 1988), the mortality of young between fledging and breeding prevented most individuals from contributing fledglings to the next generation. Depending on the season, some 45–100 % of male fledglings in the sample and 66–100 % of female fledglings died before they could breed. Of those which bred, an additional 36 % of males and 39 % of females fledged no young in their lifetimes. Overall, relatively few individuals maintained the population from one generation to the next. Among those which bred, only 17 % of males and 24 % of the females recruited young into the breeding population during their reproductive careers.

The differences between male and female lifetime reproductive output mainly related to the excess of females in the population. Females lived longer than males but started breeding at an older age and bred less frequently. Some 49 % of adult females failed to breed each year and of those which bred approximately 6 % bred in female/female pairings. Twenty-three per cent of the females formed a female–female pair bond at some stage of their reproductive lives. Males, on the other hand, did not have to compete for partners and only about 14 % failed to breed each year.

Only 8 % of Red-billed Gull eggs gave rise to young which were recruited into the breeding population (unpublished data). This is less than 12 % reported by Newton (1985) for Sparrowhawks but the same as Hötker (1988) found for Meadow Pipits *Anthus pratensis*. This relatively low success, and the finding that the population was being maintained by a small number of successful individuals, led Newton (1985) to postulate that the genetic variance in the population could be greatly reduced if it were not continually counteracted by annual and local variations in selection, and by recombination and mutation. With only a small number of individuals contributing, favourable genotypes could spread quickly through the population, particularly if the traits were inherited (Newton 1985, Hötker 1988). For Red-billed Gulls, however, comparison of the productivity of parents and their offspring over a five-year period has shown that fledging success and the rate of recruitment were not strongly inherited (Fig. 23.8). In addition, laying date and frequency of breeding, factors which accounted for a large amount of variability in lifetime fledging success, were factors which were not significantly correlated between parents and offspring.

QUALITY OF THE PARENT

A number of findings from this study draw attention to the quality of the bird and its influence on lifetime production. As well as the differences in annual breeding success between individuals, there were differences in the frequency of breeding and date of laying which ultimately affected the lifetime reproductive rate. Some females bred infrequently and experienced difficulty in attracting or keeping a partner. Heavy body weight was linked to more frequent breeding in females and more likelihood of attaining a partner, but this accounted for only 5 % and 13 % respectively of the variability. Partly as a consequence of breeding more frequently, heavier females hatched more eggs in their lifetimes and tended to fledge more young, but in each case this accounted for only about 6 % of the variability. Another factor associated with heavier female body weight, which would assist in hatching relatively more eggs, was that heavier birds tended to lay larger eggs than lighter birds (Mills 1979). Egg size is an important factor in determining survival after hatching, at least in the Herring Gull *Larus argentatus* (Parsons 1975). Heavy body weight of males was also linked with higher lifetime fledging rates but, as with females, it accounted for only 6 % of the variability.

A characteristic of the less productive individuals was that they tended to breed later in the season, presumably because they were less efficient or competitive, and this is supported by the finding that late laying females had shorter lifespans than early breeders.

In summary, lifetime production of individual Red-billed Gulls varied considerably, more so in females than males. The surplus of females in the population caused differences in production between the sexes by deferring the age of first breeding and reducing the opportunity for subsequent breeding amongst females. Mortality between fledging and attainment of breeding age prevented the majority of individuals in the population from contributing young to the next generation. The population was maintained by a small number of productive individuals, but their productivity was not strongly inherited, and nor was the frequency of breeding or laying date, factors which accounted for the greatest variability in lifetime production.

Acknowledgements

Many people have helped me over the years. Drs B. Stonehouse, E.C. Young and M.C. Crawley provided advice and support in the early stages of the study. I am particularly grateful to Dr G.R. Williams and Dr M.C. Crawley, Directors of Research in the N.Z. Wildlife Service, and to Dr R.M.S. Sadleir and Dr P.J. Moors of the Department of Conservation, for allowing me to continue with the study.

Peter Shaw, Andy Garrick, Jack Cowie, Peter Moore, Ian Flux, Jane Maxwell, Chris Petyt, Bert Rebergen and Helen Mills contributed substantially to the field work; Nina Swift, Chris Petyt, Steve Jamieson, Karen Irik and Rod Cossee assisted with coding data for the computer with great accuracy; and John Yarrall developed computer programs used in the analysis.

I am grateful to Mr Jack Van Berkel and the Zoology Department, University of Canterbury for the use of facilities at Edward Percival Field Station, Kaikoura. Drs M.C. Crawley, P.J. Moors, M.J. Williams, R.M.S. Sadleir, G. Sherley and Messrs H.A. Best and D.G. Newman made constructive comments on the manuscript, and Mrs June Bullock typed various drafts.

References

Dwight, J. 1925. The Gulls (Laridae) of the World. *Bulletin of American Museum of Natural History* **52**: 63–401.

Hötker, H. 1988. Lifetime reproductive output of male and female Meadow Pipits *Anthus pratensis*. *J. Anim. Ecol.* **57**: 109–17.

Johnston, R.E. 1982. Distribution, status and variation of the Silver Gull *Larus novaehollandiae* Stephens, with notes on the *Larus cirrocephalus* species-group. *Records of Western Australia Museum* **10**: 133–65.

Mills, J.A. 1973. The influence of age and pair-bond on the breeding biology of the Red-billed Gull, *Larus novaehollandiae scopulinus*. *J. Anim. Ecol.* **42**: 147–62.

Mills, J.A. 1979. Factors affecting the egg size of Red-billed Gulls *Larus novaehollandiae scopulinus*. *Ibis* **121**: 53–67.

Newton, I. 1985. Lifetime reproductive output of female Sparrowhawks. *J. Anim. Ecol.* **54**: 241–53.

Newton, I. 1988. Individual performance in Sparrowhawks: the ecology of two sexes. *Proc. Int. Orn. Congr.* **19**: 125–54.

Parsons, J. 1975. Seasonal variation in the breeding success of the Herring Gull: an experimental approach to pre-fledging success. *J. Anim. Ecol.* **44**: 553–73.

24. Short-tailed Shearwater

R.D. WOOLLER, J.S. BRADLEY, I.J. SKIRA & D.L.
SERVENTY

The Short-tailed Shearwater *Puffinus tenuirostris* is a medium-sized (500 g) seabird of the order Procellariiformes. Like other shearwaters, petrels and albatrosses, it is essentially pelagic and wide-ranging but clumsy on land. Shearwaters are colonial burrow-nesters and about 23 million Short-tailed Shearwaters breed annually on islands and headlands of south-eastern Australia, especially Tasmania, in colonies of up to three million burrows.

An alternative name for the species, the muttonbird, stems from its exploitation for food and oil. Muttonbirds have been harvested commercially for well over a century, mainly as fat (about 1 kg) young taken from burrows. At its peak, this harvest took about one million young annually; currently about 350,000 young are harvested each year by commercial operators.

Short-tailed Shearwaters are transequatorial migrants. Exhausted and starved birds are often washed up on beaches in Japan, the Aleutian Islands, the western sea-board of North America and eastern Australia. This led to the suggestion that they followed a figure-of-eight migration path. However, more recent studies, especially ornithological transects of the Pacific, suggest that most birds merely fly north over the western Pacific to the Arctic and return southwards over the centre of that ocean. On either route, the birds would travel about 30,000 km annually. Adults leave their breeding grounds on this migration, deserting their chicks in the burrows from early April onwards. The young follow them, unaccompanied, two to three weeks later, by which time they weigh about 600 g, little more than an adult (Lill & Baldwin 1983).

Short-tailed Shearwaters are present at their breeding sites from September/October until April/May and, on land, are almost entirely nocturnal. After arrival, established pairs are re-united or new pairs

LIFETIME REPRODUCTION IN BIRDS
ISBN 0-12-517370-9

formed, the species being essentially monogamous. Existing burrows are occupied or new ones excavated before a pre-laying exodus of about 20 days. Egg-laying occurs between 23 and 28 November in 85 % of all instances known throughout the distribution of the species and extrapolation from harvesting dates suggests a similarly consistent and narrow laying period for over a century. Only a single egg is laid, and lost or unsuccessful eggs are not replaced. This invariable clutch size and laying period, together with strong fidelity, make the Short-tailed Shearwater well-suited to studies of lifetime breeding success.

The study population

In 1947, a study began of a small colony of Short-tailed Shearwaters on Fisher Island (40° 10'S, 148° 16'E), a 0.8 ha islet just off the southern tip of Flinders Island, in Bass Strait, between Tasmania and mainland Australia. Much of the island is bare granite but some friable soil, up to 0.6 m deep, supports tussock grass *Poa poiformis* in which shearwaters burrow. Occasionally, heavy rainfall floods these burrows during the breeding season. All burrows have been numbered individually, the identities of all birds in them recorded and the presence of an egg or a chick noted (Serventy & Curry 1984).

Thus, patterns of burrow occupation, mate change or retention, laying success and annual production of young have been followed continuously over 40 years at this colony. Banding started in 1947 and, since 1950, all adults in burrows have been marked with individually numbered, durable, metal leg bands. The longevity of this species means that bands progressively wear out, but a programme of double-banding has allowed consistency in identification of breeding adults (Wooller *et al.* 1985).

All fledgling young were also banded individually. Many later returned to their natal colony to breed. It was therefore possible to examine the ages of first-breeding of recruits to the Fisher Island population and to distinguish the effects upon reproductive success of both chronological age and breeding experience. If it is assumed that, once they have first laid on Fisher Island, birds do not breed elsewhere (see later), then disappearance of marked individuals may be used to estimate annual survival rates, both in relation to age and to reproductive status.

Recruitment and age of first-breeding

The total burrow population of shearwaters on Fisher Island is about 100–200 birds. A decline in the breeding population after the start of the study was attributed to handling birds outside burrows (Serventy & Curry 1984). This practice was stopped and the population has remained relatively stable for at least 25 years. Known-age recruits, first banded as nestlings on Fisher Island, comprise about 45 % of the breeding population. Some annual cohorts of young returned at a higher rate than others but no long-term trends in return rate were apparent (Serventy & Curry 1984). Just over half the returning birds were males. Known-age birds and recruits from elsewhere did not differ significantly in demography and were combined in all analyses, except where chronological age was a factor.

Young birds bred for the first time at a wide range of ages (4–15 years), but most between 5 and 8 (Table 24.1). The mean ages at first-breeding of males (7.3 years) and females (7.0 years) did not differ significantly. Offspring showed no tendency to breed first at the same age as either of their parents had done (Wooller *et al.* 1988). Most birds (41 %) bred in the year that they were first recorded on Fisher Island, or after one year at the colony (21 %), but some spent several years prospecting. Some of these pre-breeding birds, banded on Fisher Island, were found on other islands but no bird which bred on Fisher Island has been found breeding elsewhere (Serventy & Curry 1984).

Shearwaters that started breeding at the earlier ages appeared to have a slightly lower survival rate than those that deferred breeding for some further years (Table 24.1). No comparison was possible with non-breeding birds of the same age. The reproductive success of those birds starting very young was also relatively low, especially during the first few breeding attempts. There may thus be penalties associated with breeding first at too early an age.

Survival

Because only 0.6 % of breeding shearwaters returned to Fisher Island after an absence of three years or more, for purposes of analysis, birds not seen for three years were classed as dead. In addition, birds recorded as visiting the island before 1950 were excluded to ensure a high probability that the recorded year of first-breeding was indeed the year in which an individual bred for the first time. All birds were dead 27 years after first-breeding.

On average, male shearwaters lived for 9.2 years and females for 9.4 years after breeding for the first time. However, mortality rate was not constant throughout breeding life, but varied with breeding experience, and hence with age, in both sexes. Annual mortality was about 13 % in

Table 24.1 The reproductive success and expectation of further life of known-age Short-tailed Shearwaters breeding on Fisher Island, in relation to the ages at which they bred for the first time.

Age (years) at which first bred	Mean number of young fledged per year		Median survival time (years after first bred)	Number of individuals
	First three breeding attempts	All breeding attempts		
4	0	0	2.5	3
5	0.34	0.45	8.8	24
6	0.47	0.53	9.1	52
7	0.46	0.48	11.1	45
8	0.39	0.42	9.4	35
9	0.57	0.56	12.2	16
10+	0.44	0.44	13.2	11

the year following the first breeding attempt, fell to about 5 % a few years later before rising again especially in birds with twenty or more years breeding experience (Table 24.2). Although there was no difference between males and females in their overall survival rates, females appeared to have a slightly higher survival rate than males in the earlier years but a somewhat lower rate later.

Reproductive performance

Once they had started to breed, many Short-tailed Shearwaters known to be alive (i.e. returned in later years) were absent from the colony, especially during the early years of their reproductive careers (Table 24.2). Overall, 12 % of breeding birds were absent from Fisher Island in any one year, with no difference between males and females. It is most unlikely that these individuals were missed, or that they bred elsewhere.

In addition to these absences, about 19 % of all breeding birds (i.e. those that had laid at least once before) recorded in the burrows of Fisher Island each year had no egg. Possibly some were unable to secure a mate that year or had failed after laying, but many paired birds appeared to make no attempt to lay. Again, this phenomenon seemed more pronounced in the early part of their reproductive careers and birds may have enhanced their long-term survival by not breeding every year.

Among birds that did lay, the success with which young were raised to fledging increased from 43 % at the first attempt to over 70 % about 10 years later. This improvement in performance appeared to have at least two components, one involving the total cumulative breeding experience of an individual and the other its experience breeding with a particular partner. Both had separate and additive effects, similar in both sexes (Table 24.3). Thus, on their first breeding attempt with their first mate, birds produced 0.25–0.35 young per pair but this rose rapidly to 0.60–0.67 young if the pair bred together for three more years. However, by the time that birds bred for the first time with a fourth or subsequent mate, they produced 0.52–0.58 young per pair, presumably the result of accumulated breeding experience or greater effort.

Changing mates resulted in a temporary drop in reproductive perform- ance (Table 24.3). Among birds that had completed their reproductive careers, none had more than eight mates, 15 % had three mates, 24 % two mates and 42 % had only one mate (Wooller *et al.* 1988). Nearly 60 % had partners whose age differed from theirs by two years or less and there was no tendency for one sex to be the older partner. About 8 % of birds changed mate each year following the death of their partner

Table 24.2 Mean annual mortality, breeding success, percentage attendance and percentage laying in Short-tailed Shearwaters according to breeding experience.

Years since first bred	Mean (± SE) annual percentage mortality	Percentage of eggs laid that resulted in young fledged	Percentage of breeding birds known to be alive but not present at colony	Percentage of breeding birds present at colony that did not lay	Sample size
0	12.8 ± 1.7	43	—	—	470
1	9.6 ± 1.6	49	15	25	345
2	8.3 ± 1.6	59	17	24	297
3	5.0 ± 1.3	60	17	20	268
4	5.1 ± 1.4	66	11	19	264
5	7.8 ± 1.8	60	15	16	233
6	6.0 ± 1.7	69	9	21	228
7	9.1 ± 2.2	69	11	16	201
8	8.3 ± 2.3	74	11	19	172
9	8.0 ± 2.4	71	10	12	154
10	14.9 ± 3.5	79	11	18	126
11–12	11.1 ± 3.2	72	11	13	195
13–15	10.8 ± 3.9	69	8	16	212
16–19	10.1 ± 4.8	65	9	24	168
20–27	20.6 ± 10.0	56	10	10	134

Table 24.3 The mean annual number of fledged young produced by pairs of Short-tailed Shearwaters, in relation both to the duration of the pair bond and whether it was with their first, second, third or subsequent mate, separately for males and females. Sample sizes shown in parentheses.

	Number of years breeding experience with that mate			
	---	---	---	---
	0	1	2	≥ 3
Males:				
First mate	0.35 (196)	0.47 (119)	0.55 (72)	0.67 (200)
Second mate	0.38 (118)	0.49 (64)	0.66 (53)	0.69 (98)
Third mate	0.43 (74)	0.61 (55)	0.68 (34)	0.83 (102)
≥Fourth mate	0.58 (67)	0.62 (41)	0.59 (24)	0.54 (53)
Females:				
First mate	0.25 (156)	0.48 (103)	0.62 (62)	0.60 (144)
Second mate	0.43 (106)	0.44 (63)	0.67 (47)	0.78 (121)
Third mate	0.48 (77)	0.72 (52)	0.66 (34)	0.69 (108)
≥Fourth mate	0.52 (107)	0.63 (56)	0.62 (37)	0.68 (77)

but mate retention in surviving birds appeared related to reproductive performance. Some 33 % of all pairs which failed to produce an egg in the preceding season changed partners by divorce. The divorce rate in pairs which produced an egg, even though it failed to hatch, was 23 % and the equivalent rate in birds which fledged young was 15 % (Wooller *et al.* 1988). There were no differences between the sexes in these respects.

There is an indication in Tables 24.1 and 24.3 that the reproductive performance of birds nearing the end of their breeding careers was less than during their earlier years, although it was never as low as when they started breeding. In both sexes this occurred at about the time their mortality rates began to rise, indicative, perhaps, of senescence.

Factors affecting annual breeding success

These age-related trends in fecundity occurred against a background of marked annual variations in success. Although, on average, 61 % of all eggs laid resulted in free-flying young, the annual rate ranged between 25 % and 83 % (Serventy & Curry, 1984). Most young hatched were raised successfully to fledging. Occasional heavy rain sometimes flooded the burrows but breeding success was less than 50 % in only four years since 1955. No long-term changes in breeding success were apparent during the course of the study and no consistent variations were apparent between different areas of the colony.

There is no natural predation on Fisher Island but at other colonies snakes, feral cats and large gulls are more frequent predators. Limey-bird disease is present but relatively unimportant (Mykytowycz 1963). Overall, the most frequent cause of egg loss is flooding; most chick mortality is the result of starvation and occasional flooding of burrows.

Lifetime reproductive success

Both survival and reproductive output vary with breeding experience, and hence effectively with age, in Short-tailed Shearwaters. On average, the annual mortality of first-breeders is relatively high, falling thereafter to its lowest levels in mid-career before rising again in older birds. Conversely, the number of free-flying young raised per year was lowest when birds started breeding and rose rapidly thereafter before falling again towards the end of their breeding life. Thus the numbers of fledglings produced during their lifetimes by birds which were known to have completed their reproductive careers, increased with the lengths of their breeding lifespans, probably in an approximately linear manner, although variability was high (Fig. 24.1a).

Figure 24.1 For Short-tailed Shearwaters on Fisher Island known to have completed their reproductive careers: (a) The relationship between the mean duration of an individual's completed lifespan, from the time it first bred until its final disappearance, and the mean number of free-flying young that it produced during that period ($r_{10} = + 0.979$, $P < 0.001$; $y = 0.481x + 0.047$). (b) The relationship between the mean number of free-flying young produced in a completed lifetime and the mean number of those young which returned to breed on Fisher Island ($r_7 = + 0.989$, $P < 0.001$; $y = 0.116x + 0.097$). (c) The relationship between the mean duration of an individual's completed lifespan and the mean number of reproducing offspring produced during that period ($r_{10} = + 0.991$, $P < 0.001$: $y = 0.069x - 0.024$).

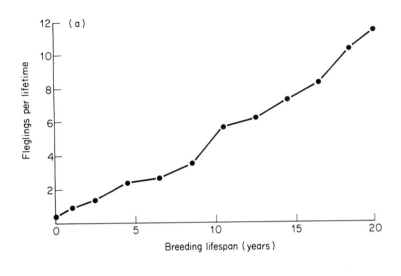

(a)
Fleglings per lifetime
12
10
8
6
4
2
0
5
10
15
20
Breeding lifespan (years)
(b)
2.0
1.0
0
5
10
15
20
Fledglings per lifetime
Reproducing offspring per lifetime
(c)
1.6
1.2
0.8
0.4
0
2
4
6
8
10
12
14
16
18
20
Breeding lifespan (years)

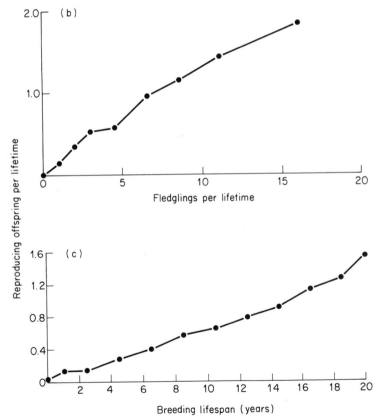

The proportion of young fledged which returned to Fisher Island to breed in their turn, increased slightly with an increase in the breeding experience of both their male and female parents (Wooller *et al.* 1988). Thus about 15 % of all young produced by parents which had bred once before, returned to breed, and this rose slowly to 19 % for the young of birds with 10 years breeding experience. There was thus a clear relationship between the number of young fledged by a bird during its lifetime and the number of reproducing offspring which it produced (Fig. 24.1b). This, in turn, meant that the number of reproducing offspring produced by a shearwater in its lifetime, one measure of fitness, was related to the length of that lifetime (Fig. 24.1c). Thus, the length of breeding lifespan was a major factor in determining lifetime success.

Discussion

The Short-tailed Shearwater showed great variation in the ages at which known-age birds bred for the first time. It was clear, however, that those which started very young were relatively unsuccessful initially. In another procellariiform seabird, the Fulmar *Fulmarus glacialis*, birds breeding for the first time at younger ages were less successful in their first breeding attempts than birds starting at older ages (Ollason & Dunnet 1978). In both species, these differences soon disappeared and, thereafter, breeding success increased with accumulated breeding experience.

Age-specific, non-linear patterns of mortality, similar to that in the Short-tailed Shearwater, have also been found in Fulmars (Dunnet & Ollason 1978) and Kittiwakes *Rissa tridactyla* (Coulson & Wooller 1976), although sample sizes of old birds were inevitably rather small in these long-lived birds. The closely related Manx Shearwater *Puffinus puffinus* showed a maximal longevity of 29 years (Harris 1966), similar to that of the Short-tailed Shearwater.

Absences from their colony by breeding birds or non-breeding years at the colony, may be more common in seabirds than has been realized. For instance, Kittiwakes are absent on 8 % of all their potential breeding occasions (Wooller & Coulson 1977), a value similar to the 12 % absences of Short-tailed Shearwaters. Indeed, there may be a continuum between such intermittent breeding in normally annual breeders and the biennial or less frequent breeding of very long-lived seabirds such as albatrosses.

The enhancement of reproductive performance by familiarity, not just with the breeding process but with a particular partner, has also been noted in Fulmars and Kittiwakes (Ollason & Dunnet 1978, Coulson & Thomas 1985). The more generalized increase in reproductive performance with experience is seen in many seabirds (Coulson 1966, Brooke 1978, Ollason & Dunnet 1978), but it is unclear whether this results from greater effort or greater efficiency (Nur 1984). It is also difficult to separate the effects of chronological age from breeding experience, since most birds start breeding at similar ages. However, there is little evidence that age *per se* has an effect except during the first few years of a bird's reproductive career.

This relationship between divorce and a failure to rear young in the preceding season is also apparent in the Manx Shearwater (Brooke 1978) and the Kittiwake (Coulson & Thomas 1985). Such failure may stem from the incompatibility of partners, particularly during incubation. Although divorce may lead to a temporary decrease in reproductive output, it is likely that, in a long-lived bird, this is outweighed by the consequent improvement in reproductive performance resulting from greater compatibility of partners.

On average, Short-tailed Shearwaters seemed to experience years of low output once every 10 years. Some, but not all, of these poor years were caused by flooding of burrows. The relatively long breeding lifespan of some members of the species may thus have reduced the impact of such poor years on their fitness. However, the relationships between survival and production of offspring may be more complex than Fig. 24.1 implies. Bradley *et al.* (1989) showed that early in a shearwater's breeding career, increasing offspring production was associated with a decreasing mortality rate. In mid-life there was no significant relationship between survival and fecundity but, later in life, higher offspring production was associated with an increasing rate of mortality. It was also clear that individuals which produced the most free-flying young during their early years lived significantly longer thereafter than less productive individuals.

Thus the common assumption that most selection occurs prior to, or at, the first breeding attempt may not be true in this species, since differential effects are apparent for several years after first breeding. Indeed there seem to be at least two separate processes operating. Early in life, some individuals, presumably more susceptible to environmental pressures, produce fewer offspring and also die earlier than others with a relatively higher reproductive output and longer subsequent life. However, towards the end of life those birds which have produced fewer young during their lives have a slight, but significant, advantage in terms of future survival over those that have produced more offspring.

The low survival and low reproductive success of birds breeding for the first time do indeed suggest that this represents a critical period in their life. There appear to be constraints on how early a shearwater can breed effectively and also on how long it can realistically defer breeding. Once this period is past, there are probably other trade-offs between survival and fecundity by which shearwaters can increase their offspring production and extend their breeding life. Intermittent breeding and mate change by divorce may represent two ways to accomplish this. However, the decline in both survival and reproductive performance of old birds suggests that senescence may set in and curtail lifespan.

Overall, 71 % of all shearwaters which had completed their reproductive careers produced no offspring which returned to breed on Fisher Island (Wooller *et al.* 1988). Thus a relatively small proportion of the breeding population was responsible for most of the next generation. However, it is not clear what characterizes the long-lived, successful shearwaters. No significant correlations with size or other morphological characteristics have yet been found. Nor is there any evidence of competition for burrows, which are always in excess, or of differential effects between areas of the colony or individual burrows. Parasitism and disease are slight, and the harvesting of young and grazing or trampling by domestic stock do not occur on Fisher Island. As with most seabirds, our knowledge of Short-tailed Shearwaters at sea is extremely limited. They feed mainly on euphausiid crustaceans and squid and their annual transequatorial migration may imply seasonal changes in this food. Some individuals may be better foragers than others, especially at times when food is short, but the roles of chance and other factors in determining which birds have long and productive lifespans have yet to be determined.

References

Bradley, J.S., Wooller, R.D., Skira, I.J. & Serventy, D.L. 1989. Age-dependent survival of breeding Short-tailed Shearwaters *Puffinus tenuirostris*. *J. Anim. Ecol.* **58**: 175–188.

Brooke, M. de L. 1978. Some factors affecting the laying date, incubation and breeding success of the Manx Shearwater *Puffinus puffinus*. *J. Anim. Ecol.* **47**: 477–95.

Coulson, J.C. 1966. The influence of the pair-bond and age on the breeding biology of the Kittiwake Gull *Rissa tridactyla*. *J. Anim. Ecol.* **35**: 269–79.

Coulson, J.C. & Thomas, C. 1985. Differences in the breeding performance of individual Kittiwake Gulls *Rissa tridactyla* (L.). In *Behavioural Ecology*, ed. R.M. Sibly & R.H. Smith, pp. 489–503. Oxford: Blackwell.

Coulson, J.C. & Wooller, R.D. 1976. Differential survival rates among breeding Kittiwake Gulls *Rissa tridactyla* (L.). *J. Anim. Ecol.* **45**: 205–13.

Dunnet, G.M. & Ollason, J.C. 1978. The estimation of survival rate in the Fulmar, *Fulmarus glacialis*. *J. Anim. Ecol.* **47**: 507–20.

Harris, M.P. 1966. Age of return to the colony, age of breeding and adult survival of Manx Shearwaters. *Bird Study* **13**: 84–95.

Lill, A. & Baldwin, J. 1983. Weight changes and the mode of depot fat accumulation in migratory Short-tailed Shearwaters. *Aust. J. Zool.* **31**: 891–902.

Mykytowycz, R. 1963. "Limey-bird disease" in chicks of the Tasmanian mutton-bird (*Puffinus tenuirostris*). *Avian Diseases* **7**: 67–79.

Nur, N. 1984. Increased reproductive success with age in the California Gull: due to increased effort or improvement of skill? *Oikos* **43**: 407–8.

Ollason, J.C. & Dunnet, G.M. 1978. Age, experience and other factors affecting the breeding success of the Fulmar, *Fulmarus glacialis*, in Orkney. *J. Anim. Ecol.* **47**: 961–76.

Serventy, D.L. & Curry, P.J. 1984. Observations on colony size, breeding success, recruitment and inter-colony dispersal in a Tasmanian colony of Short-tailed Shearwaters *Puffinus tenuirostris* over a 30-year period. *Emu* **84**: 71–9.

Wooller, R.D. & Coulson, J.C. 1977. Factors affecting the age of first breeding of the Kittiwake, *Rissa tridactyla*. *Ibis* **119**: 339–49.

Wooller, R.D., Skira, I.J. & Serventy, D.L. 1985. Band wear on Short-tailed Shearwaters *Puffinus tenuirostris*. *Corella* **9**: 121–2.

Wooller, R.D., Bradley, J.S., Serventy, D.L. & Skira, I.J. 1988. Factors contributing to reproductive success in Short-tailed Shearwaters *Puffinus tenuirostris*. *Proc. Int. Orn. Congr.* **19**: 848–856.

Part VI. General Issues

This last section consists of two chapters which examine the findings from previous chapters in relation to the wider issues of life history theory and population biology.

One of the fundamental problems of evolutionary biology is to explain life histories: why some species are short-lived and have high annual breeding rates, while others are long-lived, and have deferred maturity and low annual breeding rates. Most relevant theory assumes that current reproduction has costs, which are manifest in reduced survival and future reproduction. There is therefore a trade-off between current and future reproduction and the balance point is drawn differently in different species, supposedly according to selection resulting from species–specific circumstances. In Chapter 25, Partridge examines life-history theory in the light of recent work on age-related reproduction in birds and other animals, and discusses the problems in assessing reproductive costs reliably.

In the final chapter, Newton reviews the main findings to come from studies of LRS in birds, and discusses their relevance to demography and population ecology. One of the main general findings to emerge is the enormous individual variation in LRS found in natural populations, and another is the widespread occurrence of "senescence", involving decline in both survival and reproduction towards old age. The chapter ends with an assessment of the value of LRS as a measure of individual performance, and as a measure of biological fitness.

25. Lifetime Reproductive Success and Life-history Evolution

LINDA PARTRIDGE

Lifetime reproductive success (LRS) is the sum of the reproductive contributions from each of the ages to which an individual succeeds in surviving. Both survival probability and fertility often show age-specific variation in natural populations; young adults may differ in their demographic characteristics from individuals in the prime of life or in old age. Reproduction is a costly activity which has often been demonstrated to have negative effects on both survival and future fertility. In consequence, fertility at all ages cannot be simultaneously maximized by natural selection. There is therefore a trade-off between current and future reproduction, and much theoretical work has been devoted to devising solutions to the problem of optimal reproductive rate, and the way it changes with age. Clutch size in birds has become a classic example in this context.

As well as age-specific variation, there is often marked individual variability in lifespan and reproductive performance, leading to variation in LRS (Clutton-Brock 1988, present volume). The evolutionary implications of this variability are often unclear. It could be of purely phenotypic origin, if it arises from non-heritable variation in body condition or environment. If there is some genetic basis, it could indicate heritable variation in fitness, because LRS is related to, although not identical with, fitness (see below). Birds are again ideal subjects for investigating these issues in the field.

The aim of this chapter is to outline some evolutionary theories of age-specific variation in reproductive rate and survival probability, and to discuss the implications of individual variation for the evolution of lifetime reproductive success. The examples will be drawn mainly from birds.

LIFETIME REPRODUCTION IN BIRDS
ISBN 0-12-517370-9

What is an optimal life-history?

The life-history of an organism is the combination of age-specific survival rates and fertilities seen in a typical individual. The life-history leading to the greatest fitness is the one that maximizes the Malthusian parameter *r*, or intrinsic rate of population increase, under the prevailing ecological conditions, including population density. *r* is determined by age-specific survival and reproductive rates, and is defined by the characteristic equation:

$$1 = \int_0^w l_x m_x e^{-rx} dx$$

where *x* is age, l_x is survival probability to age *x*, m_x is fertility at age *x* and *w* is the age of last breeding. The main assumptions involved are frequency-independent weak selection, stable age distribution and a constant environment (Charlesworth 1980).

An evolutionarily ideal organism would therefore commence breeding at birth at the maximum possible rate and would continue to do so throughout its infinite lifespan. The great diversity seen in age at first breeding, reproductive rate and lifespan would make no evolutionary sense if this were possible, and most theoretical models of life-history evolution assume that reproduction incurs costs in terms of survival and future fertility. Hence, only certain combinations of survival and reproductive rates are possible in practice, and it is these that set an upper limit to LRS. The life-history favoured by selection therefore depends upon the extrinsic and intrinsic constraints on fertility and survival rates at different ages. These constraints are likely to differ between species, giving rise to different life-history strategies.

Maximizing LRS, even in the field, is not the same thing as maximizing *r* because LRS ignores the effects of reproductive timing. In an expanding population where offspring will more than replace parents, an offspring produced early in the lifespan is more valuable than one produced later, because it constitutes a higher proportion of its cohort of zygotes, while in a decreasing population the opposite is true. This point might seem irrelevant, since any population that does not go extinct must on the long-term experience a balance of periods of increase and of decrease. However, what matters here is the way that selection acts on genetic variation for life-history variables during the different parts of the cycle. For instance, many temperate zone insects during the summer months experience stable periods of population expansion, during which early breeding will be favoured. The eventual crash in numbers before the next breeding season may well fall equally on genotypes with differing ages of first reproduction, and no selection on this character will therefore occur at this time. The population will hence be selected for an early age of first breeding, despite the absence of any long-term trend in numbers. LRS is probably a good approximation to fitness for many bird populations, because they are often relatively stable in numbers (Nur in press).

An example of a model of the evolution of avian clutch size that includes the effect on the adult of reproductive costs is illustrated in Fig. 25.1. The model assumes parthenogenesis, so that offspring are genetically identical to their parents, and that offspring start breeding at age 1 and so become reproductively equivalent to their parents at that time. Offspring survival probability to age 1 declines with the size of the clutch, so that there is an intermediate most productive clutch size b_0 (Fig. 25.1a). If parental mortality is linearly related to clutch size, then optimal brood size b_0 declines, to maximize the difference between parental mortality and total offspring surviving to year 1 (Fig. 25.1b) (Charnov & Krebs 1974). This is an example of a reproductive effort model of life-history evolution (e.g. Williams 1966, Gadgil & Bossert 1970), which highlights the importance of reproductive costs.

Calow (1979) has made a broad distinction between ecological and physiological costs of reproduction. Costs are ecological in origin if breeding causes increased exposure to external sources of risk, such as predation or parasitism, while physiological costs occur if, for instance, allocation of nutrients to reproductive activities denies them to growth or somatic maintenance. The distinction is not a sharp one because, for example, denial of nutrients to somatic maintenance may make animals less able to escape from predators or to fight infection. Nonetheless, the explicit recognition of extrinsic and intrinsic sources of cost is valuable. In birds ecological costs might include an increase in predation on breeders (Slagsvold 1984, Lima 1987), while physiological costs could include effects of hormonal changes, mating, egg-production, incubation and care of young.

In the Charnov–Krebs model of reproductive rate, the relevant reproductive costs are those that increase with the number of offspring produced, and therefore would exclude costs incurred as a result of breeding at all, such as nest building or mate attraction. In contrast, models of the age of first breeding (e.g. Pianka & Parker 1975) would include these, because the decision for any age is whether to incur any reproductive costs in the light of their potential effect on survival or future fertility.

Despite their central position in life history theory, there is still some dispute both about the occurrence of costs of reproduction, and how best to measure them.

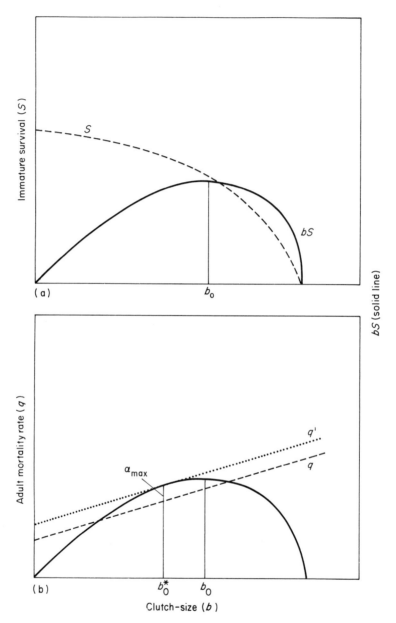

Figure 25.1 The Charnov & Krebs (1974) model of optimal clutch size. (a) Immature survival (S) declines with increasing clutch size (b) leading to a domed relationship between total clutch productivity (bS) and clutch size, with the most productive brood size at b_0. (b) Adult mortality rate (q) shows a linear increase with clutch size, and the optimal clutch size b_0^* maximizes the difference (α) between the loss of the parent and gain of offspring, leading to an optimum at a lower clutch size than b_0.

Measuring costs of reproduction

Controversy has arisen over costs of reproduction because some empirical studies have apparently failed to demonstrate them. These negative or contradictory findings usually occur in studies where individuals are allowed to reproduce at their chosen rate, and subsequent survival and fertility are monitored, while costs of reproduction are in general found in studies where reproductive rate is experimentally manipulated (Partridge & Harvey 1985, 1988; Reznick 1985). The problem with many non-manipulative studies is that important confounding variables are not controlled; some individuals may be at an advantage in both breeding and survival because of their own physical condition, or because they live in a better environment (Partridge & Harvey 1985, 1988, van Noordwijk & de Jong 1986, Partridge in press). These individuals are likely to show high rates of current and future fertility as well as high survival, so that costs of reproduction will be underestimated or appear negative. It is therefore desirable to allocate individuals randomly to experimental groups whose reproductive rates are then experimentally varied to obtain an accurate estimate of subsequent costs. If confounding variables can be controlled in the field, then correlational studies can be useful as a preliminary approach to detecting reproductive costs.

Additional discussion has centred on the importance of genetic as opposed to purely phenotypic studies of the costs of reproduction. Genetic correlations between different characters can be measured in appropriately designed breeding experiments and by artificial selection, and the few such laboratory studies of reproductive costs have in general revealed a negative effect of reproductive rate on subsequent survival and fertility (Reznick 1985). There are strong practical motives for avoiding field work on genetic correlations. Naturally occurring genetic variation results in only small changes in reproductive rate, which makes the estimation of the shapes of cost and benefit curves very difficult. In addition, both the large errors generally associated with field estimates of genetic parameters and the extreme difficulty of measuring genetic correlations adequately under field conditions make such work unattractive.

However, phenotypic manipulations are only of any evolutionary relevance if the response to them is the same as that to a genetic effect of the same magnitude. It is at present not clear if this is always or ever the case. Phenotypic studies have come to have a bad name amongst geneticists because the correlations between reproductive rate and subsequent survival or fertility revealed by *correlational* phenotypic studies often differ in sign from the genetic correlations (Rose & Charlesworth 1981a, Lande 1982). However, there have been very few cases where the results of genetic and *experimental* phenotypic studies have been compared. Birds are not ideal material for an investigation of this issue, but some of the relevant experiments have been done in *Drosophila* in the laboratory. In female *Drosophila melanogaster* the genetic correlations between early fertility and subsequent survival and fertility are negative (Rose & Charlesworth 1981a, b, Rose 1984, Lukinbill *et al.* 1984), while the phenotypic correlations are positive (Rose & Charlesworth 1981a, Partridge 1988). However, an experimental increase in the rate of egg-production or of exposure to males produces a negative effect on female survival (Partridge *et al.* 1987). More studies of this kind are needed. Meanwhile the main task for phenotypic experiments is to identify the nature of reproductive costs and hence design the appropriate experiments to measure their magnitude (Partridge & Harvey 1988, Pease & Bull 1988).

The difference in results from correlational and experimental phenotypic studies can be illustrated from the bird literature (Tables 25.1 and 25.2). These include all the field studies known to me of reproductive costs in nidicolous birds. The experimental studies all involved changes in brood size. Only those that involved direct measurements of survival or fertility were included; indirect physiological measures such as weight or metabolic rate, while of great interest for a study of mechanisms, suffer from some difficulties in interpretation, and anyway do not give information directly relevant to life-history theory (Bryant 1988). Of the 13 correlational studies (Table 25.1), seven (Kluyver 1963, Perrins 1965, Tinbergen *et al.* 1985, Bryant 1979, McGillivray 1983, Ekman & Askenmo 1986, Nur in press) found evidence of costs, two (Gustafsson & Sutherland 1988, Newton 1988) found no significant relationship, while four (Den Boer-Hazewinkel 1987, Smith 1981, Hogstedt 1981, Korpimaki 1988) found negative costs of reproduction. The 15 experimental studies (Table 25.2) give more consistent results, with 10 (Askenmo 1979, Gustaffson & Sutherland 1988, Slagsvold 1984, Smith *et al.* 1987, Tinbergen 1987, Nur 1984, 1988, Roskaft 1985, Hegener & Wingfield 1987, Reid 1987, Dijkstra 1988) showing costs, and four (de Steven 1980, Finke *et al.* 1987, Korpimaki 1988, Pettifor *et al.* 1988) showing no relationship, while none showed negative costs. This difference is not quite statistically significant ($\chi^2 = 5.166$, $2df$, $P = 0.0755$). The comparison is anyway not really valid, because there are several studies of Great Tits in both Tables, although the results are not consistent within each Table.

In all studies recapture was taken as a measure of survival, which may not always be a reasonable assumption. In addition, not all of the studies followed adults or young through to later breeding seasons. It is clear that an increase in clutch size can have a negative effect on both offspring

Table 25.1 Correlational tests of costs of reproduction in birds.

Species	Result
Sparrowhawk *Accipiter nisus*	Survival of females and brood size unrelated to brood size the previous season (Newton 1988) (0)
Tengmalm's Owl *Aegolius funereus*	Males rearing larger broods showed marginally higher survival and no reduction in fertility the next year (Korpimaki 1988) (?−)
House Martin *Delichon urbica*	Lower survival of double brooded females (Bryant 1979) (+)
Pied Flycatcher *Ficedula hypoleuca*	Survival of females unrelated to brood size (Lack 1966, but see Nur in press) (?+)
Collared Flycatcher *Ficedula albicollis*	Survival of males and females and female fertility unrelated to clutch size (Gustaffson & Sutherland 1988) (0)
Great Tit *Parus major*	Probability of second brood reduced by size of first (Kluyver 1963, Perrins 1965). Adult survival reduced (Tinbergen *et al.* 1985) or increased (Den Boer-Hazewinkel 1987) with increased brood size (3+ 1−)
Willow Tit *Parus montanus*	Survival of adults decreased with brood size. Non-reproductive males had higher survival (Ekman & Askenmo 1986) (+)
House Sparrow *Passer domesticus*	Fledglings from late broods negatively correlated with those of earlier broods (McGillivray 1983) (+)
Song Sparrow *Melospiza melodia*	Survival of females increased with brood size (Smith 1981) (−)
Magpie *Pica pica*	Survival of adults increased with brood size (Hogstedt 1981) (−)

7+ 20 4−

0 = no evidence of costs; + = costs; − = "negative" costs.

(e.g. Gustaffson & Sutherland 1988) and parental (e.g. Roskaft 1985, Nur 1988) fertility in ensuing breeding seasons. In addition, none of the experimental studies manipulated all aspects of reproductive effort. The omission of costs of egg-production may be particularly important (Winkler 1985, Nur 1986). It is easier to point to this problem than to suggest remedies, because it is precisely where egg-production is costly that birds are less likely to respond to egg-removal. A hormonal manipulation of egg number might be appropriate.

Table 25.2 Experimental tests of costs of reproduction in birds.

Species	Manipulation	Result
Kestrel *Falco tinnunculus*	Enlarge and reduce random broods	Adult survival affected (Dijkstra 1988) (+)
Glaucous winged Gull *Larus glaucescens*	Brood size randomized	Adult survival affected (Reid 1987) (+)
Tengmalm's Owl *Aegolius funereus*	Broods enlarged	Adult survival unaffected (Korpimaki 1988) (0)
Tree Swallow *Iridoprocne bicolor*	Enlarge random broods	No effect on female survival (de Steven 1980) (0)
House Wren *Troglodytes aedon*	Enlarge and reduce random broods	No effect on second broods (Finke *et al.* 1987) (0)
Pied Flycatcher *Ficedula hypoleuca*	Enlarge random broods	Male survival reduced (Askenmo 1979) (+)
Collared Flycatcher *Ficedula albicollis*	Reduce and enlarge random broods	Female fertility reduced next season (Gustaffson & Sutherland 1988) (+)
Great Tit *Parus major*	Enlarge and reduce random broods	Female fertility (Slagsvold 1984, Smith *et al.* 1987, Tinbergen 1987, Tinbergen *et al.* 1987) affected, no effect on female survival or fertility (Pettifor *et al.* 1988) (3+ 10)
Blue Tit *Passer domesticus*	Brood size randomized	Female survival and male fertility affected (Nur 1984, 1988) (+)
House Sparrow *Passer domesticus*	Enlarge and reduce random broods	Female fertility affected (Hegener & Wingfield 1987) (+)
Rook *Corvus frugilegus*	Enlarge random broods	Fertility reduced (Roskaft 1985) (+)

10+ 40

0 = no evidence of costs; + = costs; − = "negative" costs.

It is thus probably safe to conclude that costs of reproduction do exist in birds, and that the task in hand is to measure their extent in properly controlled studies.

Optimal clutch size

Models of life histories, such as the one outlined above (Charnov & Krebs 1974), make quantitative predictions. For avian clutch size, to a first approximation, the model predicts that the optimum will maximize the difference between breeding offspring raised in the current attempt and future offspring lost by the parental mortality. In fact, effects on subsequent parental fertility must also be included, so the new prediction would become that the optimal clutch size would maximize the amount by which breeding offspring gained through the current breeding attempt exceeds the breeding offspring lost through all reproductive costs to the parent. If there are no reproductive costs, then this corresponds to the most productive brood size, otherwise the optimum is likely to be lower. The model also ignores the effects of clutch size on offspring fertility, and again these should be included to complete the picture.

This prediction is in principle testable, but suitable data do not at present exist. We have seen that experimental studies have so far not manipulated all potentially costly aspects of reproduction, and many have not followed parents and offspring through to ensuing breeding seasons, to get full measures of the costs and benefits associated with different clutch sizes. In addition, conditions for parents and offspring vary between breeding seasons; in his study of Blue Tits *Parus caeruleus*, Nur (1988) found that the extent of reproductive costs varied from year to year, and the same result emerged from his reanalysis of Campbell's Pied Flycatcher *Ficedula hypoleuca* data (Nur in press), first analysed by Lack (1966). It is also clear that the most productive brood size can vary from year to year (Perrins & Moss 1975, van Noordwijk *et al.* 1981). This means that long term studies are necessary, and such annual variation also affects the theory of optimal clutch size.

TEMPORALLY VARIABLE ENVIRONMENTS

Most theoretical and empirical studies of optimal clutch size have assumed that the environment stays constant, so that the costs and benefits of a given level of reproductive allocation do not vary between breeding attempts. However, where conditions for breeding vary, the optimal clutch size can also vary. Under these circumstances, individuals that lay large clutches may do well in good years, but may also do particularly badly in poor years. Long-term fitness is then better served by achieving the optimal clutch size in bad years (when the optimum is low) than in good years (when the optimum is high), because each offspring successfully reared in a bad year forms a higher proportion of its cohort than does an offspring from a good year. Similarly, a parent that comes through a population bottle-neck has a higher relative fitness than one that succeeds in surviving in a large population. If parents cannot anticipate and lay the optimal clutch size for a given season, then the net effect is to pull the long term optimal clutch size down, and theory suggests that under these circumstances the optimal clutch size is the geometric, not the arithmetic, mean of the optimal clutch sizes in different years, because this clutch size reduces the variance in fitness experienced from year to year (Gillespie 1977, Bulmer 1985). There are some simplifying assumptions involved in reaching this conclusion, and these may in practice be a problem (Nur 1987). Boyce & Perrins (1987) have compared observed Great Tit clutch sizes with the most productive clutch size predicted by a model allowing for year to year variability, and found good agreement. However, their analysis ignored costs of reproduction, which would lower the predicted optimum still further.

INDIVIDUAL VARIATION

If individuals differ either in the effort they need to expend to rear offspring or in the costs that they incur as a result of doing so, then there will be individual variation in optimal clutch size. There is mounting evidence, some of it circumstantial, that this kind of variation occurs. For instance reproductive rate often changes with age in a manner consistent with the hypothesis (see below). In addition, Hogstedt (1980) found that variation in Magpie clutch size was associated with territory quality, although he did not investigate if the variation was adaptive, that is whether optimal brood size varied between different territory types. Perrins & Moss (1975) found that the most productive brood size for natural Great Tit broods was larger than for manipulated broods; this would be predicted if the birds best able to rear broods of a particular size laid the appropriate natural clutch size. This interpretation was confirmed by experimental work by Pettifor et al. (1988) who increased and decreased brood size, and found that the initial individual variation in clutch size was adaptive; individuals were laying their personal optimal clutch size, and deviations from this resulted in fewer offspring surviving to breed. One oddity of this study was that its findings contradicted those of another study (Boyce & Perrins 1987) using the same data set; the latter authors found a two-egg difference between mean clutch size and optimal clutch size within a season. The two studies used different methods of data analysis, and the discrepancy in findings suggests that the results obtained are sensitive to this. An absence of individual differences was reported for Blue Tits by Nur (1986), who found that the ability of parents to care for experimental broods and the costs incurred in doing so showed no association with natural clutch size.

As with studies of reproductive costs, brood size manipulations are not entirely satisfactory for examining individual differences in costs and benefits, even when subsequent effects on offspring and parents are monitored, because they do not examine all aspects of reproductive effort, particularly the costs of egg-laying. In addition, the individuals being compared for parental ability have already put differing levels of effort into egg-laying. Furthermore, the existence of individual variation points to the possibility that individuals may differ in their response to artificial broods of different sizes; if parental work rate is already to some extent determined by the time the clutch is laid, then the individual responses to the manipulation may not be optimal.

The presence of individual differences poses some theoretical problems for determining the population optimal clutch size. If individual cost and benefit curves vary, there are two ways of calculating the population optimum, but only one of them is correct. The first is to measure individual cost and benefit curves, deduce clutch size optima for individual birds, and then combine these to produce the population mean optimum. This is the correct way to proceed. However, this is not feasible, because it requires repeated measures with the same individual parent rearing broods of different sizes. What is usually done is to measure the population mean cost curve and the population mean benefit curve, and then use these to deduce the population optimal brood size. This is inappropriate, because the individual cost and benefit curves will not have additive effects on the optimum, because the curves for different individuals will probably have different shapes. This problem could be circumvented if individuals belonging to different cost-benefit curve categories could be identified in advance and then treated as a group for analysis of the data from clutch size manipulations. It is possible that initial clutch size might provide a basis on which to do this, possibly combined with other measurements on the birds themselves. What is required is a reliable predictor of the parental ability and reproductive costs of different individuals.

Constraints on optimality

Life history theory has the idea of constraints built into it, because of the assumption of reproductive costs. However, there are many other constraints on the action of natural selection which play no explicit part in the existing optimality models, and which mean that in practice the optimum may not be realized. Examples are lack of additive genetic variance for a response to selection, insufficient time for the response to occur, and an effect of starting conditions on the way a response to selection occurs. These issues have been thoroughly considered elsewhere in the context of optimality in general (Maynard Smith 1978, Gould & Lewontin 1979) and of optimal life histories in particular (Partridge & Harvey 1988). Their importance in the present context is that they suggest that a comparative approach may be more successful in detecting the effects of natural selection on life-histories than direct testing of the fit of particular optimality models. This is not to deny the importance of the models, which make assumptions explicit and therefore testable, and identify the variables to be used in comparative studies. Comparisons of populations or of species differing in critical variables, such as juvenile and adult survival rates or the costs and benefits associated with different clutch sizes, may provide the best means of detecting the effects of natural selection on life histories. Birds could provide ideal material for this kind of work and we would then be well on the way to understanding the observed diversity in avian life histories.

Changes in reproductive rate with age

Two theoretical considerations suggest that reproductive rate might be expected to change with age. The first, as we have seen, is that individual phenotype may change in a way that alters the costs or benefits of a given level of reproduction. For instance, an improvement in foraging efficiency with age could increase energy intake so that more offspring could be produced without additional cost. This type of consideration may be important in setting the age of first breeding in many birds, as well as influencing reproductive rate once breeding has commenced (Nol & Smith 1987). There is a general tendency for survival probabilities to be lowest in young age cohorts. Examples include Ring-necked Ducks *Anas collaris* (Conroy & Eberhardt 1983), Sparrowhawks *Accipiter nisus* (Newton *et al.* 1983), Black-capped Chickadees *Parus atricapillus* (Loery *et al.* 1987) and Great Tits (Clobert *et al.* 1988). The low survival probably reflects a variety of deficiencies such as low feeding skill, low social rank and hence poor access to resources and possibly incompetence with predators, which may mean that little or no reproductive effort can profitably be made. This may explain the low fertilities often found in young breeders (e.g. Coulson & Horobin 1976, Perrins 1979, Newton *et al.* 1981).

The second reason for changes in reproductive rate with age is senescence, which is the tendency for all aspects of performance, including survival probability and fertility, to decline with advancing age. Ageing has been documented in avian field studies. For instance, a decline in adult survival rates later in the lifespan has been reported in Fulmars (Dunnet & Ollason 1978), Great Tits (Webber 1975), Black-capped Chickadees (Loery *et al.* 1987), Sparrowhawks (present volume) and Short tailed Shearwaters (present volume). It must be mentioned in passing that both the collection and analysis of data on age-specific survival rates can be problematic (Lakhani & Newton 1983, Fryxell 1986, Seber 1986). Ring loss, emigration and age-specific variation in capture probabilities are obvious difficulties and, depending on the methods used, temporal variation in survival rate affecting all age classes could also be a snag. A more subtle problem, which means that the effects of ageing will nearly always be underestimated, occurs if survivors to later ages are not a random sample of the cohort from which they come; they are likely to be better quality individuals than those already dead. Fertility, which can be measured and analysed in longitudinal studies and presents less difficulty for study, often shows a similar pattern of decline later in the lifespan (e.g. Great Tits (Perrins 1979), Arctic Terns (Coulson & Horobin 1976) and Sparrowhawks (Newton *et al.* 1981)).

Senescence shows that the level of performance that can be achieved early in adult life is for some reason not maintained. One possible explanation is that ageing is a consequence of inevitable wear and tear. However, this does not explain why creatures with otherwise similar biological characteristics can have very different lifespans, not only in the field, but also in captivity where natural hazards are removed and optimal conditions provided (Williams 1957). Among birds many parrots are notably long-lived for their body size, while emus and ostriches are extremely short-lived (Comfort 1956). This finding implies that the effects of wear and tear are combatted to different extents in different species, so that ageing has evolved.

By what processes could a maladaptive character like senescence evolve? Even in a population free of ageing there will be mortality, caused by accidents, predation and disease. As age increases, there will therefore be a steadily declining number of survivors in any cohort. We can superimpose on this picture the occurrence of mutations with age-specific effects on survival probability or fertility; these are unhappily familiar in the human population in the form of certain hereditary late onset genetic diseases such as Huntingdon's chorea. Mutations that act on young individuals will be expressed in a higher proportion of their original carriers than will mutations acting on older age classes, because here many of the carriers will already be dead from other causes before the time when the effects of the mutation become apparent. The intensity of natural selection will therefore be greater on a mutation affecting survival probability or fertility earlier in the lifespan, and mutations that decrease performance will be more readily eliminated and mutations that increase it will be more readily incorporated if they act early. For this reason, this "mutation accumulation" theory predicts that survival probability and fertility will decline with advancing age in the adult lifespan (Medawar 1952, Charlesworth 1980).

There is a second theory of the evolution of ageing, which suggests that genetic variants with beneficial effects earlier in the lifespan may tend to reduce performance later on. Mutations are often pleiotropic in their effects, in other words affect more than one phenotypic characteristic. If a mutation had the effect of increasing fitness early in the lifespan but at the cost of reducing it later on, the mutation could be favoured by selection, because the early beneficial effect would be apparent in more of the carriers of the mutation than would the deleterious late effect (Williams 1957, Hamilton 1966). A cost of reproduction would be an obvious candidate for gene action of the kind proposed by this theory. If breeding results in the diversion of nutrients away from somatic repair or maintenance, then it could in addition result in an irreversible drop in somatic condition identifiable as ageing.

These two theories are not mutually exclusive, tests of them have been virtually confined to *Drosophila* and there has been no work in the field (Partridge 1987). An obvious prediction from both theories is that the effects of senescence should be more apparent in species or populations with high levels of age-independent adult mortality, because it is this that reduces the impact of the late effects of mutations. Birds would be good subjects for a comparative test. What is required is a comparison of age-independent mortality rates in the field with the timing of onset of senescence in captivity. In addition, the relative ease with which reproductive rate can be manipulated in the field might make a field test of the effects of reproductive costs on ageing feasible. For this, what is required is an investigation of the effects of manipulation of reproductive rate on survival probability and fertility late in the lifespan.

Finally, the existence of senescence can, paradoxically, lead to an increase in reproductive rate at the end of the lifespan. For instance, old

California Gulls *Larus californicus* (Pugesek 1981) and, in a longitudinal study, elderly red deer *Cervus elaphus* hinds invest more in their young (Clutton-Brock 1984) than do younger parents. Such an increase would be predicted if the animal can detect the decline in its own somatic condition, because the reduced expectation of future reproduction should tip the tradeoff towards current reproduction. Some caution is needed in interpretation of studies of this kind that use cross sectional data, because in the absence of longitudinal studies there is a possibility that individuals surviving to old ages were better parents at all ages. Ideally, longitudinal data should be gathered.

Individual variation in lifetime reproductive success

Field studies usually reveal considerable variation in LRS between individuals, and much interest has focused on the causes, both the proximate mechanisms, such as the relative contributions of variation in survival and fertility, and the ultimate significance in evolutionary terms. Although LRS is not identical with fitness, it is probably a good approximation to it for many bird populations. The existence of extensive variation for the character is therefore in some need of explanation.

Variation in LRS could result from chance or it could reflect genuine biological differences between individuals. Chance variation would not have any particular evolutionary significance. Biological differences could be a result of events during the lifetime of the individual, for instance its nutrition during development, or the quality of its territory or of its mate. Developmental effects are likely to be persistent and to some extent irreversible, while the effects of adult nutrition are likely to be more short-term. Offspring would not on the whole be expected to inherit any of these differences, although there might be some transmission during the time of parental care. This purely phenotypic variation in LRS would not cause the character to evolve, but it could nonetheless have some important evolutionary implications. Thus if individuals differ in condition in a way that affects their survival or fertility, selection may favour condition-dependent variation in allocation of effort to reproduction, as we have already seen. The nature of any such adjustment would depend upon the exact effects of condition (Nur in press). For instance, variation in predator impact could affect survival prospects for both parents and offspring, and the optimal response would depend upon the relative strength of the two effects. Even if only parents were affected, for instance by the effects of nutrition, their best response would depend upon the effects of condition on fertility and survival (Nur in press). The crucial universal prediction is that an individual with low current fertility should have a lower fertility optimum than an individual with high fertility.

Another cause of variation in LRS could be differences in genotype. There is an urgent need for more studies investigating the heritability of LRS in the field; this is no easy undertaking, but birds provide promising material. Characters closely correlated with fitness are in general expected to show low heritability, because those genetic variants increasing the character value should have already been incorporated by natural selection (Fisher 1930, Robertson 1955). Field data now provide some support for this theory. For instance the heritability of a variety of morphological and life-history traits has been measured in the field for the Collared Flycatcher, and those traits probably most closely correlated with fitness, including lifetime reproductive success and longevity, showed the lowest heritabilities, while various morphological traits showed the highest (Gustaffson 1986). Although the information would be harder to obtain, it would be most valuable to have data of this kind for long lived species, where the effects of environmental fluctuations are likely to be smaller, so that any genetic effects are more likely to be detected.

However, heritability of LRS is probably likely to be significant in some cases. We have seen that selection on reproductive rate shows temporal variation, and this type of selection can maintain genetic variation for the character (Charlesworth 1987). In addition, selection may vary geographically and in different environments, and genetic differentiation together with gene flow between areas could result in variation within a single area (Endler 1977). Even so, the vast majority of variation in LRS is likely to be non-heritable, giving little or no potential for evolutionary increase in the character.

References

Askenmo, C. 1979. Reproductive effort and return rate of male Pied Flycatchers. *Amer. Nat.* **114**: 748–53.

Boyce, M.S. & Perrins, C.M. 1987. Optimizing Great Tit clutch size in a fluctuating environment. *Ecology* **68**: 142–53.

Bryant, D.M. 1979. Reproductive costs in the House Martin *Delichon urbica*. *J. Anim. Ecol.* **48**: 655–75.

Bryant, D.M. 1988. Energy expenditure and body mass changes as measures of reproductive costs in birds. *Funct. Ecol.* **2**: 23–34.

Bulmer, M. 1985. Selection for iteroparity in a variable environment. *Amer. Nat.* **126**: 63–71.

Calow, P. 1979. The cost of reproduction—a physiological approach. *Biol. Rev.* **54**: 23–40.

Charlesworth, B. 1980. *Evolution in Age-Structured Populations.* Cambridge: University Press.

Charlesworth, B. 1987. The heritability of fitness. In *Sexual Selection: Testing the Alternatives*, ed. J.W. Bradbury & M.B. Andersson, pp. 21–40. Winchester: John Wiley.

Charnov, E.L. & Krebs, J.R. 1974. On clutch size and fitness. *Ibis* **116**: 217–19.

Clobert, J., Perrins, C.M., McCleery, R.H. & Gosler, A.G. 1988. Survival rate in the Great Tit *Parus major* in relation to sex, age, and immigration status. *J. Anim. Ecol.* **57**: 287–306.

Clutton-Brock, T.H. 1984. Reproductive effort and terminal investment in iteroparous animals. *Amer. Nat.* **123**: 212–29.

Clutton-Brock, T.H. (ed.) 1988. *Reproductive Success.* Chicago: University Press.

Comfort, A. 1956. *The Biology of Senescence.* New York: Rinhart and Co. Inc.

Conroy, M.J. & Eberhardt, R.T. 1983. Variation in survival and recovery rates of Ring-necked Ducks. *J. Wildlife Management* **47**: 127–37.

Coulson, J.C. & Horobin, J. 1976. The influence of age on the breeding biology and survival of the Arctic Tern *Sterna paradisea. J. Zool. Lond.* **178**: 247–60.

Den Boer-Hazewinkel, J. 1987. On the costs of reproduction: parental survival and production of second clutches in the Great Tit. *Ardea* **75**: 99–110.

De Steven, D. 1980. Clutch size, breeding success, and parental survival in the Tree Swallow (*Irodoprocne bicolor*). *Evolution* **34**: 278–91.

Dijkstra, C. 1988. Reproductive tactics in the Kestrel. Unpublished Ph. D. thesis, University of Groningen.

Dunnet, G.M. & Ollason, J.C. 1978. The estimation of survival rate in the Fulmar. *J. Anim. Ecol.* **47**: 507–20.

Ekman, J. & Askenmo, C. 1986. Reproductive cost, age-specific survival and a comparison of the reproductive strategy in two European tits (Genus *Parus*). *Evolution* **40**: 159–68.

Endler, J.A. 1977. *Geographic Variation, Speciation, and Clines.* Princeton: University Press.

Finke, M.A., Milinkovich, D.J. & Thompson, C.F. 1987. Evolution of clutch size: an experimental test in the House Wren (*Troglodytes aedon*). *J. Anim. Ecol.* **56**: 99–114.

Fisher, R.A. 1930. *The Genetical Theory of Natural Selection.* New York, Dover.

Fryxell, J.M. 1986. Age-specific mortality: an alternative approach. *Ecology* **67**: 1687–92.

Gadgil, M. & Bossert, W. 1970. Life historical consequences of natural selection. *Amer. Nat.* **104**: 1–24.

Gillespie, J.H. 1977. Natural selection for variance in offspring numbers: a new evolutionary principle. *Amer. Nat.* **111**: 1010–14.

Gould, S.J. & Lewontin, R.C. 1979. The spandrels of San Marco and the panglossian paradigm: a critique of the adaptationist programme. *Proc. Roy. Soc. Lond.B.* **205**: 581–98.

Gustaffson, L. 1986. Lifetime reproductive success and heritability: empirical support for Fisher's fundamental theorem. *Amer. Nat.* **128**: 761–4.

Gustaffson, L. & Sutherland, W.J. 1988. The cost of reproduction in the Collared Flycatcher *Ficedula albicollis. Nature* **335**: 813–15.

Hamilton, W.D. 1966. The moulding of senescence by natural selection. *J. Theor. Biol.* **12**: 12–45.

Hegener, R.E. & Wingfield, J.C. 1987. Effects of brood size manipulations on parental investment, breeding success, and reproductive endocrinology of House Sparrows. *Auk* **104**: 470–80.

Hogstedt, G. 1980. Evolution of clutch size in birds: adaptive variation in relation to territory quality. *Science* **210**: 1148–50.

Hogstedt, G. 1981. Should there be a positive or negative correlation between survival of adults in a bird population and their clutch size? *Amer. Nat.* **118**: 568–71.

Kluyver, H.N. 1963. The determination of reproductive rates in Paridae. *Proc. Int. Orn. Congr.* **13**: 706–16.

Korpimaki, E. 1988. Costs of reproduction and success of manipulated broods under varying food conditions in Tengmalm's Owl. *J. Anim. Ecol.* **57**: 1027–39.

Lack, D. 1966. *Population Studies of Birds*. Oxford University Press.

Lakhani, K.H. & Newton, I. 1983. Estimating age-specific bird survival rates from ring recoveries - can it be done? *J. Anim. Ecol.* **52**: 83–91.

Lande, R. 1982. A quantitative genetic theory of life history evolution. *Ecology* **63**: 607–15.

Lima, S.L. 1987. Clutch size in birds: a predation perspective. *Ecology* **68**: 1062–70.

Loery, G., Pollock, K.H., Nichols, J.D. & Hines, J.E. 1987. Age-specificity of Black-capped Chickadee survival rates: analysis of capture-recapture data. *Ecology* **68**: 1038–44.

Luckinbill, L.S., Arking, R., Clare, M.J., Cirocco, W.C. & Buck, S.A. 1984. Selection for delayed senescence in *Drosophila melanogaster*. *Evolution* **38**: 996–1003.

Maynard Smith, J. 1978. Optimization theory in evolution. *Ann. Rev. Ecol. Syst.* **9**: 31–56.

McGillivray, W.B. 1983. Intraseasonal reproductive costs for the House Sparrow (*Passer domesticus*). *Auk* **100**: 25–32.

Medawar, P.B. 1952. *An Unsolved Problem of Biology*. London: H.K. Lewis.

Newton, I. 1988. Age and reproduction in the sparrowhawk. In *Reproductive Success*, ed. T.H. Clutton-Brock, pp. 201–219. Chicago: University Press.

Newton, I., Marquiss, M. & Moss, D. 1981. Age and breeding in sparrowhawks. *J. Anim. Ecol.* **50**: 839–53.

Newton, I., Marquiss, M. & Rothery, P. 1983. Age structure and survival in a Sparrowhawk population. *J. Anim. Ecol.* **52**: 591–602.

Nol, E. & Smith, J.N.M. 1987. Effects of age and breeding experience on seasonal reproductive success in the Song Sparrow. *J. Anim. Ecol.* **56**: 301–13.

Nur, N. 1984. The consequences of brood size for breeding Blue Tits: I. Adult survival, weight change and the cost of reproduction. *J. Anim. Ecol.* **53**: 479–96.

Nur, N. 1986. Is clutch size variation in the Blue Tit (*Parus caeruleus*) adaptive? An experimental study. *J. Anim. Ecol.* **55**: 983–99.

Nur, N. 1987. Population growth rate and the measurement of fitness: a critical reflection. *Oikos* **48**: 338–41.

Nur, N. 1988. The consequences of brood size for breeding Blue Tits. III. Measuring the cost of reproduction: survival, future fecundity, and differential dispersal. *Evolution* **42**: 351–62.

Nur, N. in press. The cost of reproduction in birds: an examination of the evidence. *Ardea*.

Partridge, L. 1987. Is accelerated senescence a cost of reproduction? *Funct. Ecol.* **1**: 317–20.

Patridge, L. 1988. Lifetime reproductive success in Drosophilia. In Reproductive Success ed. T.H. Clutton-Brock, pp. 11–23. Chicago: University Press.

Partridge, L. In press. An experimentalist's approach to evolutionary ecology. In *Proceedings of the British Ecological Society's Centenary Symposium.*

Partridge, L. & Harvey, P.H. 1985. Costs of reproduction. *Nature* **316**: 20–1.

Partridge, L. & Harvey, P.H. 1988. The ecological context of life history evolution. *Science* **241**: 1449–1455.

Partridge, L., Green, A. & Fowler, K. 1987. Effect of egg-production and of exposure to males on female survival in *Drosophila melanogaster. J. Insect Physiol.* **33**: 745–9.

Pease, C.M. & Bull, J.J. 1988. A critique of methods for measuring life history trade-offs. *J. Evol. Biol.* **1**: 293–303.

Perrins, C.M. 1965. Population fluctuations and clutch size in the Great Tit *Parus major* L. *J. Anim. Ecol.* **34**: 601–47.

Perrins, C.M. 1979. *British Tits.* Collins, London.

Perrins, C.M. & Moss, D. 1975. Reproductive rates in the Great Tit. *J. Anim. Ecol.* **44**: 659–706.

Pettifor, R.A., Perrins, C.M. & McCleery, R.H. 1988. Individual optimization of clutch size in Great Tits. *Nature* **336**: 160–2.

Pianka, E.R. & Parker, W.S. 1975. Age-specific reproductive tactics. *Amer. Nat.* **109**: 453–64.

Pugesek, B.H. 1981. Increased reproductive effort with age in the California Gull (*Larus californicus*). *Science* **212**: 822–3.

Reid, W.V. 1987. The cost of reproduction in the Glaucous-winged Gull. *Oecologia* **74**: 458–67.

Reznick, D. 1985. Costs of reproduction: an evaluation of the empirical evidence. *Oikos* **44**: 257–67.

Robertson, A. 1955. Selection in animals: synthesis. *Cold Spring Harbor Symp. Quant. Biol.* **20**: 225–9.

Rose, M.R. 1984. Laboratory evolution of postponed senescence in *Drosophila melanogaster. Evolution* **38**: 1004–10.

Rose, M., & Charlesworth, B. 1981a. Genetics of life history in *Drosophila melanogaster.* I. Sib analysis of adult females. *Genetics* **97**: 173–86.

Rose, M., & Charlesworth, B. 1981b. Genetics of life history in *Drosophila melanogaster.* II. Exploratory selection experiments. *Genetics* **97**: 187–96.

Roskaft, E. 1985. The effect of enlarged brood-size on the future reproductive potential of the Rook. *J. Anim. Ecol.* **54**: 255–60.

Seber, G.A.F. 1986. A review of estimating animal abundance. *Biometrics* **42**: 267–92.

Slagsvold, T. 1984. Clutch size variation of birds in relation to nest predation: on the cost of reproduction. *J. Anim. Ecol.* **53**: 945–53.

Smith, H.G., Kallander, H. & Nilsson, J.A. 1987. Effects of experimentally altered brood size on frequency and timing of second clutches in the Great Tit. *Auk* **104**: 700–6.

Smith, J.N.M. 1981. Does high fecundity reduce survival in Song Sparrows? *Evolution* **35**: 1142–8.

Tinbergen, J.M. 1987. Costs of reproduction in the Great Tit; interseasonal costs associated with brood size. *Ardea* **75**: 111–22.

Tinbergen, J.M., van Balen, J.H. & Van Eck, H.M. 1985. Density-dependent survival in an isolated Great Tit population: Kluyver's data reanalysed. *Ardea* **73**: 38–48.

Tinbergen, J.M., van Balen, J.H., Drent, P.J., Cave, A.J., Mertens, J.A.L. & den Boer-Hazewinkel. 1987. Population dynamics and cost-benefit analysis. *Neth. J. Zool.* **37**: 180–213.

Van Noordwijk, A.J. & de Jong, G. 1986. Acquisition and allocation of resources: their influence on variation in life history tactics. *Amer. Nat.* **128**: 137–42.

Van Noordwijk, A.J. van Balen, J.H. & Scharloo, W. 1981. Genetic and environmental variation in clutch size of Great Tit (*Parus major*). *Neth. J. Zool.* **31**: 342–72.

Webber, M.I. 1975. Some aspects of the non-breeding population dynamics of the Great Tit (*Parus major*). Unpublished D. Phil. thesis, University of Oxford.

Williams, G.C. 1957. Pleiotropy, natural selection and the evolution of senescence. *Evolution* **11**: 398–411.

Williams, G.C. 1966. Natural selection, the costs of reproduction, and a refinement of Lack's Principle. *Amer. Nat.* **100**: 687–90.

Winkler, D.W. 1985. Factors determining a clutch size reduction of California Gulls (*Larus californicus*): A multi-hypothesis approach. *Evolution* **39**: 667–77.

26. Synthesis

IAN NEWTON

The importance of lifetime reproductive success (LRS), as a measure of individual performance, is two-fold. First it reveals more than any other measure the full extent of individual variation in reproductive success; and secondly, it provides a better basis for estimating biological fitness than any other measure yet available. These points are true for birds, whether LRS is measured as numbers of fledglings or of recruits to future breeding populations, as both measures are correlated. In this final chapter, I shall attempt to review the main patterns of LRS in birds, and the factors that influence LRS, and then discuss the use of LRS as a measure of individual performance and of biological fitness. I shall refer mainly to studies in this book, but also to others where relevant. Most studies suffer to some degree from gaps in records resulting from movements and incompletely recorded lifespans, and some also suffer from uncertainties over parentage (Chapter 1).

Individual variation in LRS

From the preceding chapters, the following generalizations can be made: (a) a large fraction of all the fledglings that are produced by a bird population die before they can breed; (b) not all the individuals which survive to attempt breeding subsequently produce offspring; and (c) successful individuals vary greatly in productivity. The most successful individuals raise far more young than are needed to replace themselves, and hence they contribute disproportionately to the next generation. Moreover, in any population which remains demographically stable, the proportion of contributing individuals is likely to remain more or less constant through time. But this proportion may of course alter during a period of population change.

LIFETIME REPRODUCTION IN BIRDS
ISBN 0-12-517370-9

In a stable breeding population, pre-breeding mortality is largely determined by the difference between adult mortality, which governs the number of openings available for new breeders each year, and the reproductive rate, which governs the number of young produced each year, and from which breeders are ultimately drawn. The proportion of fledglings which die without making a breeding attempt can be estimated at 42–86 % in the different species considered here (Table 26.1). This proportion shows no obvious relationship to life history type, being large both in short-lived passerines with high reproductive rates, and in long-lived species with long-deferred first breeding.

The proportion of breeders which make at least one nesting attempt, but fail to produce young, varied between 5 % and 35–49 % in species considered here (Table 26.1). The proportion was small in short-lived hole-nesting passerines which experienced high nest success, and in long-lived species in which most breeders had more than one nesting attempt. Conversely, the proportion tended to be large in those short-lived passerines which experienced high predation of nest contents, and in those long-lived species which bred only in certain years. A marked sexual difference in the proportion of unproductive adults was associated with polygyny or with greater survival in one sex than the other.

Taking both types of unproductive individual into account, the total non-contributing individuals can be estimated at 62–87 % of all fledglings in different species (Table 26.1). If losses at the egg stage are included too, then the proportion of eggs (zygotes) which fail to give rise to breeders is of course even higher, estimated at 92 % in Meadow Pipit (Chapter 8), 92 % in Red-billed Gull (Chapter 23), 91 % in Common Gull *Larus canus* (Rattiste & Lilleleht 1987), 90 % in Kingfisher (Chapter 7), 89 % in Splendid Fairy-wren (Chapter 15), 88 % in Sparrowhawk (Newton 1985) and 85 % in Osprey (Chapter 18). So, in a wide range of species, roughly one in 10 eggs can be expected to produce breeding adults.

Among the productive individuals of a population, the distibution of LRS values is highly skewed, with most individuals producing small numbers of young, and a few producing many young (Fig. 26.1). This contributes to the tendency for a small fraction of individuals in one generation to produce a large proportion of the next. In the Blue Tit, for example, only 3 % of individuals in one generation of fledglings (or 30 % of breeders) produce 50 % of the next generation of fledglings. Equivalent figures for some other species are given in Table 26.2, calculated from the data in previous chapters.

Table 26.1 Percentage of individuals which produced no young during their lives. M = male, F = female, B = both sexes.

	% of fledglings which died before they could breed			% of breeders which produced no young			Total % of fledglings which were non-productive			Source
	M	F	B	M	F	B	M	F	B	
Blue Tit	78	76	86	<1	2	5			87	Dhondt, Chapter 2
Pied Flycatcher				3	6		78	78		Sternberg, Chapter 4
House Martin			81							Bryant, Chapter 6
Kingfisher			>72	22	(6)[a]					Bunzel & Druke, Chapter 7
Meadow Pipit				12	12		78	74		Hötker, Chapter 8
Song Sparrow										Hochachka & Smith, Chapter 9
Indigo Bunting[b]				27,37	17,36					Payne, Chapter 10
Red-winged Blackbird				23	37					Orians & Beletsky, Chapter 11
Magpie				43	41					Birkhead & Goodburn, Chapter 12
Green Woodhoopoe	78	79	78							Ligon & Ligon, Chapter 14
Splendid Fairy Wren				25	43		84	87		Rowley & Russell, Chapter 15
Screech Owl		72			14				73	Gehlbach, Chapter 19
Ural Owl		72			3					Saurola, Chapter 20
Sparrowhawk					16			76	73	Newton, Chapter 17
Osprey			71	12	22					Postupalsky, Chapter 18
Barnacle Goose			42	35	49		62	70		Owen & Black, Chapter 21
Snow Goose			70							Cooke & Rockwell 1988
Red-billed Gull				36	39					Mills, Chapter 23
Kittiwake	63	56		3	2		68	59		Coulson 1988

[a] Small sample.
[b] Two different areas.

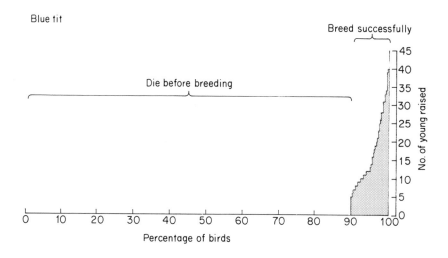

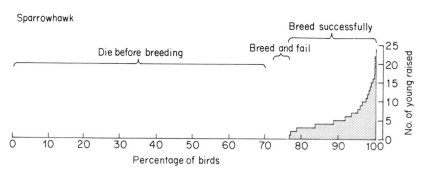

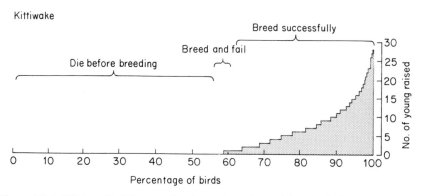

Figure 26.1 Lifetime fledgling productions of a cohort of female fledglings in three bird species, showing the markedly skewed pattern in lifetime reproductive success. Calculated from the data in Chapters 2 (Plot c, Dhondt), 17 (Newton) and Coulson (1988).

Table 26.2 Percentage of breeders which produce 50 % of all offspring.[a] M = male, F = female, B = both sexes.

	% of fledglings which produce 50% of offspring			% of breeders which produce 50% of offspring			Source
	M	F	B	M	F	B	
Blue Tit	5		3			30	Dhondt, Chapter 2
Meadow Pipit		9		18	30		Hotker, Chapter 8
House Martin		6		20	28		Bryant, Chapter 6
Kingfisher	6			30	31		Bunzel & Drüke, Chapter 7
Indigo Bunting[b]				14–18	19–20		Payne, Chapter 10
Sparrowhawk		5			20		Newton, Chapter 17
Osprey	7	5		24	16		Postupalsky, Chapter 18
Screech Owl					21		Gehlbach, Chapter 19
Ural Owl		6			23		Saurola, Chapter 20
Barnacle Goose							Owen & Black, Chapter 21
Red-billed Gull			9			15	Mills, Chapter 23
Mute Swan					15	22	Bacon & Andersen-Harild, Chapter 22

[a] In some cases the figures have been calculated from data given in the relevant chapters.
[b] Two different areas.

Comparison between male and female LRS

As expected, in monogamous species in which the sexes show similar survival, the mean and variance in LRS is equal in males and females (e.g. Scrub Jay, Chapter 13). Deviations from this pattern in some monogamous species can be attributed to occasional polygyny, which gives wider variance in male LRS, or to differential survival between the sexes, which sometimes gives greater variance in LRS of the longer-lived sex. In almost all species considered here, a sex difference in survival was apparent during part of the lifespan, leading in turn to a difference in sex ratio among birds of breeding age. The result was that, in the longer lived sex, a greater proportion of individuals remained unpaired compared with the shorter-lived sex, the mean age of first breeding was greater, intermittent breeding was more frequent and the maximum recorded LRS was greater. Not all these differences were found in every species, however.

In most species with a sex difference in survival males were longer lived than females, but in four species (Woodhoopoe, Osprey, Sparrowhawk, Red-billed Gull) females lived longer. In the two raptors the difference was associated with reversed sexual size dimorphism, in which females are larger than males.

In species where males are regularly polygynous, greater variance in male than in female LRS would be expected, because of the variable numbers of mates which males (but not females) can have. The difference in variance is especially marked in annual reproduction, but less so in LRS. This is because the males typically breed over a shorter part of the lifespan than do females, and the females themselves may build up considerable variance in LRS through variations in lifespan (see also Clutton-Brock 1988). In the three species in this book where polygyny was regular, the variance in male success was much greater in the Redwinged Blackbird than in the *Ficedula* flycatchers, partly because of the greater number of females that each blackbird male could mate with. Some males of this species had up to 14 (partly sequential) mates in a season, sired up to 29 fledglings in a season, and up to 159 in a lifetime. This compares with a maximum of 24 raised by a female blackbird in her lifetime (Chapter 12). Male Pied Flycatchers normally had up to two mates at a time, and fathered up to 12 young in a year, or up to 37 in a lifetime, little different from the maximum of 36 raised by females of this species (Chapter 4).

Sexual selection "depends on the advantage which certain individuals have over others of the same sex and species solely in respect of reproduction" (Darwin 1871). Hence, it is often assumed, the greater the sex difference in variance of LRS, the greater the opportunity for sexual selection (Wade & Arnold 1980). Interestingly, in the studies reviewed here, the extent of the difference in variance in LRS correlated with the degree of sexual dimorphism in different species. The Red-winged Blackbird was by far the most dimorphic of the species reviewed here, both in size and colour, followed by the *Ficedula* flycatchers. Species showing virtually no dimorphism, such as the Scrub Jay, showed similar variance in LRS between the sexes. Hence the common correlation between polygyny and marked sexual dimorphism in birds (Payne 1984) presumably occurs because selection pressures in males and females are most divergent in polygynous species. An obvious next step is to investigate the extent to which variations in male phenotype within polygynous species affect LRS.

While in polygynous bird species, the numbers of mated females per male influenced the LRS of males or of both sexes (Chapter 4), in polyandrous species the numbers of mated males per female is presumably important. This has been demonstrated for annual breeding success (Davies 1985, Faaborg 1986), but not, to my knowledge, for lifetime success.

Components of LRS

For an individual breeder, total LRS is limited by the numbers of breeding years, the numbers of broods per year and the numbers of young per brood. The two latter are in turn limited by egg production (i.e. fecundity) and by subsequent offspring survival. But which of these components contributes most importantly to variance in LRS?

In all species studied so far, breeding lifespan, in turn dependent on total lifespan minus pre-breeding lifespan (up to age of first breeding), emerged as the major demographic determinant of LRS, where LRS was measured as fledgling production. Using regression analysis, variance in lifespan (or breeding lifespan) within species accounted for 30–86 % of the variance in fledgling production (Table 26.3). No tendency was apparent for lifespan to contribute less to the variance in LRS of short-lived species than of long-lived ones. Rather, lifespan was least important in those species in which annual production was highly variable, as in males of polygynous species, in multi-brooded species with variable nest success, or in long-lived species which breed intermittently (Barnacle Goose, Chapter 20) or over variable parts of a potential breeding lifespan (Red-billed Gull, Chapter 23).

Table 26.3 Percentage of variance in lifetime fledgling production accounted for by variation in lifespan. M = male parent, F = female parent, B = both parents.

Species	LRS measured by numbers of Fledglings			Recruits			Source
	M	F	B	M	F	B	
Blue Tit	86	77					Dhondt, Chapter 2
Great Tit			57			11	McCleery & Perrins 1988
Pied Flycatcher	46	56					Sternberg, Chapter 4
Collared Flycatcher				25	16		Gustafsson, Chapter 5
Meadow Pipit	52	48					Hötker, Chapter 8
Song Sparrow	64	61		48	32		Hochachka & Smith, Chapter 9
Indigo Bunting	40	66					Payne, Chapter 10
Red-winged Blackbird	83	78					Orians & Beletsky, Chapter 11
House Martin							Bryant 1988
Florida Scrub Jay[a]	78	81		36	36		Fitzpatrick & Woolfenden 1988
Splendid Fairy-wren	74	43		44	66		Rowley & Russell, Chapter 15
Sparrowhawk							Newton, Chapter 17
Osprey	64	67					Postupalsky, Chapter 18
Screech Owl		65					Gehlbach, Chapter 19
Ural Owl		67					Saurola, Chapter 20
Barnacle Goose	29	30					Owen & Black, Chapter 21
Mute Swan							Bacon & Andersen-Harild, Chapter 22
Red-billed Gull	31	29		5	6		Mills, Chapter 23
Fulmar[a]	61	54					Ollason & Dunnet 1988

Note: in all cases variance calculated by regression analysis of lifetime production (y) against lifespan (x).

[a] Breeding lifespan, not total lifespan.

Second to the effects of lifespan, offspring survival between the egg and fledgling stages contributed importantly to variation in LRS, while lifetime fecundity (total eggs laid) contributed least importantly to variance in LRS, once the effect of lifespan was allowed for (Chapters 6, 13; Clutton-Brock 1988). However, the exact contribution of these components varied greatly between species.

Where LRS was measured by the number of recruits to local breeding populations, rather than number of fledglings, lifespan contributed much less to variance in LRS, and the survival of young between fledging and recruitment emerged as important. This was especially so in species in which offspring survival varied greatly between years. In the short-lived Great Tit, in which most individuals breed in only one or two years, hitting a year with a good beech crop had a major effect on subsequent recruitment (van Noordwijk & van Balen 1987, McCleery & Perrins 1988; Chapter 3). In general, therefore, the later the stage at which offspring numbers are measured, the more variable the survival of offspring appears and the more this variance contributes to variance in the LRS of parents. In the Great Tit, variance in post-fledging offspring survival over-rode the effect of lifespan in its contribution to variance in LRS.

Age of first breeding

Comparing different species of birds, the general trend is for age of first breeding to rise with increase in potential lifespan (Lack 1968). While short-lived bird species usually begin nesting in their first year of life, long-lived ones wait for up to several years. In many of the species in Table 26.4, the age of first breeding in the latest individuals exceeds the mean adult life expectancy. This accounts for the fact that, even with high annual survival, a large proportion of fledglings in long-lived species dies without reproducing. Within species, individuals show considerable variation in age of first breeding: 1–5 years in short-lived Pied Flycatchers to 6–19 years in long-lived Fulmars (Table 26.4). Individual variation in age of first breeding contributes importantly to variations in length of breeding life, and hence to variance in LRS.

The figures in Table 26.4 may be biased in favour of early-breeding individuals. This is because not all the studies have continued long enough to enable all late starters to be recorded, and also because observers are often reluctant to include very late starters for fear that they may have bred somewhere else before. However, in some of the species listed, this is an unlikely bias because individuals show extreme site fidelity and can often be seen to be present in the breeding areas as non-breeders.

Table 26.4 Age of first breeding and mean adult life expectancy in various bird species. M = male, F = female.

Species	Age (years) of first breeding		Mean expectation of further life (years)	Reference
	Mean	Range		
Passerines				
Pied Flycatcher M	1.9	1–5	1.5	Sternberg, Chapter 4
F	1.5	1–5	1.5	
Red-winged Blackbird M	2.4	1–6	2.0	Orians & Beletsky, Chapter 11
Florida Scrub Jay M	2.9	1–7	4.3	Fitzpatrick & Woolfenden 1988
F	2.4	1–5	4.3	
Arabian Babbler				Zahavi, Chapter 16
Magpie M	1.8	1–3	3.5	Birkhead & Goodburn, Chapter 10
F	1.6	1–3	2.0	
Raptors				
Sparrowhawk F	2.0	1–4	2.3	Newton 1985
Osprey[a]	3.1	3–6	6.2	Postupalsky, Chapter 18
Peregrine F[a]	2.0	1–5	9.5	Mearns & Newton 1984
Red Kite[a]		2–7	19.5	Newton, Davis & Davis 1989
Ural Owl F	4.0	1–9	6.2	Saurola, Chapter 20

Water-birds

Oyster-catcher M	6.6	4–14	9.7	Saffriel et al. 1985, M.P. Harris, pers. comm.
F	4.5	3–6	9.7	
White Stork M	3.4	2–6	3.7	Zink 1967
F	3.9	2–6	3.7	
Barnacle Goose M[a]	6.7		9.7	Owen & Black, Chapter 21
F[a]	6.0	2–10	9.7	
Snow Goose F	2.2	2–4	4.8	Cooke & Rockwell 1988
Mute Swan[a]	6.0	2–15	5.8	Bacon & Andersen-Harild, Chapter 22

Sea-birds

Kittiwake M[a]	4.7		4.8	Wooller & Coulson 1977, Coulson 1988
F[a]	5.1		6.6	
Fulmar	8.0[b]	6–19	30.3	Ollason & Dunnet 1988
Short-tailed Shearwater	7.0	4–15	9.7	Wooller, Chapter 24
Red-billed Gull M[a]	3.3	2–5	5.8	Mills, Chapter 23
F[a]	4.2	2–6	8.6	
Adélie Penguin M	6.2	4–8	8.9	Ainley et al. 1983
F	4.9	3–7	8.9	
Wandering Albatross M	12.1	9–16	22.2	Weimerskirch & Jouventin 1987
F	11.2	7–16	12.3	

[a] Calculated from an increasing population in which age of first breeding may be lower, and adult life expectancy higher than in a stable population of that species.
[b] Modal.

The earliest age at which individuals of a particular species could breed, given ideal conditions, is presumably dependent on the age of physiological maturity, when the gonads first become functional. In long-lived species physiological maturity probably occurs at an earlier age than that at which most individuals first breed, so that other factors are responsible for the delay. Constraints are likely to include adequate food-supplies, mates and nesting places, all of which can be difficult to get or in short supply. Indeed, recruitment to a breeding population is often a competitive process, in which foraging skills, experience and social status are all involved.

As evidence for this view, the mean age of first breeding in any given species can vary with conditions (Chapter 16, 20). In those monogamous bird species in which the sex ratio among potential breeders is unequal, the surplus sex starts, on average, at a greater age than the other (see above). In territorial birds of prey, breeding at a young age is more frequent when territories are freely available, as in depleted populations, or when territories (and nest sites) are made available by the experimental removal of the previous occupants (Newton 1979, Village 1983). Some raptor species can breed successfully while still in "immature plumage", so physiological maturity precedes the age at which the definitive adult dress is acquired. In geese and other species subject to annual variations in food-supplies, young individuals are more likely to breed in good food-years than in poor ones. It seems reasonable to assume that individuals which delayed their breeding in poor food-years, would have nested in those years if they had been able to acquire sufficient nutrients to do so (Cooke & Rockwell 1988). However, the fact that, at any given time, some individuals start much younger than others implies that age is not the only factor giving precedence in recruitment.

Non-breeding by established breeders

Non-breeding years among established breeders are clearly frequent in species subject to annual fluctuations in conditions, such as Ural Owl and Barnacle Goose (Chapters 18, 20). They are also regular in long-lived species subject to more stable conditions, such as Short-tailed Shearwater (at least 12 % of experienced adults per year) and Kittiwake (2 % of males and 5 % of females), especially early in the breeding life (Chapter 24, Wooller & Coulson 1977). Non-breeding years have been linked with loss or change of mate (Coulson 1988, Scott 1988, Ollason & Dunnet 1988), but often have no obvious cause, except perhaps temporary lapses in body condition or territory ownership. It is also possible that some very old individuals cease breeding some years before they die, but are seldom recorded once they disappear from breeding populations (Chapter 7).

Factors affecting LRS

The LRS of any individual could result from the action of chance events, environmental factors (including social factors), phenotype (and underlying genotype), or some combination of these. Even if all individuals in a population had identical probabilities of survival and annual breeding success, then some would be more productive than others purely by chance, and the distribution of LRS values in a population would fit a predictable pattern. So far, at least two attempts have been made to compare the observed pattern of LRS values with that expected by chance, given the mean annual survival, ages of first breeding and brood-sizes prevalent in the population. These species were Scrub Jay (Chapter 13) and Sparrowhawk (Newton 1988b), and in neither did the observed and expected patterns of LRS values differ significantly. This indicates that chance alone, with all individuals having identical probabilities of annual survival and reproductive success, could not be excluded as the factor determining the observed distribution of LRS values. But nor does it confirm that the pattern was indeed due entirely to chance, partly because the method does not predict which particular individuals are the most productive and which the least. Analyses of this type may in fact prove of little value because, if the observed LRS values did deviate significantly from random expectation, this might be because one of the parameters was wrongly estimated. The whole question of chance in LRS may be best side-stepped, by checking the extent to which variations in LRS can be related to environmental and other factors. Given that the environment might itself impose a certain pattern of LRS values, by limiting reproduction and survival, it becomes more important to ask which types of individuals succeed in prevailing conditions and which do not, and which aspects of environment, upbringing and phenotype (or genotype) predispose success.

ENVIRONMENTAL FACTORS

Environmental factors affect LRS through their influence on annual reproduction or survival. By definition these are factors over which the individual has no control. Some such factors could affect individuals entirely at random and regardless of phenotype: for example, death by lightning or volcanic eruption. Others might affect phenotype, but with varying severity, at certain times and places but not at others. For example, in a good year all individuals in a population may breed well, while in a poor year all may do badly, but some phenotypes worse than others. The performance of a given phenotype thus varies with individual circumstances. Environmental fluctuations in time and space emerged as a major factor affecting LRS in most of the species discussed in this book. Temporal variations included annual and longer-term changes in conditions (Chapters 3, 7, 11, 15, 18, 21, 22), while spatial variations applied to regions (Chapter 21), study areas in the same region (Chapters 2, 8, 9, 17, 22), and territories in the same study area (Chapters 6, 7, 13, 14, 15, 19).

Under temporal fluctuations, the exact calendar years in which a bird lived were crucial. Those individuals whose breeding lives fell within a period of good food-supplies (Chapter 14), or of low or increasing population (Chapter 7), generally had higher LRS than did other individuals whose lives fell in less fortunate times. Some studies fell largely within a period of population increase (Mute Swan, Barnacle Goose) or decline (Sparrowhawk), so the observed LRS values would be expected to differ from those in stable populations.

In long-lived species exposed to marked annual fluctuations in conditions, such as Barnacle Goose, LRS depended less on total breeding lifespan than on the number of good years included (Chapter 20). In short-lived species, entire lifetimes could fall wholly within good or bad periods. Thus, where weather or food-supply differed between years, the LRS of once-only breeders depended largely on which year they happened to hit, giving large differences in LRS between cohorts (Chapters 7, 11). In some studies, rare catastrophes, such as a volcanic eruption or an epidemic, curtailed the reproductive careers of many individuals alive at the time, but those which survived the population crash did particularly well under reduced competition.

In many other bird species LRS is likely to be time-dependent. This will be true of any species whose numbers fluctuate greatly, either on a regular (cyclic) or irregular pattern (Lack 1954, Keith 1963, Krebs 1978). Individuals that breed during a population increase may have higher average lifetime success, but contribute relatively few of their genes to the next cycle (Chitty 1967). Conversely, individuals that breed successfully during population declines may contribute greatly to subsequent generations, even though their average per capita reproductive success may be lower. It is not the absolute contribution of an individual genotype to subsequent generations that matters in terms of biological fitness, but its contribution relative to that of other genotypes.

SOCIAL FACTORS

In birds, as in most other animals, reproduction involves a coalition between at least two individuals of opposite sex, which collaborate for part or all of the breeding cycle. An individual's LRS can thus depend as much on the attributes of its mate (or breeding group) as on its own attributes. For instance, in Kittiwakes the number of young fledged by different pairs is influenced by the laying date of the female, which is affected by the age of the male (Thomas & Coulson 1988). Similarly, in Sparrowhawks the age of the male affects the clutch size and nest success of the female (Newton *et al.* 1981, Newton 1988a). In Great Tits, female age predicts clutch size and laying date, while male age predicts fledging success and reruitment (McCleery & Perrins 1988). These effects apply far beyond the annual breeding attempt. Thus, in Bewick's Swans, male characteristics, particularly size, predicted the long term breeding success of pairs more accurately than did female characteristics (Scott 1988). Hence, in many birds mate quality contributes to individual LRS, and selection would be expected to favour choosiness in both sexes (Clutton-Brock 1988).

The effect of social environment on LRS is most obvious in group-living species. The social group typically consists of the dominant pair, plus several subordinates, often offspring from previous years, which help to raise the young. The size of the group can affect the success of all its members, while changes in group membership can affect survival and fecundity (Chapters 13–16). The breeding success of the dominant pair is often enhanced by helpers, while the opportunity for helpers to reproduce depends largely on their rank within the group, and on breeding vacancies arising in their own or neighbouring groups. At any one time, therefore, a bird low in the hierarchy of a group will usually have to wait longer to breed than a bird high in the hierarchy, and many individuals which live for several years die without ever getting to breed. In the meantime, however, their help has contributed to the success of their parents, and hence to their own inclusive fitness.

Group-living species illustrate how the behaviour of an individual depends on circumstances, as the same bird may be a helper at one stage in its life and a breeder at another. Being a breeder is the most productive, but both roles are adaptive, as the bird is in each case taking the best option for itself in the circumstances prevailing. This is important in assessing the adaptive significance of behaviour, because at first sight, the fitness of breeders would seem to be so much higher than that of non-reproductive helpers that helping seems no more than altruism. Until breeding opportunities become available, however, the only way subordinates can increase their own fitness is by helping kin. They may also benefit from experience gained while helping, and from the rise in social status that helping brings (Chapter 16).

Wherever it could be studied, social rank emerged as a correlate of breeding success in both sexes. It influenced access to prime resources, such as territories, food or mates. As it was related to age or size, however, any independent effects are not always easy to discern (but see Chapter 15). Because of competition, many individuals are prevented from pursuing the option most likely to promote high LRS. They may be relegated to poor breeding sites, have fewer or poorer mates, or (in co-operative breeders) have to spend their whole lives as helpers. It is through competition that population density (Chapter 7) and sex ratio influence LRS. Indeed shortage of possible mates probably lowers the LRS of some individuals in all monogamous species in which the sex ratio among potential breeders is unequal (see above). While the performance of an individual in the face of competition may depend partly on its phenotype, the same individual might clearly do better in some circumstances than in others. So yet again external factors influence LRS over and above phenotype and underlying genotype.

PHENOTYPIC FACTORS

The full effects of phenotype on LRS are impossible to assess, partly because only certain aspects of phenotype are ever studied. However, correlations between some phenotypic character and LRS indicate that phenotype may influence LRS, which is not then solely a product of chance and circumstance. The extent to which some phenotypic character is under genetic control, and hence susceptible to natural or sexual selection, is of course a further question.

An obvious aspect of phenotype is morphology, but most observers of bird populations record only measures of body size, such as wing length and weight. This is a pity because it limits investigation into the extent to which variation in LRS can be related to phenotype. Some species show no relationship between LRS and body size (Chapters 2, 20), but in others larger individuals are the most productive (Chapters 11, 19, Scott 1988), partly because they have better access to resources controlling reproduction. Thus large size is associated with longer life and more productive mates in male House Martins (Chapter 6), with earlier age of first breeding in female Sparrowhawks (Chapter 17), and with better winter foraging sites in male Bewick's Swans (Scott 1988). In other bird species female body size is commonly correlated with components of reproductive success, such as egg size or laying date (Chapter 3), the effects of which may be swamped by other factors in LRS values. For assessing the adaptive value of certain characters under genetic control, LRS may be too all-encompassing a measure of fitness to use. Nonetheless it puts the effects of particular attributes in the context of overall fitness.

Greater LRS in large breeders need not imply consistent selection for increased body size. Within each species variation in size may be largely environmental in origin, perhaps reflecting feeding conditions in early life (Chapter 6). Selection on the genetic component of variance might act in different directions at different times, or at different stages in the life cycle (Grant 1986, Smith 1988). The costs of large body size may be manifest mainly in the pre-breeding stage, for example if it is associated with faster growth rates or longer growth periods, making larger individuals more susceptible to starvation. If large size were inherited and favoured by selection, it would be difficult to explain why modern birds are no bigger than their predecessors preserved in museums, or in short-lived species why successive generations did not become larger during the course of a study (Chapter 6).

EARLY EXPERIENCE AND RECRUITMENT CHANCES

The above sections were concerned primarily with the LRS of those individuals which breed. But because many individuals die early, pre-breeding survival is an important component of fitness. Little is known of the factors involved, but in some bird-species an early fledging date within the year favoured survival and subsequent recruitment (Chapters 3, 5, 10, 19). In addition, weight at fledging emerged as important in some species, notably Great Tit (Perrins 1979), but not in others, such as Sparrowhawk (Newton & Moss 1986). In species whose environmental conditions varied greatly between years, the year of birth influenced phenotype (Chapters 5, 6), recruitment chances (Chapters 3, 4, 7, 18, 22), and hence the LRS of both parents and offspring.

These observations emphasize the role of early experience on subsequent performance. An individual which is hatched early in the season and grows during good conditions is much more likely to perform well subsequently than one raised in poor conditions: the "silver spoon" effect of Grafen (1988), which can influence phenotype or act independently of it. The important point is that the LRS values of breeders can be affected by environmental conditions early in life, as well as by those prevailing at the time of breeding.

Age effects on survival and reproduction

Both survival and breeding success are age-dependent in birds, usually improving in the early years of life, and, at least in some species, deteriorating thereafter. This terminal decline is difficult to demonstrate because in any population few individuals reach old age, so samples are inevitably small. In some long-lived bird species, there is also a problem of ring loss, which can create the illusion of a decline in survival rate or exaggerate any decline that might occur. Only if individuals are re-caught frequently, and worn rings replaced (as in many recent studies) can this problem be avoided.

Initial improvements in breeding and survival can be attributed to physical maturation, to increased experience of foraging, breeding or mate, to age and experience of mate, or to improvement of breeding site and social status. It has also been suggested that young breeders may show reproductive restraint in order to enhance their subsequent performance (Curio 1983), but there is little evidence for this. Deterioration of performance in later life, in the face of greater experience, can only be put down to senescence: to general wear and tear, which reduces efficiency and social status. Such deterioration is not just a feature of long-lived bird species, but is evident in short-lived ones too. In contrast to some mammals, however, it seems not to be associated with an obvious decline in physical condition.

SURVIVAL

Ring recoveries long ago revealed that survival chances improve with age in birds, but only recently, with larger samples of older age groups, has it emerged that in some species survival declines again in later life. This contrasts to the earlier and widespread view that, after a certain age, survival rate remains constant and independent of age until the very end of the potential lifespan (Nice 1937, Lack 1954). In short-lived species improvements early in life are usually apparent only between the first and subsequent years, but may continue for several successive years in longer-lived species (up to age 4 in Sparrowhawk, Chapter 17; up to age 10 in Short-tailed Shearwater, Chapter 24). Significant declines in survival later in life have now been reliably demonstrated in several bird species, both small and large (Fig. 26.2). The pattern of change in survival with

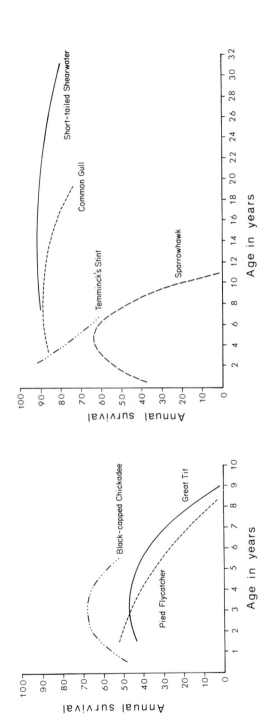

Figure 26.2 Survival in relation to age for seven bird species, based on the annual identification of individuals in intensively studied local populations. The above curves were calculated from whatever polynomial regression gave the best fit to the annual data points. In some species survival estimates were based only on birds ringed as nestlings, and hence of precisely known ages. But in Temmincks Stint, Common Gull and Short-tailed Shearwater, survival was estimated from year of first breeding, assuming that all individuals started at the mean age for their population. Based on data in Perrins (1979) for Great Tit *Parus major* (see also Chapter 3); Loery *et al.* (1987) for Black-capped Chickadee *P. atricapillus*; Sternberg (this volume) for Pied Flycatcher *Ficedula hypoleuca*; Hildén (1978) for Temminck's Stint *Calidris temminckii*; Newton (Chapter 17) for Sparrowhawk *Accipiter nisus*; Rattiste & Lilleleht (1987) for Common Gull *Larus canus*; and Wooller *et al.* (Chapter 24) for Short-tailed Shearwater *Puffinus teniurostris*.

age varies greatly, however: in some species survival rises to a peak around the middle of life and then declines again, while in others survival declines steadily from the year of first breeding on (Fig. 26.2). Moreover, in some species survival changes only slightly with age, but in others much more so.

Terminal declines may well be detected in increasing numbers of species, as records for older individuals accumulate and the problem of ring loss in long-lived species is overcome (Botkin & Miller 1974). The main interest will be in finding the exact shape of the survival curve in different species, and when during the potential lifespan, decline sets in. There is no reason to expect such survival curves to be species-specific, however, as they may well vary within a species, depending on conditions.

REPRODUCTION

In examining breeding at different ages, the usual procedure is to pool all data for each age class, and then find how mean performance alters from one age class to another. Improvements in mean performance may be apparent over only one or a few years in short-lived species, such as tit species (Chapter 2, Perrins 1979), but up to 9 years after first breeding in long-lived species, such as Fulmar (Ollason & Dunnet 1988). Declines in performance in later life, extending over several years, have been noted in Great Tit (Perrins 1979), Sparrowhawk (Chapter 17), Arctic Tern *Sterna paradisaea* (Coulson & Horobin 1976), Tengmalm's Owl *Aegolius funereus* (Korpimaki 1988) and others. They may well emerge in other species as studies continue, and more data become available for birds in the older age groups. Changes are commonly apparent in certain components of reproduction, such as laying date, egg size and clutch size, but are less often evident in production of young (but see Thomas & Coulson 1988).

The procedure of comparing successive age groups does not allow separation of the different ways in which changes in mean performance might arise, namely by changes in the year-to-year performance of the same individuals as they age, or by the progressive appearance or disappearance of distinct classes of individuals in the samples. In some species, those individuals that breed most successfully also have higher survival, perhaps linked with phenotype or territory quality (Potts 1969, de Steven 1980, Hogstedt 1981, Thomas & Coulson 1988), so the progressive disappearance of poor performers from a population would inevitably result in an improvement with age in the mean performance of the remaining individuals. However, work on Snow Geese and Sparrowhawks has shown that this is not the only cause of change, and that individuals do indeed perform progressively better in each succeeding year of early life (Hamann & Cooke 1987, Newton unpublished). In other long-lived species, such as Fulmar, the turnover in individuals is so slow that changes in mean performance with age may be assumed to result largely from changes in individual performance.

Interpretation

The improvements in both survival and reproduction early in life can be attributed in part to improvement in foraging efficiency, which must presumably improve survival chances, and increase the amount of energy which can be allocated to reproduction without incurring additional survival costs. Improvements in feeding efficiency, noted in several species between the first and later years of life (Orians 1969, Recher & Recher 1969, Dunn 1972), extend well beyond the fourth year in Herring Gull *Larus argentatus* (Coulson 1988). Such considerations presumably influence the age of first breeding in particular species, as well as success after breeding has started. Experience of breeding *per se*, irrespective of age, has emerged as important to success in certain species (Harvey *et al.* 1988), as has experience of a particular mate (Scott 1988). In some long-lived species, re-pairing with a new mate is sufficient to preclude breeding for a year or more, or lead to lowered success in otherwise experienced individuals (Chapter 21, Scott 1988). This may be why divorce is rare in such species, and why most mate changes result from deaths. In other species, in which there is no obvious penalty in change of territory or mate, changes occur most commonly after breeding failures (Brooke 1978, Newton & Marquiss 1982), thus enhancing the prospects for future success. Social status also rises with age in many species, and may improve the outcome of competitive interactions in a way that benefits reproduction and survival.

There remains the possibility that birds might invest more in reproduction as they age, but suffer additional costs, in terms of reduced survival. Such a strategy would be favoured where mean survival probability declined with age regardless of reproduction, so that the residual reproductive value of individuals also declined (Clutton-Brock 1984). Pugesek (1981) used this hypothesis of increasing effort to explain an improvement in the breeding success of California Gulls *Larus californicus* in later life. Unfortunately without more study this explanation is not separable from the alternative, that high surviving/high reproducing birds achieved greater prominence in the samples with increasing age.

The possibility that individuals might increase both the effort and the resulting cost of breeding as they age, in line with their declining survival chances, is likely to prove extremely resistant to study. One reason is that it will need large numbers of old individuals, and will entail the manipulation of reproductive costs. An experimental approach is required to allow for individual variation in reproductive potential, and because good reproduction and good survival are often associated, as mentioned above (Chapter 25, Partridge & Harvey 1985, Reznick 1985). The main question will be to what extent senescence (as reflected in declining survival) occurs as a result of reproductive effort, and to what extent it occurs regardless of reproduction.

Clearly, the extent of changes in both reproduction and survival through the lifespan varies between species and explaining such differences may provide a fruitful field for future work. Moreover, the exact shape of the reproductive value curve, which reflects the residual reproductive potential of the average individual at different ages (Fisher 1930), varies greatly with the extent of senescent decline. The study of age-related survival and reproduction is thus highly relevant to the further development of life history theory.

Individual consistency in breeding performance

Despite changes with age, some studies have pointed to consistency in the breeding performance of individuals in successive years, notably in laying dates, egg and clutch sizes. Some birds do consistently better in these respects than others of their age class, while others do consistently worse (Koskimies 1957, Brooke 1978, Findlay & Cooke 1982, Thomas & Coulson 1988). Differences of this nature can arise from a number of causes. They may have a mainly genetic basis (van Noordwijk *et al.* 1981), or they may represent the lasting effects of conditions experienced in early life, as on body size (Bryant & Westerterp 1982), or they may arise from variation in territory quality and food-supply (Hogstedt 1980; Newton 1989), and perhaps also from debilitating factors, such as disease.

What does LRS measure?

To what extent does LRS, as a measure of reproductive achievement, reflect true biological fitness? The answer to this question depends partly on what is meant by "fitness". If we were to define fitness as the contribution of an individual of distinct genotype to some subsequent (say the next) generation, relative to that of other concurrent individuals in the population, then LRS is a good measure of fitness, especially when expressed in terms of recruits. Its measurement is open to error, however, arising either from gaps in records or from those various behavioural events (extra-pair copulation, egg dumping and "helping") whose effects are hard to measure, but which influence the true productivity of a given individual (Chapter 1). However, such problems beset any measure of breeding success which is based on observation alone.

It is when the term "fitness" is meant to reflect the adaptive qualities of an individual's genotype that care is needed. Individual differences in LRS are significant in evolution only insofar as they result from differential selection on genotypes. But LRS values are influenced not only by individual attributes (which may be partly influenced by genotype), but also by chance events and by individual circumstances in which "accidents of birth" are important, as preceding chapters show. This means that an individual LRS value does not necessarily reflect accurately the adaptive potential of a particular genotype. Like any other phenotypic character, LRS values are open to variations beyond genetic control.

For studies of adaptation, therefore, "fitness" is best regarded not so much as a property of an individual, but as a property of design (Williams 1966, Grafen 1988). Then, while LRS is the number of offspring that an individual happens to produce, Darwinian fitness is reflected in the number of offspring that a given design of animal can on average expect to produce. An individual of optimal design will not always produce the largest number of offspring possible because some chance factor may curtail its reproductive career at an early stage. Conversely, given favourable conditions and luck, an individual of inferior design could sometimes be highly prolific. Hence, while the LRS of an individual can often be measured precisely, fitness is an abstraction, which can best be estimated from samples of individuals of similar design. Consider, for example, an attempt to assess the adaptive value of body size. Plotting mean LRS against body size for a range of individuals will provide good estimates of the Darwinian fitness of different sized animals. But the LRS of a given individual will approximate Darwinian fitness in this respect to a degree that depends on how close its own LRS is to the mean of its size class. In reality there may be difficulties in including all relevant individuals in any such analysis, because many may die young before their size can be measured (the "invisible fraction" of Grafen 1988), but this is another problem.

Whether, in birds, it is best to use numbers of offspring raised to fledging, to breeding or to some subsequent stage to assess fitness, depends—as in other organisms—on the character under investigation, and when in the life cycle it is expressed. There is no point in assesssing, by production to the fledgling stage, the adaptive value of a character which influences post-fledging survival, or the sex ratio of grandchildren.

In conclusion, LRS provides a useful measure of individual performance in particular environmental circumstances; but like any other measure of breeding success, it is open to error. In normal outbred populations, in which each individual genotype is unique, it can be used as a measure of individual fitness in the circumstances prevailing. However, to allow for the action of chance (external) factors in LRS, the adaptive value of particular attributes is best assessed by averaging the LRS values for different classes of individuals of similar design. This helps to reduce the role of factors which act wholly or partly regardless of genotype on the breeding and survival of individuals.

Evolutionary implications of variation in LRS

This leads to the question of how much LRS is itself subject to selection, and able to change during evolution. Because most of the variation in LRS is apparently due to chance factors, which act to a large extent regardless of phenotype (and genotype), such variation will be non-heritable and of no evolutionary significance. Other variation could be due to biological factors, with no inheritance. Some of these factors, such as quality of territory and mate, may depend on the outcome of competition, and hence again depend largely on the particular circumstances confronting the individual at the time.

The third source of variation, genetic, would be expected to be of negligible importance to LRS, for otherwise a feature so strongly related to fitness would have been optimized long ago by selection (Fisher 1930, Robertson 1955). The three studies of heritability in LRS, on Collared Flycatcher (Chapter 5), Song Sparrow (Smith 1988) and Sparrowhawk (Chapter 17), confirm an extremely low (statistically insignificant) heritability (in all cases $h^2 < 0.07$), giving little or no potential for improvement in the character. Similarly, in Meadow Pipit and Red-billed Gull there was no correlation between the productivity of parents and offspring. A low heritability of LRS does not of course preclude the possibility that certain components of LRS might show high heritability, an obvious example being clutch size (van Noordwijk & van Balen 1988).

Conclusions

Work on lifetime reproduction in birds has served to highlight the enormous variation in the performance of concurrent individuals in a population. The majority of birds which fledge or achieve independence die without producing young themselves, and those that do produce young vary greatly in their contributions. In both sexes lifespan (or breeding lifespan) is the major demographic determinant of LRS, but variations in egg production and offspring survival are also important. Much of the variation in LRS seems unrelated to genotype, as it results from chance circumstances which act on survival or reproduction to a large extent regardless of phenotype. In this respect, "accidents of birth", including the years and location of breeding are especially important, as these largely determine the conditions under which the individual lives. The social environment is also important, as this influences the range of options open to an individual. Future work might usefully concentrate on separating the relative importance of these different factors in producing the observed distribution of LRS values. More emphasis could be given to the role of phenotype in LRS, and more especially to the role of genotype, which has so far been largely neglected. In addition, more work is needed on pre-breeding survival, and on the aspects of early experience which favour the recruitment of particular individuals to breeding populations. Much less is known about the role of early experience on phenotype and subsequent performance in birds than in mammals.

Long-term studies have also highlighted the extent of age-related variation in reproduction and survival. In contrast to earlier findings, senescence has emerged as a significant factor in avian life histories, affecting both reproduction and survival. Such studies have reinforced the need for further work on life histories, and of the relationship between reproduction and survival at different ages. This should in turn promote the development of life history theory and the understanding of senescence.

References

Ainley, D.G., Le Resche, R.E & Sladen, J.C. 1983. *Breeding biology of the Adélie Penguin*. Los Angeles: University of California Press.
Botkin, D.B. & Miller, R.S. 1974. Mortality rates and survival of birds. *Amer. Nat.* **108**: 181–92

Brooke, M. De L. 1978. Some factors affecting the laying date, incubation and breeding success of the Manx Shearwater *Puffinus puffinus*. *J. Anim. Ecol.* **47**: 477–95.

Bryant, D.M. 1988. Lifetime reproductive success of House Martins. In *Reproductive Success*, ed. T.H. Clutton-Brock, pp. 173–88. Chicago: University Press.

Bryant, D.M. & Westerterp, K.R. 1980. The energy budget of the House Martin *Delichon urbica*. *Ardea* **68**: 91–102.

Chitty, D. 1967. The natural selection of self-regulatory behaviour in animal populations. *Proc. Ecol. Soc. Aust.* **2**: 51–78.

Clutton-Brock, T.H. 1983. Selection in relation to sex. In *'Evolution from Molecules to Men'*, ed. D.S. Bendall. pp. 457–81. Cambridge: University Press.

Clutton-Brock, T.H. 1984. Reproductive effort and terminal investment in iteroparous animals. *Amer. Nat.* **123**: 212–19.

Clutton-Brock, T.H. (ed.) 1988. *Reproductive Success*. Chicago: University Press.

Cooke, F. & Rockwell, R.F. 1988. Reproductive success in a Lesser Snow Goose Population. In *Reproductive Success*, ed. T.H. Clutton-Brock, pp. 237–50. Chicago: University Press.

Coulson, J.C. 1984. The population dynamics of the Eider Duck *Somateria mollissima* and evidence of extensive non-breeding by adult ducks. *Ibis* **126**: 525–43.

Coulson, J.C. & Horobin, J. 1976. The influence of age on the breeding biology & survival of the Arctic Tern *Sterna paradisaea*. *J. Zool., Lond.* **178**: 247–60.

Coulson, J.C. & Thomas, C. 1985. Differences in the breeding performance of individual Kittiwake Gulls, *Rissa tridactyla* (L.). In *Behavioural Ecology*, ed. R.M. Sibly & R.H. Smith, pp. 489–503. 25th Symp. Brit. Ecol. Soc. Oxford: Blackwell Scientific Publications.

Curio, E. 1983. Why do young birds produce less well? *Ibis* **121**: 400–4.

Darwin, C. 1871. *The Descent of man and Selection in Relation to Sex*. London: John Murray.

Davies, N.B. 1985. Cooperation and conflict among Dunnocks *Prunella modularis* in a variable mating system. *Anim. Behav.* **33**: 628–48.

De Steven, D. 1980. Clutch size, breeding success and parental survival in the Tree Swallow *(Iridoprocne bicolour)*. *Evolution* **34**: 278–91.

Dunn, E.K. 1972. Effect of age on the fishing ability of Sandwich Terns *Sterna sandvicensis*. *Ibis* **114**: 360–6.

Faaborg, J. 1986. Reproductive success and survivorship of the Galapagos Hawk *Buteo galapagoensis:* potential costs and benefits of co-operative polyandry. *Ibis* **128**: 337–47.

Findlay, C.S. & Cooke, F. 1982. Breeding synchrony in the Lesser Snow Goose *(Anser caerulescens caerulescens)*. 1. Genetic and environmental components of hatch date variability and their effects on hatch synchrony. *Evolution* **36**: 342–51.

Fisher, R.A. 1930. *The Genetical Theory of Natural Selection*. Oxford: University Press.

Fitzpatrick, J.W. & Woolfenden, G.E. 1988. Components of lifetime reproductive success in the Florida Scrub Jay. In *Reproductive Success*, ed. T.H. Clutton-Brock, pp. 305–32. Chicago: University Press.

Grafen, A. 1988. On the uses of data on lifetime reproductive success. In *Reproductive Success*, ed. T.H. Clutton-Brock, pp. 454–71. Chicago: University Press.

Grant, P.R. 1986. *Ecology and Evolution of Darwin's Finches*. Princeton: University Press.

Hamann, J. & Cooke, F. 1987. Age effects on clutch size and laying dates of individual female Lesser Snow Geese *Anser caerulescens*. *Ibis* **129**: 527–32.

Harvey, P.H., Stenning, M.J. & Campbell, B. 1988. Factors influencing reproductive success in the Pied Flycatcher. In *Reproductive Success*, ed. T.H. Clutton-Brock, pp. 189–200. Chicago: University Press.

Hildén, O. 1978. Population dynamics in Temminck's Stint *Calidris temminckii*. *Oikos* **30**: 17–28.

Hogstedt, G. 1980. Evolution of clutch-size in birds: adaptive variation in relation to territory quality. *Science, N.Y.* **210**: 1148–50.

Hodgstedt, G. 1981. Should there be a positive or negative correlation between survival of adults in a bird population and their clutch-size? *Amer. Nat.* **118**: 568–71.

Keith, L.B. 1963. *Wildlife's Ten-year Cycle*. Madison: University of Wisconsin Press.

Korpimaki, E. 1988. Factors promoting polygyny in European birds of prey – a hypothesis. *Oecologia* **77**: 278–85.

Koskimies, J. 1957. Polymorphic variability in clutch size and laying date of the Velvet Scoter, *Melanitta fusca*. *Orn. Fenn.* **34**: 118–28.

Krebs, C.J. 1978. A review of the Chitty hypothesis of population regulation. *Can. J. Zool.* **56**: 2463–80.

Lack, D. 1954. *The Natural Regulation of Animal Numbers*. Oxford: University Press.

Lack. D. 1968. *Ecological Adaptations for Breeding in Birds*. London: Methuen.

Loery, G., Pollock, K.H., Nichols, J.D. & Hines, J.E. 1987. Age-specificity of Black-capped Chickadee survival rates: analysis of capture–recapture data. *Ecology* **68**: 1038–44.

Mearns, R. & Newton, I. 1984. Turnover and dispersal in a Peregrine population. *Ibis* **126**: 347–55.

Newton, I. 1979. *Population Ecology of Raptors*. Berkhamsted: Poyser.

Newton, I. 1985. Lifetime reproductive output of female Sparrowhawks. *J. Anim. Ecol.* **54**: 241–53.

Newton, I. 1988a. Age and reproduction in the Sparrowhawk. In *Reproductive Success*, ed. T.H. Clutton-Brock, pp. 201–19. Chicago: University Press.

Newton, I. 1988b. Individual performance in Sparrowhawks: the ecology of two sexes. *Proc. Int. Orn. Congr.* **19**: 125–154.

Newton, I. & Marquiss, M. 1982. Fidelity to breeding area and mate in Sparrowhawks *Accipiter nisus*. *J. Anim. Ecol.* **51**: 327–41.

Newton, I. & Moss, D. 1986. Post-fledging survival of Sparrowhawks *Accipiter nisus* in relation to mass, brood size and brood composition at fledging. *Ibis* **128**: 73–80.

Newton, I. Marquiss, M. & Moss, D. 1981. Age and breeding in Sparrowhawks. *J. Anim. Ecol.* **50**: 839–53.

Newton, I., Davis, P.E. & Davis, J.E. 1989. Age of first breeding, dispersal and survival of Red Kites *Milvus milvus* in Wales. *Ibis* **131**: 16–21.

Nice, M.M. 1937. Studies in the life history of the Song Sparrow, Vol. 1. *Trans.*

Linn. Soc. New York **4**: 1–247.

Noordwijk, A.J. van, Balen, J.H. van, & Scharloo, W. 1981. Genetic variation in the timing of reproduction in the Great Tit. *Oecologia* **49**: 158–66.

Noordwijk, A.J. van & Balen, J.H. van 1988. The Great Tit, *Parus major*. In *Reproductive Success*, ed. T.H. Clutton-Brock, pp. 119–35. Chicago: University Press.

Ollason, J.C. & Dunnet, G.M. 1988. Variation in breeding success in Fulmars. In *Reproductive Success*, ed. T.H. Clutton-Brock, pp. 263–78. Chicago: University Press.

Orians, G.H. 1969. Age and hunting success in the Brown Pelican (*Pelecanus occidentalis*). *Anim. Behav.* **17**: 316–19.

Partridge, L. & Harvey, P.H. 1985. Costs of reproduction. *Nature* **316**: 20–1.

Payne, R.B. 1984. Sexual selection, and arena behaviour and sexual size dimorphism in birds. *Ornithol. Monogr.* **33**: 1–52.

Perrins, C.M. 1979. *British Tits*. London: Collins.

Potts, G.R. 1969. The influence of eruptive movements, age, population size and other factors on the survival of the Shag *(Phalacrocorax aristotelis)* (L). *J. Anim. Ecol.* **38**: 53–102.

Pugesek, B.H. 1981. Increased reproductive effort with age in the California Gull (*Larus californicus*). *Science* **212**: 822–3.

Rattiste, K. & Lilleleht, V. 1987. Population ecology of the Common Gull *Larus canus* in Estonia. *Ornis Fenn.* **64**: 25–6.

Recher, H.F. & Recher, J.A. 1969. Comparative foraging efficiency of adult and immature Little Blue Herons (*Florida caerulea*). *Anim. Behav.* **17**: 320–2.

Reznick, D. 1985. Costs of reproduction: an evalution of the empirical evidence. *Oikos* **44**: 257–67.

Robertson, A. 1955. Selection in animals: synthesis. *Cold Spring Harbor Symp. Quant. Biol.* **20**: 225–9.

Saffriel, U.N., Harris, M.P., Brooke, M. De L. & Britton, C.K. 1984. Survival of breeding Oystercatchers *Haematopus ostralegus*. *J. Anim. Ecol.* **53**: 867–77.

Scott, D.H. 1988. Reproductive success in Bewick's Swans. In *Reproductive Success*, ed. T.H. Clutton-Brock, pp. 220–36. Chicago: University Press.

Smith, J.N.M. 1988. Determinants of lifetime reproductive success in the Song Sparrow. In *Reproductive Success*, ed. T.H. Clutton-Brock, pp. 154–72. Chicago: University Press.

Thomas, C.S. & Coulson, J.C. 1988. Reproductive success of Kittiwake Gulls, *Rissa tridactyla*. In *Reproductive Success*, ed. T.H. Clutton-Brock, pp. 251–62. Chicago: University Press.

Village, A. 1983. The role of nest-site availability & territorial behaviour in limiting the breeding density of Kestrels. *J. Anim. Ecol.* **52**: 635–45.

Wade, M.J. & Arnold, S.J. 1980. The intensity of sexual selection in relation to male behaviour, female choice, and sperm precedence. *Anim. Behav.* **28**: 446–61.

Weimerskirch. H. & Jouventin, P. 1987. Population dynamics of the Wandering Albatross, *Diomedea exulans*, of the Crozet Islands: causes and consequences of the population decline. *Oikos* **49**: 315–22.

Williams, G.C. 1986. Natural selection, the costs of reproduction and a refinement of Lack's principle. *Amer. Nat.* **100**: 687–90.

Wooller, R.D. & Coulson, J.C. 1977. Factors affecting the age of first breeding of the Kittwake *Rissa tridactyla*. *Ibis* **119**: 339–49.

Zink, G. 1967. Populationsdynamik des Weissen Storchs in Mitteleuropa. *Proc. Int. Orn. Congr.* **14**: 191–215.

Index